GAME THEORY. Edition 2

Volume 1: Basic Concepts.

Kindle Direct Publishing (formerly CreateSpace Independent Publishing);
2nd edition (January 2018)

https://www.amazon.com/Game-Theory-1-Basic-Concepts/dp/1983604631

ISBN-13: 978-1983604638
ISBN-10: 1983604631

Giacomo Bonanno is Professor of Economics at the
University of California, Davis
http://faculty.econ.ucdavis.edu/faculty/bonanno/

Preface to the first edition, 2015

After teaching game theory (at both the undergraduate and graduate level) at the University of California, Davis for 25 years, I decided to organize all my teaching material in a textbook. There are many excellent textbooks in game theory and there is hardly any need for a new one. However, there are two distinguishing features of this textbook: (1) it is open access and thus free,[1] and (2) it contains an unusually large number of exercises with complete and detailed answers.

I tried to write the book in such a way that it would be accessible to anybody with minimum knowledge of mathematics (high-school level algebra and some elementary notions of probability) and no prior knowledge of game theory. However, the book is intended to be rigorous and it includes several proofs. I believe it is appropriate for an advanced undergraduate class in game theory and also for a first-year graduate-level class.

I expect that there will be some typos and (hopefully minor) mistakes. If you come across any typos or mistakes, I would be grateful if you could inform me: I can be reached at gfbonanno@ucdavis.edu. I will maintain an updated version of the book on my web page at

http://www.econ.ucdavis.edu/faculty/bonanno/

I also intend to add, some time in the future, a further collection of exercises and exam questions with detailed answers. Details will appear on my web page.

I am very grateful to Elise Tidrick for meticulously going through each chapter of the book and for suggesting numerous improvements. Her insightful and constructive comments have considerably enhanced this book.

I would also like to thank Nicholas Bowden, Lester Lusher, Burkhard Schipper, Matthieu Stigler, Sukjoon Lee, Minfei Xu and Pedro Paulo Funari for pointing out typos.

[1]There may be several other free textbooks on game theory available. The only one I am aware of is the excellent book by Ariel Rubinstein and Martin Osborne, MIT Press, 1994, which can be downloaded for free from Ariel Rubinstein's web page: `http://arielrubinstein.tau.ac.il`. In my experience this book is too advanced for an undergraduate class in game theory.

Preface to the second edition, 2018

This second edition introduces

- ◇ substantial reformatting (having translated into LaTeX the previous file written in Microsoft Word),
- ◇ the addition of 15 new exercises (bringing the total to 180),
- ◇ improved exposition, with additional text and examples, and enhanced topics,
- ◇ the availability of a printed version of the book, split into two volumes (the first covering the basic concepts and the second dealing with advanced topics).

I am indebted to Elise Tidrick for the enormous amount of work she put into helping me with this new edition. Not only did she design the covers of the two volumes with original artwork, but she spent hours helping me translate the original file into LaTeX code, taking care of reformatting the figures and nudging and instructing me on how to modify the LaTeX output so as to ensure a smooth flow in the reading of the book.

I would like to thank Dr. Chula Kooanantkul for pointing out several typos and Mathias Legrand for making the latex template used for this book available for free (the template was downloaded from `http://www.latextemplates.com/template/the-legrand-orange-book`).

v.8 10-22

Contents

1. Introduction

The discipline of game theory was pioneered in the early 20th century by mathematicians Ernst Zermelo (1913) and John von Neumann (1928). The breakthrough came with John von Neumann and Oskar Morgenstern's book, *Theory of Games and Economic Behavior*, published in 1944. This was followed by important work by John Nash (1950-51) and Lloyd Shapley (1953). Game theory had a major influence on the development of several branches of economics (industrial organization, international trade, labor economics, macroeconomics, etc.). Over time the impact of game theory extended to other branches of the social sciences (political science, international relations, philosophy, sociology, anthropology, etc.) as well as to fields outside the social sciences, such as biology, computer science, logic, etc. In 1994 the Nobel Memorial prize in economics was given to three game theorists, John Nash, John Harsanyi and Reinhard Selten, for their theoretical work in game theory which was very influential in economics. At the same time, the US Federal Communications Commission was using game theory to help it design a $7-billion auction of the radio spectrum for personal communication services (naturally, the bidders used game theory too!). The Nobel Memorial prize in economics was awarded to game theorists three more times: in 2005 to Robert Aumann and Thomas Schelling, in 2007 to Leonid Hurwicz, Eric Maskin and Roger Myerson and in 2012 to Lloyd Shapley and Alvin Roth.

Game theory provides a formal language for the representation and analysis of *interactive situations*, that is, situations where several "entities", called *players*, take actions that affect each other. The nature of the players varies depending on the context in which the game theoretic language is invoked: in evolutionary biology (see, for example, John Maynard Smith, 1982) players are non-thinking living organisms;[1] in computer science

[1]Evolutionary game theory has been applied not only to the analysis of animal and insect behavior but also to studying the "most successful strategies" for tumor and cancer cells (see, for example, Gerstung *et al.*, 2011).

(see, for example, Shoham-Leyton-Brown, 2008) players are artificial agents; in behavioral game theory (see, for example, Camerer, 2003) players are "ordinary" human beings, etc. Traditionally, however, game theory has focused on interaction among intelligent, sophisticated and rational individuals. For example, Robert Aumann describes game theory as follows:

> "Briefly put, game and economic theory are concerned with the interactive behavior of *Homo rationalis* – rational man. *Homo rationalis* is the species that always acts both purposefully and logically, has well-defined goals, is motivated solely by the desire to approach these goals as closely as possible, and has the calculating ability required to do so." (Aumann, 1985, p. 35.)

This book is concerned with the traditional interpretation of game theory.

Game theory is divided into two main branches. The first is *cooperative game theory*, which assumes that the players can communicate, form coalitions and sign binding agreements. Cooperative game theory has been used, for example, to analyze voting behavior and other issues in political science and related fields.

We will deal exclusively with the other main branch, namely *non-cooperative game theory*. Non-cooperative game theory models situations where the players are either unable to communicate or are able to communicate but cannot sign binding contracts. An example of the latter situation is the interaction among firms in an industry in an environment where antitrust laws make it illegal for firms to reach agreements concerning prices or production quotas or other forms of collusive behavior.

The book is divided into five parts. The printed version of the book is split into two volumes. **Volume 1** covers the basic concepts and encompasses Chapters 1-7 (Parts I and II), while **Volume 2** is devoted to advanced topics, encompassing Chapters 8-16 (Parts III to V).

Part I deals with games with *ordinal* payoffs, that is, with games where the players' preferences over the possible outcomes are only specified in terms of an ordinal ranking (outcome o is better than outcome o' or o is just as good as o'). Chapter 2 covers strategic-form games, Chapter 3 deals with dynamic games with perfect information and Chapter 4 with general dynamic games with (possibly) imperfect information.

Part II is devoted to games with *cardinal* payoffs, that is, with games where the players' preferences extend to uncertain prospects or lotteries: players are assumed to have a consistent ranking of the set of lotteries over basic outcomes. Chapter 5 reviews the theory of expected utility, Chapter 6 discusses the notion of mixed strategy in strategic-form games and of mixed-strategy Nash equilibrium, while Chapter 7 deals with mixed strategies in dynamic games.

Parts III, IV and V cover a number of advanced topics.

Part III deals with the notions of knowledge, common knowledge and belief. Chapter 8 explains how to model what an individual knows and what she is uncertain about and how to extend the analysis to the interactive knowledge of several individuals (e.g. what Individual 1 knows about what Individual 2 knows about some facts or about the state of knowledge of Individual 1). The chapter ends with the notion of common knowledge. Chapter 9 adds probabilistic beliefs to the knowledge structures of the previous chapter and discusses the notions of Bayesian updating, belief revision, like-mindedness and the possibility of "agreeing to disagree". Chapter 10 uses the interactive knowledge-belief structures of the previous two chapters to model the players' state of mind in a possible play of a given game and studies the implications of common knowledge of rationality in strategic-form games.

Part IV focuses on dynamic (or extensive-form) games and on the issue of how to refine the notion of subgame-perfect equilibrium (which was introduced in Chapters 4 and 7). Chapter 11 introduces a simple notion, called weak sequential equilibrium, which achieves some desirable goals (such as the elimination of strictly dominated choices) but fails to provide a refinement of subgame-perfect equilibrium. Chapter 12 explains the more complex notion of sequential equilibrium, which is extensively used in applications of game theory. That notion, however, leaves much to be desired from a practical point of view (it is typically hard to show that an equilibrium is indeed a sequential equilibrium) and also from a conceptual point of view (it appeals to a topological condition, whose interpretation is not clear). Chapter 13 introduces an intermediate notion, called perfect Bayesian equilibrium, whose conceptual justification is anchored in the so called AGM theory of belief revision, extensively studied in philosophy and computer science, which was pioneered by Carlos Alchourrón (a legal scholar), Peter Gärdenfors (a philosopher) and David Makinson (a logician) in 1985. In Chapter 13 we also provide an alternative characterization of sequential equilibrium based on the notion of perfect Bayesian equilibrium, which is free of topological conditions.

Part V deals with the so-called "theory of games of incomplete information", which was pioneered by John Harsanyi (1967-68). This theory is usually explained using the so-called "type-space" approach suggested by Harsanyi. However, we follow a different approach: the so-called "state-space" approach, which makes use of the interactive knowledge-belief structures developed in Part III. We find this approach both simpler and more elegant. For completeness, in Chapter 16 we explain the commonly used type-based structures and show how to convert a state-space structure into a type-space structure and *vice versa* (the two approaches are equivalent). Chapter 14 deals with situations of incomplete information that involve static (or strategic-form) games, while Chapter 15 deals with situations of incomplete information that involve dynamic (or extensive-form) games.

At the end of each section of each chapter the reader is invited to try the exercises for that section. All the exercises are collected in the penultimate section of the chapter, followed by a section containing complete and detailed answers for each exercise. For each chapter, the set of exercises culminates in a "challenging question", which is more difficult and more time consuming than the other exercises. In game theory, as in mathematics in general, it is essential to test one's understanding of the material by attempting to solve exercises and problems. The reader is encouraged to attempt solving exercises after the introduction of every new concept.

The spacing in this book does not necessarily follow conventional formatting standards. Rather, it is the editor's intention that each step is made plain in order for the student to easily follow along and quickly discover where he/she may grapple with a complete understanding of the material.

I

Games with Ordinal Payoffs

2. Ordinal Games in Strategic Form

2.1 Game frames and games

Game theory deals with interactive situations where two or more individuals, called players, make decisions that jointly determine the final outcome. To see an example point your browser to the following video:[1]

`https://www.youtube.com/watch?v=tBtr8-VMj0E`.

In this video each of two players, Sarah and Steve, has to pick one of two balls: inside one ball appears the word 'split' and inside the other the word 'steal' (each player is first asked to secretly check which of the two balls in front of him/her is the split ball and which is the steal ball). They make their decisions simultaneously. The possible outcomes are shown in Figure 2.1, where each row is labeled with a possible choice for Sarah and each column with a possible choice for Steven. Each cell in the table thus corresponds to a possible pair of choices and the resulting outcome is written inside the cell.

		Steven	
		Split	Steal
Sarah	Split	Sarah gets \$50,000 Steven gets \$50,000	Sarah gets nothing Steven gets \$100,000
	Steal	Sarah gets \$100,000 Steven gets nothing	Sarah gets nothing Steven gets nothing

Figure 2.1: The *Golden Balls* "game"

What should a rational player do in such a situation? It is tempting to reason as follows.

[1]The video shows an excerpt from *Golden Balls*, a British daytime TV game show. If you search for 'Split or Steal' on youtube.com you will find several instances of this game.

Let us focus on Sarah's decision problem. She realizes that her decision alone is not sufficient to determine the outcome; she has no control over what Steven will choose to do. However, she can envision two scenarios: one where Steven chooses *Steal* and the other where he chooses *Split*.

- If Steven decides to *Steal*, then it does not matter what Sarah does, because she ends up with nothing, no matter what she chooses.
- If Steven picks *Split*, then Sarah will get either $50,000 (if she also picks *Split*) or $100,000 (if she picks *Steal*).

Thus Sarah should choose *Steal*.

The above argument, however, is not valid because it is based on an implicit and unwarranted assumption about how Sarah ranks the outcomes; namely, it assumes that Sarah is selfish and greedy, which may or may not be true. Let us denote the outcomes as follows:

o_1 : Sarah gets $50,000 and Steven gets $50,000.
o_2 : Sarah gets nothing and Steven gets $100,000.
o_3 : Sarah gets $100,000 and Steven gets nothing.
o_4 : Sarah gets nothing and Steven gets nothing.

Table 2.1: Names for the outcomes shown in Figure 2.1

If, indeed, Sarah is *selfish and greedy* – in the sense that, in evaluating the outcomes, she focuses exclusively on what she herself gets and prefers more money to less – then her ranking of the outcomes is as follows: $o_3 \succ o_1 \succ o_2 \sim o_4$ (which reads 'o_3 is better than o_1, o_1 is better than o_2 and o_2 is just as good as o_4'). But there are other possibilities. For example, Sarah might be *fair-minded* and view the outcome where both get $50,000 as better than all the other outcomes. For instance, her ranking could be $o_1 \succ o_3 \succ o_2 \succ o_4$; according to this ranking, besides valuing fairness, she also displays *benevolence* towards Steven, in the sense that – when comparing the two outcomes where she gets nothing, namely o_2 and o_4 – she prefers the one where at least Steven goes home with some money. If, in fact, Sarah is fair-minded and benevolent, then the logic underlying the above argument would yield the opposite conclusion, namely that she should choose *Split*.

Thus we cannot presume to know the answer to the question "What is the rational choice for Sarah?" if we don't know what her preferences are. It is a common mistake (unfortunately one that even game theorists sometimes make) to reason under the assumption that players are selfish and greedy. This is, typically, an unwarranted assumption. Research in experimental psychology, philosophy and economics has amply demonstrated that many people are strongly motivated by considerations of fairness. Indeed, fairness seems to motivate not only humans but also primates, as shown in the following video:[2] `http://www.ted.com/talks/frans_de_waal_do_animals_have_morals`.

The situation illustrated in Figure 2.1 is not a game as we have no information about the preferences of the players; we use the expression *game-frame* to refer to it. In the case

[2]Also available at `https://www.youtube.com/watch?v=GcJxRqTs5nk`

where there are only two players and each player has a small number of possible choices (also called *strategies*), a game-frame can be represented – as we did in Figure 2.1 – by means of a table, with as many rows as the number of possible strategies of Player 1 and as many columns as the number of strategies of Player 2; each row is labeled with one strategy of Player 1 and each column with one strategy of Player 2; inside each cell of the table (which corresponds to a pair of strategies, one for Player 1 and one for Player 2) we write the corresponding outcome.

Before presenting the definition of game-frame, we remind the reader of what the Cartesian product of two or more sets is. Let S_1 and S_2 be two sets. Then the Cartesian product of S_1 and S_2, denoted by $S_1 \times S_2$, is the set of *ordered pairs* (x_1, x_2) where x_1 is an element of S_1 ($x_1 \in S_1$) and x_2 is an element of S_2 ($x_2 \in S_2$). For example, if $S_1 = \{a,b,c\}$ and $S_2 = \{D,E\}$ then

$$S_1 \times S_2 = \{(a,D),(a,E),(b,D),(b,E),(c,D),(c,E)\}.$$

The definition extends to the general case of n sets ($n \geq 2$): an element of $S_1 \times S_2 \times ... \times S_n$ is an ordered n-tuple $(x_1, x_2, ..., x_n)$ where, for each $i = 1, \ldots, n$, $x_i \in S_i$.

The definition of game-frame is as follows

Definition 2.1.1 A *game-frame in strategic form* is a list of four items (a quadruple) $\langle I, (S_1, S_2, ..., S_n), O, f \rangle$ where:

- $I = \{1, 2, \ldots, n\}$ is a set of *players* ($n \geq 2$).
- $(S_1, S_2, \ldots, S_n)$ is a list of sets, one for each player. For every Player $i \in I$, S_i is the set of *strategies* (or possible choices) of Player i. We denote by S the Cartesian product of these sets: $S = S_1 \times S_2 \times \cdots \times S_n$; thus an element of S is a list $s = (s_1, s_2, \ldots, s_n)$ consisting of one strategy for each player. We call S the set of *strategy profiles*.
- O is a set of *outcomes*.
- $f : S \to O$ is a function that associates with every strategy profile s an outcome $f(s) \in O$.

Using the notation of Definition 2.1.1, the situation illustrated in Figure 2.1 is the following game-frame in strategic form:

- $I = \{1, 2\}$ (letting Sarah be Player 1 and Steven Player 2),
- $(S_1, S_2) = (\{Split, Steal\}, \{Split, Steal\})$; thus $S_1 = S_2 = \{Split, Steal\}$, so that the set of strategy profiles is
  $S = \{(Split, Split), (Split, Steal), (Steal, Split), (Steal, Steal)\}$,
- O is the set of outcomes listed in Table 2.1,
- f is the following function:

$s:$	$(Split, Split)$	$(Split, Steal)$	$(Steal, Split)$	$(Steal, Steal)$
$f(s):$	o_1	o_2	o_3	o_4

  (that is, $f(Split, Split) = o_1$, $f(Split, Steal) = o_2$, etc.).

From a game-frame one obtains a *game* by adding, for each player, her preferences over (or ranking of) the possible outcomes. We use the notation shown in Table 2.2. For example, if M denotes 'Mexican food' and J denotes 'Japanese food', then $M \succ_{Alice} J$ means that Alice prefers Mexican food to Japanese food and $M \sim_{Bob} J$ means that Bob is indifferent between the two.

Notation	Interpretation
$o \succsim_i o'$	Player i considers outcome o to be *at least as good as* o' (that is, either better than or just as good as)
$o \succ_i o'$	Player i considers outcome o to be *better than* o' (that is, she prefers o to o')
$o \sim_i o'$	Player i considers outcome o to be *just as good as* o' (that is, she is indifferent between o and o')

Table 2.2: Notation for preference relations

R The "at least as good" relation $\succsim$ is sufficient to capture also strict preference $\succ$ and indifference $\sim$. In fact, starting from $\succsim$, one can define strict preference as follows: $o \succ o'$ if and only if $o \succsim o'$ and $o' \not\succsim o$ and one can define indifference as follows: $o \sim o'$ if and only if $o \succsim o'$ and $o' \succsim o$.

We will assume throughout this book that the "at least as good" relation $\succsim_i$ of Player i – which embodies her preferences over (or ranking of) the outcomes – is complete (for every two outcomes o_1 and o_2, either $o_1 \succsim_i o_2$ or $o_2 \succsim_i o_1$, or both) and transitive (if $o_1 \succsim_i o_2$ and $o_2 \succsim_i o_3$ then $o_1 \succsim_i o_3$).[3]

There are (at least) four ways of representing, or expressing, a complete and transitive preference relation over (or ranking of) a set of outcomes. For example, suppose that $O = \{o_1, o_2, o_3, o_4, o_5\}$ and that we want to represent the following ranking (expressing the preferences of a given individual): o_3 is better than o_5, which is just as good as o_1, o_1 is better than o_4, which, in turn, is better than o_2 (thus, o_3 is the best outcome and o_2 is the worst outcome). We can represent this ranking in one of the following ways:

- As a subset of $O \times O$ (the interpretation of $(o, o') \in O \times O$ is that o is at least as good as o'):

$$\begin{array}{l}\{(o_1,o_1),(o_1,o_2),(o_1,o_4),(o_1,o_5)\\(o_2,o_2),\\(o_3,o_1),(o_3,o_2),(o_3,o_3),(o_3,o_4),(o_3,o_5),\\(o_4,o_2),(o_4,o_4),\\(o_5,o_1),(o_5,o_2),(o_5,o_4),(o_5,o_5)\}\end{array}$$

- By using the notation of Table 2.2: $o_3 \succ o_5 \sim o_1 \succ o_4 \succ o_2$.

[3]Transitivity of the "at least as good" relation implies transitivity of the indifference relation (if $o_1 \sim o_2$ and $o_2 \sim o_3$ then $o_1 \sim o_3$) as well as transitivity of the strict preference relation (not only in the sense that (1) if $o_1 \succ o_2$ and $o_2 \succ o_3$ then $o_1 \succ o_3$, but also that (2) if $o_1 \succ o_2$ and $o_2 \sim o_3$ then $o_1 \succ o_3$ and (3) if $o_1 \sim o_2$ and $o_2 \succ o_3$ then $o_1 \succ o_3$).

- By listing the outcomes in a column, starting with the best at the top and proceeding down to the worst, thus using the convention that if outcome o is listed above outcome o' then o is preferred to o', while if o and o' are written next to each other (on the same row), then they are considered to be just as good:

best	o_3
	o_1, o_5
	o_4
worst	o_2

- By assigning a number to each outcome, with the convention that *if the number assigned to o is greater than the number assigned to o'* then o is preferred to o', and if two outcomes are assigned the same number then they are considered to be just as good. For example, we could choose the following numbers: $\begin{array}{ccccc} o_1 & o_2 & o_3 & o_4 & o_5 \\ 6 & 1 & 8 & 2 & 6 \end{array}$. Such an assignment of numbers is called a *utility function*. A useful way of thinking of utility is as an "index of satisfaction": the higher the index the better the outcome; however, this suggestion is just to aid memory and should be taken with a grain of salt because a utility function *does not measure anything* and, furthermore, as explained below, the actual numbers used as utility indices are completely arbitrary.[4]

Definition 2.1.2 Given a complete and transitive ranking $\succsim$ of a finite set of outcomes O, a function $U : O \to \mathbb{R}$ (where $\mathbb{R}$ denotes the set of real numbers)[a] is said to be an *ordinal utility function that represents the ranking* $\succsim$ if, for every two outcomes o and o', $U(o) > U(o')$ if and only if $o \succ o'$ and $U(o) = U(o')$ if and only if $o \sim o'$. The number $U(o)$ is called the *utility of outcome* o.[b]

[a]The notation $f : X \to Y$ is used to denote a function which associates with every $x \in X$ an element $y = f(x)$ with $y \in Y$.

[b]Thus, $o \succsim o'$ if and only if $U(o) \geq U(o')$.

R Note that the statement "for Alice the utility of Mexican food is 10" is in itself a meaningless statement; on the other hand, what would be a meaningful statement is "for Alice the utility of Mexican food is 10 and the utility of Japanese food is 5" , because such a statement conveys the information that she prefers Mexican food to Japanese food. However, the two numbers 10 and 5 have no other meaning besides the fact that 10 is greater than 5: for example, we **cannot** infer from these numbers that she considers Mexican food twice as good as Japanese food. The reason for this is that we could have expressed the same fact, namely that she prefers Mexican food to Japanese food, by assigning utility 100 to Mexican food and -25 to Japanese food, or with any other two numbers (as long as the number assigned to Mexican food is larger than the number assigned to Japanese food).

[4]Note that assigning a utility of 1 to an outcome o does not mean that o is the "first choice". Indeed, in this example a utility of 1 is assigned to the worst outcome: o_2 is the worst outcome because it has the lowest utility (which happens to be 1, in this example).

It follows from the above remark that there is an infinite number of utility functions that represent the same ranking. For instance, the following are equivalent ways of representing the ranking $o_3 \succ o_1 \succ o_2 \sim o_4$ (f, g and h are three out of the many possible utility functions):

$outcome \rightarrow$ $utility\ function \downarrow$	o_1	o_2	o_3	o_4
f:	5	2	10	2
g:	0.8	0.7	1	0.7
h:	27	1	100	1

Utility functions are a particularly convenient way of representing preferences. In fact, by using utility functions one can give a more condensed representation of games, as explained in the last paragraph of the following definition.

Definition 2.1.3 An *ordinal game in strategic form* is a quintuple $\langle I, (S_1, \ldots, S_n), O, f, (\succsim_1, \ldots, \succsim_n) \rangle$ where:

- $\langle I, (S_1, \ldots, S_n), O, f \rangle$ is a game-frame in strategic form (Definition 2.1.1) and
- for every Player $i \in I$, $\succsim_i$ is a complete and transitive ranking of the set of outcomes O.

If we replace each ranking $\succsim_i$ with a utility function U_i that represents it, and we assign, to each strategy profile s, Player i's utility of $f(s)$ (recall that $f(s)$ is the outcome associated with s) then we obtain a function $\pi_i : S \rightarrow \mathbb{R}$ called Player i's *payoff function*. Thus $\pi_i(s) = U_i(f(s))$.[a] Having done so, we obtain a triple $\langle I, (S_1, \ldots, S_n), (\pi_1, \ldots, \pi_n) \rangle$ called a *reduced-form ordinal game in strategic form* ('reduced-form' because some information is lost, namely the specification of the possible outcomes).

[a]Note that, in this book, the symbol π is not used to denote the irrational number used to compute the circumference and area of a circle, but rather as the Greek letter for 'p' which stands for 'payoff'.

For example, take the game-frame illustrated in Figure 2.1, let Sarah be Player 1 and Steven Player 2 and name the possible outcomes as shown in Table 2.1. Let us add the information that both players are selfish and greedy (that is, Player 1's ranking is $o_3 \succ_1 o_1 \succ_1 o_2 \sim_1 o_4$ and Player 2's ranking is $o_2 \succ_2 o_1 \succ_2 o_3 \sim_2 o_4$) and let us represent their rankings with the following utility functions (note, again, that the choice of numbers 2, 3 and 4 for utilities is arbitrary: any other three numbers would do):

$outcome \rightarrow$ $utility\ function \downarrow$	o_1	o_2	o_3	o_4
U_1 (Player 1):	3	2	4	2
U_2 (Player 2):	3	4	2	2

Then we obtain the reduced-form game shown in Figure 2.2, where in each cell the first number is the payoff of Player 1 and the second number is the payoff of Player 2.

		Player 2 (Steven)	
		Split	Steal
Player 1 (Sarah)	Split	3 3	2 4
	Steal	4 2	2 2

Figure 2.2: One possible game based on the game-frame of Figure 2.1

On the other hand, if we add to the game-frame of Figure 2.1 the information that Player 1 is fair-minded and benevolent (that is, her ranking is $o_1 \succ_1 o_3 \succ_1 o_2 \succ_1 o_4$), while Player 2 is selfish and greedy and represent these rankings with the following utility functions:

outcome → *utility function* ↓	o_1	o_2	o_3	o_4
U_1 (Player 1):	4	2	3	1
U_2 (Player 2):	3	4	2	2

then we obtain the reduced-form game shown in Figure 2.3.

		Player 2 (Steven)	
		Split	Steal
Player 1 (Sarah)	Split	4 3	2 4
	Steal	3 2	1 2

Figure 2.3: Another possible game based on the game-frame of Figure 2.1

In general, a player will act differently in different games, even if they are based on the same game-frame, because her incentives and objectives (as captured by her ranking of the outcomes) will be different. For example, one can argue that in the game of Figure 2.2 a rational Player 1 would choose *Steal*, while in the game of Figure 2.3 the rational choice for Player 1 is *Split*.

Test your understanding of the concepts introduced in this section, by going through the exercises in Section 2.9.1 at the end of this chapter.

2.2 Strict and weak dominance

In this section we define two relations on the set of strategies of a player. Before introducing the formal definition, we shall illustrate these notions with an example. The first relation is called "strict dominance". Let us focus our attention on one player, say Player 1, and select two of her strategies, say a and b. We say that a **strictly** dominates b (for Player 1) if, for every possible strategy profile of the other players, strategy a of Player 1, in conjunction with the strategies selected by the other players, yields a payoff for Player 1 which is **greater than** the payoff associated with strategy b (in conjunction with the strategies selected by the other players). For example, consider the following two-player game, where only the payoffs of Player 1 are shown:

		Player 2		
		E	F	G
Player 1	A	3 ...	2 ...	1 ...
	B	2 ...	1 ...	0 ...
	C	3 ...	2 ...	1 ...
	D	2 ...	0 ...	0 ...

Figure 2.4: A game showing only the payoffs of Player 1

In this game for Player 1 strategy A strictly dominates strategy B:

- if Player 2 selects E then A in conjunction with E gives Player 1 a payoff of 3, while B in conjunction with E gives her only a payoff of 2,
- if Player 2 selects F then A in conjunction with F gives Player 1 a payoff of 2, while B in conjunction with F gives her only a payoff of 1,
- if Player 2 selects G then A in conjunction with G gives Player 1 a payoff of 1, while B in conjunction with G gives her only a payoff of 0.

In the game of Figure 2.4 we also have that A strictly dominates D and C strictly dominates D; however, it is not the case that B strictly dominates D because, in conjunction with strategy E of Player 2, B and D yield the same payoff for Player 1.

The second relation is called "weak dominance". The definition is similar to that of strict dominance, but we replace 'greater than' with 'greater than or equal to' while insisting on at least one strict inequality: a **weakly** dominates b (for Player 1) if, for every possible strategy profile of the other players, strategy a of Player 1, in conjunction with the strategies selected by the other players, yields a payoff for Player 1 which is **greater than or equal to** the payoff associated with strategy b (in conjunction with the strategies selected by the other players) and, furthermore, there is at least one strategy profile of the other players against which strategy a gives a larger payoff to Player 1 than strategy b. In the example of Figure 2.4, we have that, while it is not true that B strictly dominates D, it is true that B weakly dominates D:

- if Player 2 selects E, then B in conjunction with E gives Player 1 the same payoff as D in conjunction with E (namely 2),
- if Player 2 selects F, then B in conjunction with F gives Player 1 a payoff of 1, while D in conjunction with F gives her only a payoff of 0,
- if Player 2 selects G then B in conjunction with G gives Player 1 the same payoff as D in conjunction with G (namely 0).

In order to give the definitions in full generality we need to introduce some notation. Recall that S denotes the set of strategy profiles, that is, an element s of S is an ordered list of strategies $s = (s_1, ..., s_n)$, one for each player. We will often want to focus on one player, say Player i, and view s as a pair consisting of the strategy of Player i and the remaining strategies of all the other players. For example, suppose that there are three players and the strategy sets are as follows: $S_1 = \{a,b,c\}$, $S_2 = \{d,e\}$ and $S_3 = \{f,g\}$. Then one possible strategy profile is $s = (b,d,g)$ (thus $s_1 = b$, $s_2 = d$ and $s_3 = g$). If we focus on, say, Player 2 then we will denote by s_{-2} the sub-profile consisting of the strategies of the players *other than* 2: in this case $s_{-2} = (b,g)$. This gives us an alternative way of denoting s, namely as (s_2, s_{-2}). Continuing our example where $s = (b,d,g)$, letting $s_{-2} = (b,g)$, we can denote s also by (d, s_{-2}) and we can write the result of replacing Player 2's strategy d with her strategy e in s by (e, s_{-2}); thus $(d, s_{-2}) = (b,d,g)$ while $(e, s_{-2}) = (b,e,g)$. In general, given a Player i, we denote by S_{-i} the set of strategy profiles of the players other than i (that is, S_{-i} is the Cartesian product of the strategy sets of the other players; in the above example we have that $S_{-2} = S_1 \times S_3 = \{a,b,c\} \times \{f,g\} = \{(a,f),(a,g),(b,f),(b,g),(c,f),(c,g)\}$. We denote an element of S_{-i} by s_{-i}.

Definition 2.2.1 Given an ordinal game in strategic form, let i be a Player and a and b two of her strategies ($a, b \in S_i$). We say that, for Player i,

- *a strictly dominates b* (or *b is strictly dominated by a*) if, in every situation (that is, no matter what the other players do), a gives Player i a payoff which is *greater* than the payoff that b gives. Formally: for every $s_{-i} \in S_{-i}$, $\pi_i(a, s_{-i}) > \pi_i(b, s_{-i})$.[a]
- *a weakly dominates b* (or *b is weakly dominated by a*) if, in every situation, a gives Player i a payoff which is *greater than or equal to* the payoff that b gives and, furthermore, there is at least one situation where a gives a greater payoff than b. Formally: for every $s_{-i} \in S_{-i}$, $\pi_i(a, s_{-i}) \geq \pi_i(b, s_{-i})$ and there exists an $\bar{s}_{-i} \in S_{-i}$ such that $\pi_i(a, \bar{s}_{-i}) > \pi_i(b, \bar{s}_{-i})$.[b]
- *a is equivalent to b* if, in every situation, a and b give Player i the *same* payoff. Formally: for every $s_{-i} \in S_{-i}$, $\pi_i(a, s_{-i}) = \pi_i(b, s_{-i})$.[c]

[a] Or, stated in terms of rankings instead of payoffs, $f(a, s_{-i}) \succ_i f(b, s_{-i})$ for every $s_{-i} \in S_{-i}$.

[b] Or, stated in terms of rankings, $f(a, s_{-i}) \succsim_i f(b, s_{-i})$, for every $s_{-i} \in S_{-i}$, and there exists an $\bar{s}_{-i} \in S_{-i}$ such that $f(a, \bar{s}_{-i}) \succ_i f(b, \bar{s}_{-i})$.

[c] Or, stated in terms of rankings, $f(a, s_{-i}) \sim_i f(b, s_{-i})$, for every $s_{-i} \in S_{-i}$.

For example, in the game of Figure 2.5 (which reproduces Figure 2.4), we have that

- A strictly dominates B.
- A and C are equivalent.
- A strictly dominates D.
- B is strictly dominated by C.
- B weakly (but not strictly) dominates D.
- C strictly dominates D.

Player 1 \ Player 2	E	F	G
A	3 ...	2 ...	1 ...
B	2 ...	1 ...	0 ...
C	3 ...	2 ...	1 ...
D	2 ...	0 ...	0 ...

Figure 2.5: Copy of Figure 2.4

R Note that if strategy a strictly dominates strategy b then it also satisfies the conditions for weak dominance, that is, 'a strictly dominates b' implies 'a weakly dominates b'. Throughout the book the expression 'a *weakly* dominates b' will be interpreted as 'a dominates b *weakly but not strictly*'.

The expression 'a dominates b' can be understood as 'a is better than b'. The next term we define is 'dominant' which can be understood as 'best'. Thus one cannot meaningfully say "a dominates" because one needs to name another strategy that is dominated by a; for example, one would have to say "a dominates b". On the other hand, one *can* meaningfully say "a is dominant" because it is like saying "a is best", which means "a is better than every other strategy".

Definition 2.2.2 Given an ordinal game in strategic form, let i be a Player and a one of her strategies ($a \in S_i$). We say that, for Player i,

- a is a *strictly dominant* strategy if a strictly dominates every other strategy of Player i.
- a is a *weakly dominant* strategy if, for every other strategy x of Player i, one of the following is true: either (1) a weakly dominates x or (2) a is equivalent to x.

For example, in the game shown in Figure 2.5, A and C are both weakly dominant strategies for Player 1. Note that if a player has two or more strategies that are weakly dominant, then any two of those strategies must be equivalent. On the other hand, there can be at most one strictly dominant strategy.

R The reader should convince herself/himself that the definition of weakly dominant strategy given in Definition 2.2.2 is equivalent to the following: $a \in S_i$ is a weakly dominant strategy for Player i if and only if, for every $s_{-i} \in S_{-i}$, $\pi_i(a, s_{-i}) \geq \pi_i(s_i, s_{-i})$ for every $s_i \in S_i$.[5]

In accordance with the convention established earlier, the expression 'a is a weakly dominant strategy' will have the default interpretation 'a is a *weakly but not strictly dominant strategy*'.

Note: if you claim that, for some player, "strategy x is (weakly or strictly) dominated" then you ought to name another strategy of that player that dominates x. Saying "x is dominated" is akin to saying "x is worse": worse than what? On the other hand, claiming that strategy x is weakly dominant is akin to claiming that it is best, that is, better than, or just as good as, any other strategy.

Definition 2.2.3 Given an ordinal game in strategic form, let $s = (s_1, ..., s_n)$ be a strategy profile. We say that

- s is a *strictly dominant-strategy equilibrium* if, for every Player i, s_i is a strictly dominant strategy.
- s is a *weakly dominant-strategy equilibrium* if, for every Player i, s_i is a weakly dominant strategy and, furthermore, for at least one Player j, s_j is not a strictly dominant strategy.

If we refer to a strategy profile as a dominant-strategy equilibrium, without qualifying it as weak or strict, then the default interpretation will be 'weak'.

In the game of Figure 2.6 (which reproduces Figure 2.2), *Steal* is a weakly dominant strategy for each player and thus (*Steal*,*Steal*) is a weakly dominant-strategy equilibrium.

		Player 2 (Steven)			
		Split		Steal	
Player 1 (Sarah)	Split	3	3	2	4
	Steal	4	2	2	2

Figure 2.6: Copy of Figure 2.2

		Player 2 (Steven)			
		Split		Steal	
Player 1 (Sarah)	Split	4	3	2	4
	Steal	3	2	1	2

Figure 2.7: Copy of Figure 2.3

In the game of Figure 2.7 (which reproduces Figure 2.3), *Split* is a strictly dominant strategy for Player 1, while *Steal* is a weakly (but not strictly) dominant strategy for Player 2 and thus (*Split*,*Steal*) is a weakly dominant-strategy equilibrium.

[5]Or, stated in terms of rankings, for every $s_{-i} \in S_{-i}$, $f(a, s_{-i}) \succsim_i f(s_i, s_{-i})$ for every $s_i \in S_i$..

The *Prisoner's Dilemma* is an example of a game with a strictly dominant-strategy equilibrium. For a detailed account of the history of this game and an in-depth analysis of it see `http://plato.stanford.edu/entries/prisoner-dilemma` or `http://en.wikipedia.org/wiki/Prisoner's_dilemma`.
An instance of the Prisoner's Dilemma is the following situation. Doug and Ed work for the same company and the annual party is approaching. They know that they are the only two candidates for the best-worker-of-the-year prize and at the moment they are tied; however, only one person can be awarded the prize and thus, unless one of them manages to outperform the other, nobody will receive the prize. Each chooses between exerting *Normal* effort or *Extra* effort (that is, work overtime) before the party. The corresponding game-frame is shown in Figure 2.8.

		Player 2 (Ed)	
		Normal effort	Extra effort
Player 1 (Doug)	Normal effort	o_1	o_2
	Extra effort	o_3	o_4

o_1 : nobody gets the prize and nobody sacrifices family time
o_2 : Ed gets the prize and sacrifices family time, Doug does not
o_3 : Doug gets the prize and sacrifices family time, Ed does not
o_4 : nobody gets the prize and both sacrifice family time

Figure 2.8: The Prisoner's Dilemma game-frame

Suppose that both Doug and Ed are willing to sacrifice family time to get the prize, but otherwise value family time; furthermore, they are envious of each other, in the sense that they prefer nobody getting the prize to the other person's getting the prize (even at the personal cost of sacrificing family time). That is, their rankings are as follows: $o_3 \succ_{Doug} o_1 \succ_{Doug} o_4 \succ_{Doug} o_2$ and $o_2 \succ_{Ed} o_1 \succ_{Ed} o_4 \succ_{Ed} o_3$. Using utility functions with values from the set $\{0, 1, 2, 3\}$ we can represent the game in reduced form as shown in Figure 2.9. In this game exerting extra effort is a strictly dominant strategy for every player; thus (*Extra effort, Extra effort*) is a strictly dominant-strategy equilibrium.

Definition 2.2.4 Given an ordinal game in strategic form, let o and o' be two outcomes. We say that o is *strictly Pareto superior* to o' if every player prefers o to o' (that is, if $o \succ_i o'$, for every Player i). We say that o is *weakly Pareto superior* to o' if every player considers o to be at least as good as o' and at least one player prefers o to o' (that is, if $o \succsim_i o'$, for every Player i and there is a Player j such that $o \succ_j o'$).
In reduced-form games, this definition can be extended to strategy profiles as follows. If s and s' are two strategy profiles, then s is *strictly Pareto superior* to s' if $\pi_i(s) > \pi_i(s')$ for every Player i and s is *weakly Pareto superior* to s' if $\pi_i(s) \geq \pi_i(s')$ for every Player i and, furthermore, there is a Player j such that $\pi_j(s) > \pi_j(s')$.

		Player 2 (Ed)	
		Normal effort	Extra effort
Player 1 (Doug)	Normal effort	2 2	0 3
	Extra effort	3 0	1 1

Figure 2.9: The Prisoner's Dilemma game

For example, in the Prisoner's Dilemma game of Figure 2.9, outcome o_1 is strictly Pareto superior to o_4 or, in terms of strategy profiles, (*Normal effort*, *Normal effort*) is strictly Pareto superior to (*Extra effort*, *Extra effort*).

When a player has a strictly dominant strategy, it would be irrational for that player to choose any other strategy, since she would be guaranteed a lower payoff in every possible situation (that is, no matter what the other players do). Thus in the Prisoner's Dilemma individual rationality leads to (*Extra effort, Extra effort*) despite the fact that both players would be better off if they both chose *Normal effort*. It is obvious that if the players could reach a *binding* agreement to exert normal effort then they would do so; however, the underlying assumption in non-cooperative game theory is that such agreements are not possible (e.g. because of lack of communication or because such agreements are illegal or cannot be enforced in a court of law, etc.). Any non-binding agreement to choose *Normal effort* would not be viable: if one player expects the other player to stick to the agreement, then he will gain by cheating and choosing *Extra effort* (on the other hand, if a player does not believe that the other player will honor the agreement then he will gain by deviating from the agreement herself). The Prisoner's Dilemma game is often used to illustrate a conflict between individual rationality and collective rationality: (*Extra effort, Extra effort*) is the individually rational outcome while (*Normal effort, Normal effort*) would be the collectively rational one.

Test your understanding of the concepts introduced in this section, by going through the exercises in Section 2.9.2 at the end of this chapter.

2.3 Second-price auction

The *second-price auction*, or *Vickrey auction*, is an example of a game that has a weakly dominant-strategy equilibrium. It is a "sealed-bid" auction where bidders submit bids without knowing the bids of the other participants in the auction. The object which is auctioned is then assigned to the bidder who submits the highest bid (the winner), but the winner pays not her own bid but rather the second-highest bid, that is the highest bid among the bids that remain after removing the winner's own bid. Tie-breaking rules must be specified for selecting the winner when the highest bid is submitted by two or more bidders (in which case the winner ends up paying her own bid, because the second-highest bid is equal to the winner's bid). We first illustrate this auction with an example:

Two oil companies bid for the right to drill a field. The possible bids are \$10 million, \$20 million, ..., \$50 million. In case of ties the winner is Player 2 (this was decided earlier by tossing a coin). Let us take the point of view of Player 1. Suppose that Player 1 ordered a geological survey and, based on the report, concludes that the oil field would generate a profit of \$30 million. Suppose also that Player 1 is indifferent between any two outcomes where the oil field is given to Player 2 and prefers to get the oil field herself if and only if it has to pay not more than \$30 million for it; furthermore, getting the oil field for \$30 million is just as good as not getting it. Then we can take as utility function for Player 1 the net gain to Player 1 from the oil field (defined as profits from oil extraction minus the price paid for access to the oil field) if Player 1 wins, and zero otherwise.

		Player 2				
		\$10M	\$20M	\$30M	\$40M	\$50M
	\$10M	0	0	0	0	0
	\$20M	20	0	0	0	0
Player 1 (value \$30M)	\$30M	20	10	0	0	0
	\$40M	20	10	0	0	0
	\$50M	20	10	0	−10	0

Figure 2.10: A second-price auction where, in case of ties, the winner is Player 2

In Figure 2.10 we have written inside each cell only the payoff of Player 1. For example, why is Player 1's payoff 20 when it bids \$30M and Player 2 bids \$10M? Since Player 1's bid is higher than Player 2's bid, Player 1 is the winner and thus the drilling rights are assigned to Player 1; hence Player 1 obtains something worth \$30M and pays, not its own bid of \$30M, but the bid of Player 2, namely \$10M; it follows that Player 1's net gain is $\$(30-10)\text{M} = \$20\text{M}$.

It can be verified that for Player 1 submitting a bid equal to the value it assigns to the object (namely, a bid of \$30M) is a weakly dominant strategy: it always gives Player 1 the largest of the payoffs that are possible, given the bid of the other player. This does not imply that it is the only weakly dominant strategy; indeed, in this example bidding \$40M is also a weakly dominant strategy for Player 1 (in fact, it is equivalent to bidding \$30M).

Now we can describe the second-price auction in more general terms. Let $n \geq 2$ be the number of bidders. We assume that all non-negative numbers are allowed as bids and that the tie-breaking rule favors the player with the lowest index among those who submit the highest bid: for example, if the highest bid is \$250 and it is submitted by Players 5, 8 and 10, then the winner is Player 5. We shall denote the possible outcomes as pairs (i, p), where i is the winner and p is the price that the winner has to pay. Finally we denote by b_i the bid of Player i. We start by describing the case where there are only two bidders and then generalize to the case of an arbitrary number of bidders. We denote the set of non-negative numbers by $[0, \infty)$.

The case where $n = 2$: in this case we have that $I = \{1,2\}$, $S_1 = S_2 = [0,\infty)$, $O = \{(i,p) : i \in \{1,2\}, p \in [0,\infty)\}$ and $f : S \to O$ is given by

$$f((b_1,b_2)) = \begin{cases} (1,b_2) & \text{if } b_1 \geqslant b_2 \\ (2,b_1) & \text{if } b_1 < b_2 \end{cases}.$$

The case where $n \geq 2$: in the general case the second-price auction is the following game-frame:

- $I = \{1,\ldots,n\}$.
- $S_i = [0,\infty)$ for every $i = 1,\ldots,n$. We denote an element of S_i by b_i.
- $O = \{(i,p) : i \in I, p \in [0,\infty)\}$.
- $f : S \to O$ is defined as follows. Let $H(b_1,\ldots,b_n) \subseteq I$ be the set of bidders who submit the highest bid: $H(b_1,\ldots,b_n) = \{i \in I : b_i \geq b_j \text{ for all } j \in I\}$ and let $\hat{\imath}(b_1,\ldots,b_n)$ be the smallest number in the set $H(b_1,\ldots,b_n)$, that is, the winner of the auction. Finally, let $b^{\max}$ denote the maximum bid and $b^{second}(b_1,\ldots,b_n)$ the second-highest bid,[6] that is,

$$b^{\max}(b_1,\ldots,b_n) = Max\{b_1,\ldots,b_n\}$$

$$b^{second}(b_1,\ldots,b_n) = Max(\{b_1,\ldots,b_n\} \setminus \{b^{\max}(b_1,\ldots,b_n)\}).$$

Then $f(b_1,\ldots,b_n) = \left(\hat{\imath}(b_1\ldots,b_n)\,,\ b^{second}(b_1,\ldots,b_n)\right)$.

How much should a player bid in a second-price auction? Since what we have described is a game-frame and not a game, we cannot answer the question unless we specify the player's preferences over the set of outcomes O. Let us say that Player i in a second-price auction is *selfish and greedy* if she only cares about whether or not she wins and – conditional on winning – prefers to pay less; furthermore, she prefers winning to not winning if and only if she has to pay less than the true value of the object for her, which we denote by v_i, and is indifferent between not winning and winning if she has to pay v_i. Thus the ranking of a selfish and greedy player is as follows (together with everything that follows from transitivity):

$$\begin{array}{ll} (i,p) \succ_i (i,p') & \text{if and only if } p < p' \\ (i,p) \succ_i (j,p') & \text{for all } j \neq i \text{ and for all } p', \text{ if and only if } p < v_i \\ (i,v_i) \sim_i (j,p') & \text{for all } j \neq i \text{ and for all } p' \\ (j,p) \sim_i (k,p') & \text{for all } j \neq i, k \neq i \text{ and for all } p \text{ and } p'. \end{array}$$

An ordinal utility function that represents these preferences is:[7]

$$U_i(j,p) = \begin{cases} v_i - p & \text{if } i = j \\ 0 & \text{if } i \neq j \end{cases}$$

[6] For example, if $n = 5, b_1 = \$10, b_2 = \$14, b_3 = \$8, b_4 = \$14$ and $b_5 = \$14$ then $H(\$10,\$14,\$8,\$14,\$14) = \{2,4,5\}$, $\hat{\imath}(\$10,\$14,\$8,\$14,\$14) = 2$, $b^{\max}(\$10,\$14,\$8,\$14,\$14) = \$14$ and $b^{second}(\$10,\$14,\$8,\$14,\$14) = \$14$.

[7] Of course there are many more. For example, also the following utility function represents those preferences: $U_i(j,p) = \begin{cases} 2^{(v_i - p)} & \text{if } i = j \\ 1 & \text{if } i \neq j \end{cases}$

Using this utility function we get the following payoff function for Player i:

$$\pi_i(b_1,...,b_n) = \begin{cases} v_i - b^{second}(b_1,...,b_n) & \text{if } i = \hat{\imath}(b_1,...,b_n) \\ 0 & \text{if } i \neq \hat{\imath}(b_1,...,b_n) \end{cases}$$

We can now state the following theorem. The proof is given in Section 2.8.

Theorem 2.3.1 — **Vickrey, 1961.** In a second-price auction, if Player i is selfish and greedy then it is a weakly dominant strategy for Player i to bid her true value, that is, to choose $b_i = v_i$.

Note that, for a player who is not selfish and greedy, Theorem 2.3.1 is not true. For example, if a player has the same preferences as above for the case where she wins, but, conditional on not winning, prefers the other player to pay as much as possible (she is spiteful) or as little as possible (she is generous), then bidding her true value is no longer a dominant strategy.

Test your understanding of the concepts introduced in this section, by going through the exercises in Section 2.9.3 at the end of this chapter.

2.4 The pivotal mechanism

An article in the *Davis Enterprise* (the local newspaper in Davis, California) on January 12, 2001 started with the following paragraph:

> "By consensus, the Davis City Council agreed Wednesday to order a communitywide public opinion poll to gauge how much Davis residents would be willing to pay for a park tax and a public safety tax."

Opinion polls of this type are worthwhile only if there are reasons to believe that the people who are interviewed will respond honestly. But will they? If I would like more parks and believe that the final tax I will have to pay is independent of the amount I state in the interview, I would have an incentive to overstate my willingness to pay, hoping to swing the decision towards building a new park. On the other hand, if I fear that the final tax will be affected by the amount I report, then I might have an incentive to understate my willingness to pay.

The *pivotal mechanism*, or *Clarke mechanism*, is a game designed to give the participants an incentive to report their true willingness to pay.

A public project, say to build a park, is under consideration. The cost of the project is $\$C$. There are n individuals in the community. If the project is carried out, individual i ($i = 1, \dots, n$) will have to pay $\$c_i$ (with $c_1 + c_2 + \cdots + c_n = C$); these amounts are specified as part of the project. Note that we allow for the possibility that some individuals might have to contribute a larger share of the total cost C than others (e.g. because they live closer to the projected park and would therefore benefit more from it). Individual i has an initial wealth of $\$\overline{m}_i > 0$. If the project is carried out, individual i receives benefits from it that she considers equivalent to receiving $\$v_i$. Note that for some individual i, v_i could be negative, that is, the individual could be harmed by the project (e.g. because she likes peace and quiet and a park close to her home would bring extra traffic and noise). We assume that individual i has the following utility-of-wealth function:

$$U_i(\$m) = \begin{cases} m & \text{if the project is not carried out} \\ m + v_i & \text{if the project is carried out} \end{cases}$$

The socially efficient decision is to carry out the project if and only if $\sum_{i=1}^{n} v_i > C$ (recall that $\sum$ is the summation sign: $\sum_{i=1}^{n} v_i$ is a short-hand for $v_1 + v_2 + ... + v_n$).
For example, suppose that $n = 2$, $\overline{m}_1 = 50$, $\overline{m}_2 = 60$, $v_1 = 19$, $v_2 = -15$, $C = 6$, $c_1 = 6$, $c_2 = 0$. In this case $\sum_{i=1}^{n} v_i = 19 - 15 = 4 < C = 6$ hence the project should not be carried out. To see this consider the following table:

	If the project is not carried out	If the project is carried out
Utility of Individual 1	50	$50 + 19 - 6 = 63$
Utility of Individual 2	60	$60 - 15 = 45$

If the project is carried out, Individual 1 has a utility gain of 13, while Individual 2 has a utility loss of 15. Since the loss is greater than the gain, we have a Pareto inefficient situation. Individual 2 could propose the following alternative to Individual 1: let us not carry out the project and I will pay you \$14. Then Individual 1's wealth and utility would be $50 + 14 = 64$ and Individual 2's wealth and utility would be $60 - 14 = 46$ and thus they would both be better off.

Thus Pareto efficiency requires that the project be carried out if and only if $\sum_{i=1}^{n} v_i > C$.
This would be a simple decision for the government if it knew the v_i's. But, typically, these values are private information to the individuals. Can the government find a way to induce the individuals to reveal their true valuations? It seems that in general the answer is No: those who gain from the project would have an incentive to overstate their potential gains, while those who suffer would have an incentive to overstate their potential losses.

Influenced by Vickrey's work on second-price auctions, Clarke suggested the following procedure or game. Each individual i is asked to submit a number w_i which will be interpreted as the gross benefit (if positive) or harm (if negative) that individual i associates with the project. Note that, in principle, individual i can lie and report a value w_i which is different from the true value v_i. Then the decision will be:

$$\text{Carry out the project?} \begin{cases} Yes & if \sum_{j=1}^{n} w_j > C \\ No & if \sum_{j=1}^{n} w_j \leq C \end{cases}$$

However, this is not the end of the story. Each individual will be classified as either not pivotal or pivotal.

$$\text{Individual } i \text{ is } \textbf{not} \text{ pivotal if } \begin{cases} either & \left(\sum_{j=1}^{n} w_j > C \text{ and } \sum_{j \neq i} w_j > \sum_{j \neq i} c_j\right) \\ or & \left(\sum_{j=1}^{n} w_j \leq C \text{ and } \sum_{j \neq i} w_j \leq \sum_{j \neq i} c_j\right) \end{cases}$$

and she is pivotal otherwise. In other words, individual i is pivotal if the decision about the project that would be made in the restricted society resulting from removing individual i is different from the decision that is made when she is included. If an individual is not pivotal then she has to pay no taxes. If individual i is pivotal then she has to pay a tax in the amount of

$$\left|\sum_{j \neq i} w_j - \sum_{j \neq i} c_j\right|, \text{ that is, the absolute value of } \sum_{j \neq i} w_j - \sum_{j \neq i} c_j$$

(recall that the absolute value of a is equal to a, if a is positive, and to $-a$, if a is a negative; for instance, $|4| = 4$ and $|-4| = -(-4) = 4$).
For example, let $n = 3,\ C = 10,\ c_1 = 3,\ c_2 = 2,\ c_3 = 5$.
Suppose that they state the following benefits/losses (which may or may not be the true ones): $w_1 = -1, w_2 = 8, w_3 = 3$.
Then $\sum_{i=1}^{3} w_i = 10 = C$.
Thus the project will not be carried out. Who is pivotal? The answer is provided in Figure 2.11.

Individual $i =$	$\sum w_j$ (including i)	$\sum c_j$ (including i)	Decision	$\sum w_j$ $j \neq i$ (without i)	$\sum c_j$ $j \neq i$ (without i)	Decision	Pivotal?	Tax
1	10	10	No	$8+3=11$	$2+5=7$	Yes	Yes	$11-7=4$
2	10	10	No	$-1+3=2$	$3+5=8$	No	No	0
3	10	10	No	$-1+8=7$	$3+2=5$	Yes	Yes	$7-5=2$

Figure 2.11: Example of pivotal mechanism

It may seem that, since it involves paying a tax, being pivotal is a bad thing and one should try to avoid it. It is certainly possible for individual i to make sure that she is not pivotal: all she has to do is to report $w_i = c_i$; in fact, if $\sum_{j \neq i} w_j > \sum_{j \neq i} c_j$ then adding c_i to both sides yields $\sum_{j=1}^{n} w_j > C$ and if $\sum_{j \neq i} w_j \leq \sum_{j \neq i} c_j$ then adding c_i to both sides yields $\sum_{j=1}^{n} w_j \leq C$. It is not true, however, that it is best to avoid being pivotal. The following example shows that one can gain by being truthful even if it involves being pivotal and thus having to pay a tax. Let $n = 4, C = 15, c_1 = 5, c_2 = 0, c_3 = 5$ and $c_4 = 5$.

Suppose that $\overline{m}_1 = 40$ and $v_1 = 25$.
Imagine that you are Individual 1 and, for whatever reason, you expect the following reports by the other individuals: $w_2 = -40, w_3 = 15$ and $w_4 = 20$.
If you report $w_1 = c_1 = 5$ then you ensure that you are not pivotal.
In this case $\sum_{j=1}^{4} w_j = 5 - 40 + 15 + 20 = 0 < C = 15$ and thus the project is not carried out and your utility is equal to $\overline{m}_1 = 40$. If you report truthfully, that is, you report $w_1 = v_1 = 25$ then $\sum_{j=1}^{4} w_j = 25 - 40 + 15 + 20 = 20 > C = 15$ and the project is carried out; furthermore, you are pivotal and have to pay a tax t_1 equal to

$$\left| \sum_{j=2}^{4} w_j - \sum_{j=2}^{4} c_j \right| = |(-40 + 15 + 20) - (0 + 5 + 5)| = |-15| = 15$$

and your utility will be $\overline{m}_1 + v_1 - c_1 - t_1 = 40 + 25 - 5 - 15 = 45$; hence you are better off. Indeed, the following theorem states that no individual can ever gain by lying.

The proof of Theorem 2.4.1 is given in Section 2.8.

Theorem 2.4.1 — **Clarke, 1971.** In the pivotal mechanism (under the assumed preferences) truthful revelation (that is, stating $w_i = v_i$) is a weakly dominant strategy for every Player i.

Test your understanding of the concepts introduced in this section, by going through the exercises in Section 2.9.4 at the end of this chapter.

2.5 Iterated deletion procedures

If in a game a player has a (weakly or strictly) dominant strategy then the player ought to choose that strategy: in the case of strict dominance, choosing any other strategy guarantees that the player will do worse and in the case of weak dominance, no other strategy can give a better outcome, no matter what the other players do. Unfortunately, games that have a dominant-strategy equilibrium are not very common. What should a player do when she does not have a dominant strategy? We shall consider two iterative deletion procedures that can help solve some games.

2.5.1 IDSDS

The Iterated Deletion of Strictly Dominated Strategies (IDSDS) is the following procedure or algorithm. Given a finite ordinal strategic-form game G, let G^1 be the game obtained by removing from G, for every Player i, those strategies of Player i (if any) that are strictly dominated in G by some other strategy; let G^2 be the game obtained by removing from G^1, for every Player i, those strategies of Player i (if any) that are strictly dominated in G^1 by some other strategy, and so on. Let G^∞ be the output of this procedure. Since the initial game G is finite, G^∞ will be obtained in a finite number of steps.

Figure 2.12 illustrates this procedure. If G^{∞}contains a single strategy profile (this is **not** the case in the example of Figure 2.12), then we call that strategy profile the *iterated strictly dominant-strategy equilibrium*. If G^{∞} contains two or more strategy profiles then we refer to those strategy profiles merely as the *output of the IDSDS procedure*. For example, in the game of Figure 2.12 the output of the IDSDS procedure is the set of strategy profiles $\{(A,e),(A,f),(B,e),(B,f)\}$.

What is the significance of the output of the IDSDS procedure? Consider game G of Figure 2.12. Since, for Player 2, h is strictly dominated by g, if Player 2 is rational she will not play h. Thus, if Player 1 believes that Player 2 is rational then he believes that Player 2 will not play h, that is, he restricts attention to game G^1; since, in G^1, D is strictly dominated by C for Player 1, if Player 1 is rational he will not play D. It follows that if Player 2 believes that Player 1 is rational and that Player 1 believes that Player 2 is rational, then Player 2 restricts attention to game G^2 where rationality requires that Player 2 not play g, etc. It will be shown in a later chapter that if there is common knowledge of rationality,[8] then only strategy profiles that survive the IDSDS procedure can be played; the converse is also true: any strategy profile that survives the IDSDS procedure is compatible with common knowledge of rationality.

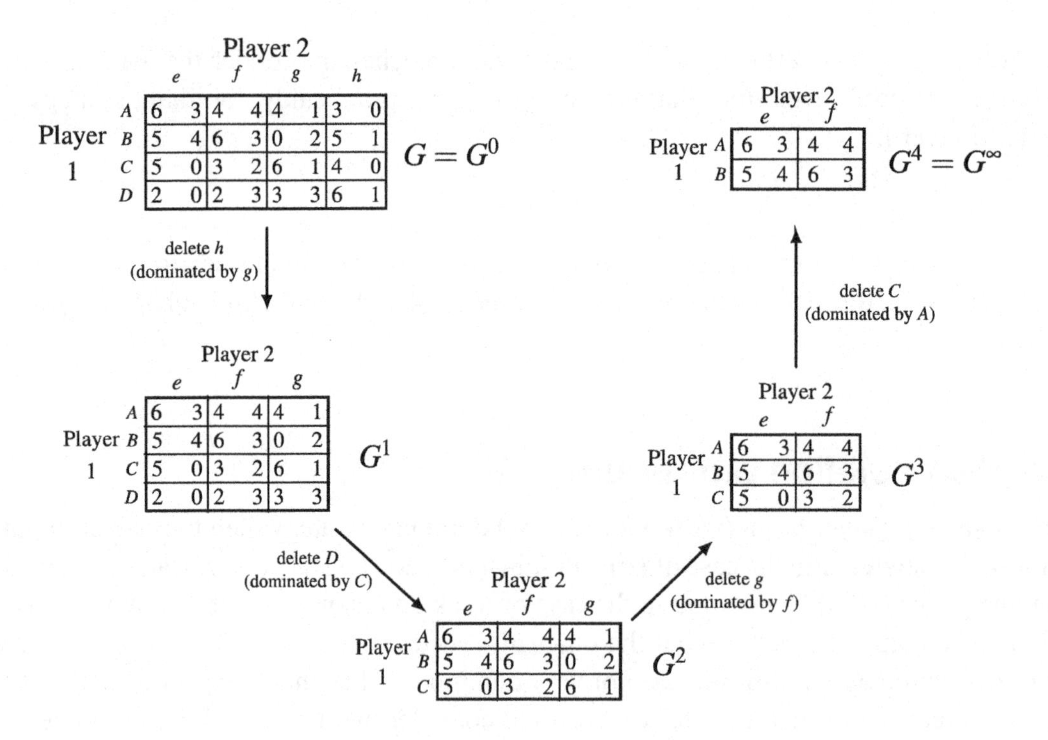

Figure 2.12: An example of the IDSDS procedure

R In finite games, the order in which strictly dominated strategies are deleted is irrelevant, in the sense that any sequence of deletions of strictly dominated strategies leads to the same output.

[8] An event E is commonly known if everybody knows E and everybody knows that everybody knows E and everybody knows that everybody knows that everybody knows E, and so on.

2.5.2 IDWDS

The Iterated Deletion of Weakly Dominated Strategies (IDWDS) is a weakening of IDSDS in that it allows the deletion also of *weakly* dominated strategies. However, this procedure has to be defined carefully, since in this case the order of deletion *can* matter. To see this, consider the game shown in Figure 2.13.

		Player 2	
		L	R
Player 1	A	4 0	0 0
	T	3 2	2 2
	M	1 1	0 0
	B	0 0	1 1

Figure 2.13: A strategic-form game with ordinal payoffs

Since M is strictly dominated by T for Player 1, we can delete it and obtain the reduced game shown in Figure 2.14

		Player 2	
		L	R
Player 1	A	4 0	0 0
	T	3 2	2 2
	B	0 0	1 1

Figure 2.14: The game of Figure 2.13 after deletion of strategy M

Now L is weakly dominated by R for Player 2. Deleting L we are left with the reduced game shown in Figure 2.15.

		Player 2
		R
Player 1	A	0 0
	T	2 2
	B	1 1

Figure 2.15: The game of Figure 2.14 after deletion of strategy L

Now A and B are strictly dominated by T. Deleting them we are left with $\boxed{(T,R)}$, with corresponding payoffs (2,2).

Alternatively, going back to the game of Figure 2.13, we could note that B is strictly dominated by T; deleting B we are left with

		Player 2	
		L	R
	A	4 0	0 0
Player 1	T	3 2	2 2
	M	1 1	0 0

Figure 2.16: The game of Figure 2.13 after deletion of strategy B

Now R is weakly dominated by L for Player 2. Deleting R we are left with the reduced game shown in Figure 2.17.

		Player 2
		L
	A	4 0
Player 1	T	3 2
	M	1 1

Figure 2.17: The game of Figure 2.16 after deletion of strategy R

Now T and M are strictly dominated by A and deleting them leads to $\boxed{(A,L)}$ with corresponding payoffs (4,0). Since one order of deletion leads to (T,R) with payoffs (2,2) and the other to (A,L) with payoffs (4,0), the procedure is not well defined.

Definition 2.5.1 — IDWDS. In order to avoid the problem illustrated above, the IDWDS procedure is defined as follows: *at every step identify, for every player, all the strategies that are weakly (or strictly) dominated and then delete all such strategies in that step.* If the output of the IDWDS procedure is a single strategy profile then we call that strategy profile the *iterated weakly dominant-strategy equilibrium* (otherwise we just use the expression 'output of the IDWDS procedure').

For example, the IDWDS procedure when applied to the game of Figure 2.13 leads to the set of strategy profiles shown in Figure 2.18.[9]

Player 2

Player 1		L		R	
	A	4	0	0	0
	T	3	2	2	2

Figure 2.18: The output of the IDWDS procedure applied to the game of Figure 2.13

Hence the game of Figure 2.13 does not have an iterated weakly dominant-strategy equilibrium.

The interpretation of the output of the IDWDS procedure is not as simple as that of the IDSDS procedure: certainly common knowledge of rationality is not sufficient. In order to delete weakly dominated strategies one needs to appeal not only to rationality but also to some notion of caution: a player should not completely rule out any of her opponents' strategies. However, this notion of caution is in direct conflict with the process of deletion of strategies. In this book we shall not address the issue of how to justify the IDWDS procedure.

> Test your understanding of the concepts introduced in this section, by going through the exercises in Section 2.9.5 at the end of this chapter.

2.6 Nash equilibrium

Games where either the IDSDS procedure or the IDWDS procedure leads to a unique strategy profile are not very common. How can one then "solve" games that are not solved by either procedure? The notion of Nash equilibrium offers a more general alternative. We first define Nash equilibrium for a two-player game.

> **Definition 2.6.1** Given an ordinal game in strategic form with two players, a strategy profile $s^* = (s_1^*, s_2^*) \in S_1 \times S_2$ is a Nash equilibrium if the following two conditions are satisfied:
> 1. for every $s_1 \in S_1$, $\pi_1(s_1^*, s_2^*) \geq \pi_1(s_1, s_2^*)$ (or stated in terms of outcomes and preferences, $f(s_1^*, s_2^*) \succsim_1 f(s_1, s_2^*)$), and
> 2. for every $s_2 \in S_2$, $\pi_2(s_1^*, s_2^*) \geq \pi_2(s_1^*, s_2)$ (or, $f(s_1^*, s_2^*) \succsim_2 f(s_1^*, s_2)$).

[9]Note that, in general, the output of the IDWDS procedure is a subset of the output of the IDSDS procedure (not necessarily a proper subset). The game of Figure 2.13 happens to be a game where the two procedures yield the same outcome.

For example, in the game of Figure 2.19 there are two Nash equilibria: (T,L) and (B,C).

		Player 2		
		L	C	R
	T	3 2	0 0	1 1
Player 1	M	3 0	1 5	4 4
	B	1 0	2 3	3 0

Figure 2.19: A strategic-form game with ordinal payoffs

There are several possible interpretations of this definition:

"No regret" interpretation: s^* is a Nash equilibrium if there is no player who, after observing the opponent's choice, regrets his own choice (in the sense that he could have done better with a different strategy of his, given the observed strategy of the opponent).

"Self-enforcing agreement" interpretation: imagine that the players are able to communicate before playing the game and reach a non-binding agreement expressed as a strategy profile s^*; then no player will have an incentive to deviate from the agreement (if she believes that the other player will follow the agreement) if and only if s^* is a Nash equilibrium.

"Viable recommendation" interpretation: imagine that a third party makes a public recommendation to each player on what strategy to play; then no player will have an incentive to deviate from the recommendation (if she believes that the other players will follow the recommendation) if and only if the recommended strategy profile is a Nash equilibrium.

"Transparency of reason" interpretation: if players are all "equally rational" and Player 2 reaches the conclusion that she should play y, then Player 1 must be able to duplicate Player 2's reasoning process and come to the same conclusion; it follows that Player 1's choice of strategy is not rational unless it is a strategy x that is optimal against y. A similar argument applies to Player 2's choice of strategy (y must be optimal against x) and thus (x,y) is a Nash equilibrium.

It should be clear that all of the above interpretations are mere rewording of the formal definition of Nash equilibrium in terms of the inequalities given in Definition 2.6.1.

The generalization of Definition 2.6.1 to games with more than two players is straightforward.

Definition 2.6.2 Given an ordinal game in strategic form with n players, a strategy profile $s^* \in S$ is a Nash equilibrium if the following n inequalities are satisfied: for every Player $i = 1, \ldots, n$,

$$\pi_i(s^*) \geq \pi_i(s_1^*, ..., s_{i-1}^*, s_i, s_{i+1}^*, ..., s_n^*) \text{ for all } s_i \in S_i.$$

The reader should convince himself/herself that a (weak or strict) dominant strategy equilibrium is a Nash equilibrium and the same is true of a (weak or strict) iterated dominant-strategy equilibrium.

Definition 2.6.3 Consider an ordinal game in strategic form, a Player i and a strategy profile $\bar{s}_{-i} \in S_{-i}$ of the players other than i. A strategy $s_i \in S_i$ of Player i is a *best reply* (or *best response*) to $\bar{s}_{-i}$ if $\pi_i(s_i, \bar{s}_{-i}) \geq \pi_i(s_i', \bar{s}_{-i})$, for every $s_i' \in S_i$.

For example, in the game of Figure 2.20, for Player 1 there are two best replies to L, namely M and T, while the unique best reply to C is B and the unique best reply to R is M; for Player 2 the best reply to T is L, the best reply to M is C and the best reply to B is C.

		Player 2					
		L		C		R	
	T	3	2	0	0	1	1
Player 1	M	3	0	1	5	4	4
	B	1	0	2	3	3	0

Figure 2.20: A strategic-form game with ordinal payoffs

R Using the notion of best reply, an alternative definition of Nash equilibrium is as follows: $\bar{s} \in S$ is a Nash equilibrium if and only if, for every Player i, $\bar{s}_i \in S_i$ is a best reply to $\bar{s}_{-i} \in S_{-i}$.

A quick way to find the Nash equilibria of a two-player game is as follows: in each column of the table underline the largest payoff of Player 1 in that column (if there are several instances, underline them all) and in each row underline the largest payoff of Player 2 in that row; if a cell has both payoffs underlined then the corresponding strategy profile is a Nash equilibrium. Underlining of the maximum payoff of Player 1 in a given column identifies the best reply of Player 1 to the strategy of Player 2 that labels that column and similarly for Player 2. This procedure is illustrated in Figure 2.21, where there is a unique Nash equilibrium, namely (B,E).

Player 1 \ Player 2	E	F	G	H
A	4 0	3 2	2 3	4 1
B	4 2	2 1	1 2	0 2
C	3 6	5 5	3 1	5 0
D	2 3	3 2	1 2	3 3

Figure 2.21: A strategic-form game with ordinal payoffs

Exercise 2.3 in Section 2.9.1 explains how to represent a three-player game by means of a set of tables. In a three-player game the procedure for finding the Nash equilibria is the same, with the necessary adaptation for Player 3: in each cell underline the payoff of Player 3 if and only if her payoff is the largest of all her payoffs in the same cell across different tables. This is illustrated in Figure 2.22, where there is a unique Nash equilibrium, namely (B,R,W).

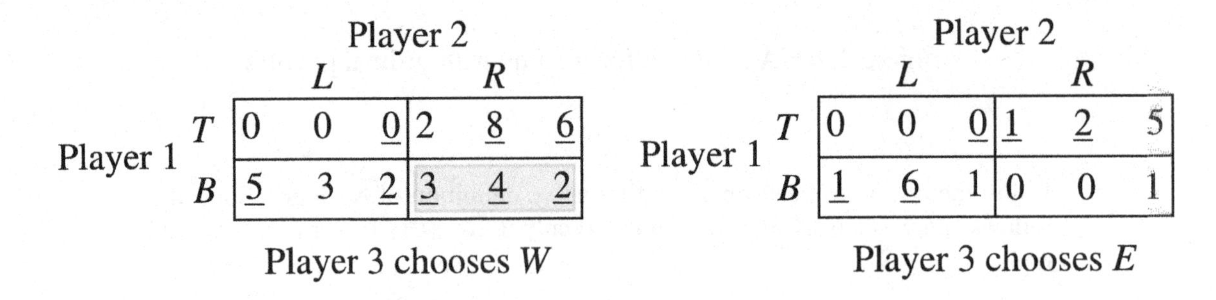

Player 3 chooses W

Player 1 \ Player 2	L	R
T	0 0 0	2 8 6
B	5 3 2	3 4 2

Player 3 chooses E

Player 1 \ Player 2	L	R
T	0 0 0	1 2 5
B	1 6 1	0 0 1

Figure 2.22: A three-player game with ordinal payoffs

Unfortunately, when the game has too many players or too many strategies – and it is thus impossible or impractical to represent it as a set of tables – there is no quick procedure for finding the Nash equilibria: one must simply apply the definition of Nash equilibrium. This is illustrated in the following example.

■ **Example 2.1** There are 50 players. A benefactor asks them to simultaneously and secretly write on a piece of paper a request, which must be a multiple of \$10 up to a maximum of \$100 (thus the possible strategies of each player are \$10, \$20, ..., \$90, \$100). He will then proceed as follows: if not more than 10% of the players ask for \$100 then he will grant every player's request, otherwise every player will get nothing. Assume that every player is selfish and greedy (only cares about how much money she gets and prefers more money to less). What are the Nash equilibria of this game? There are several:

- every strategy profile where 7 or more players request \$100 is a Nash equilibrium (everybody gets nothing and no player can get a positive amount by unilaterally changing her request, since there will still be more than 10% requesting \$100; on the other hand, convince yourself that a strategy profile where exactly 6 players request \$100 is not a Nash equilibrium),
- every strategy profile where exactly 5 players request \$100 and the remaining players request \$90 is a Nash equilibrium.

Any other strategy profile is not a Nash equilibrium: (1) if fewer than 5 players request \$100, then a player who requested less than \$100 can increase her payoff by switching to a request of \$100, (2) if exactly 5 players request \$100 and among the remaining players there is one who is not requesting \$90, then that player can increase her payoff by increasing her request to \$90. ■

We conclude this section by noting that, since so far we have restricted attention to ordinal games, there is no guarantee that an arbitrary game will have at least one Nash equilibrium. An example of a game that has no Nash equilibria is the *Matching Pennies* game. This is a simultaneous two-player game where each player has a coin and decides whether to show the Heads face or the Tails face. If both choose H or both choose T then Player 1 wins, otherwise Player 2 wins. Each player strictly prefers the outcome where she herself wins to the alternative outcome. The game is illustrated in Figure 2.23.

		Player 2	
		H	T
Player 1	H	1 0	0 1
	T	0 1	1 0

Figure 2.23: The matching pennies game

Test your understanding of the concepts introduced in this section, by going through the exercises in Section 2.9.6 at the end of this chapter.

2.7 Games with infinite strategy sets

Games where the strategy set of one or more players is infinite cannot be represented using a table or set of tables. However, all the concepts introduced in this chapter can still be applied. In this section we will focus on the notion of Nash equilibrium. We start with an example.

Example 2.2 There are two players. Each player has to write down a real number (not necessarily an integer) greater than or equal to 1; thus the strategy sets are $S_1 = S_2 = [1, \infty)$. Payoffs are as follows (π_1 is the payoff of Player 1, π_2 the payoff of Player 2, x is the number written by Player 1 and y the number written by Player 2):

$$\pi_1(x,y) = \begin{cases} x-1 & \text{if } x < y \\ 0 & \text{if } x \geq y \end{cases} \quad \text{and} \quad \pi_2(x,y) = \begin{cases} y-1 & \text{if } x > y \\ 0 & \text{if } x \leq y \end{cases}$$

What are the Nash equilibria of this game? ■

There is only one Nash equilibrium, namely (1,1) with payoffs (0,0). First of all, we must show that (1,1) is indeed a Nash equilibrium.
If Player 1 switched to some $x > 1$ then her payoff would remain 0: $\pi_1(x,1) = 0$, for all $x \in [1, \infty)$ and the same is true for Player 2 if he unilaterally switched to some $y > 1$: $\pi_2(1,y) = 0$, for all $y \in [1, \infty)$.
Now we show that no other pair (x,y) is a Nash equilibrium.
Consider first an arbitrary pair (x,y) with $x = y > 1$. Then $\pi_1(x,y) = 0$, but if Player 1 switched to an $\hat{x}$ strictly between 1 and x ($1 < \hat{x} < x$) her payoff would be $\pi_1(\hat{x},y) = \hat{x} - 1 > 0$ (recall that, by hypothesis, $x = y$).
Now consider an arbitrary (x,y) with $x < y$. Then $\pi_1(x,y) = x - 1$, but if Player 1 switched to an $\hat{x}$ strictly between x and y ($x < \hat{x} < y$) her payoff would be $\pi_1(\hat{x},y) = \hat{x} - 1 > x - 1$. The argument for ruling out pairs (x,y) with $y < x$ is similar.
Note the interesting fact that, for Player 1, $x = 1$ is a weakly dominated strategy: indeed it is weakly dominated by any other strategy: $x = 1$ guarantees a payoff of 0 for Player 1, while any $\hat{x} > 1$ would yield a positive payoff to Player 1 in some cases (against any $y > \hat{x}$) and 0 in the remaining cases. The same is true for Player 2. Thus in this game there is *a unique Nash equilibrium where the strategy of each player is weakly dominated*!

[Note: the rest of this section makes use of calculus. The reader who is not familiar with calculus should skip this part.]

We conclude this section with an example based on the analysis of competition among firms proposed by Augustine Cournot in a book published in 1838. In fact, Cournot is the one who invented what we now call 'Nash equilibrium', although his analysis was restricted to a small class of games. Consider $n \geq 2$ firms which produce an identical product. Let q_i be the quantity produced by Firm i ($i = 1, \ldots n$). For Firm i the cost of producing q_i units of output is $c_i q_i$, where c_i is a positive constant. For simplicity we will restrict attention to the case of two firms ($n = 2$) and identical cost functions: $c_1 = c_2 = c$. Let Q be total industry output, that is, $Q = q_1 + q_2$. The price at which each firm can sell each unit of output is given by the inverse demand function $P = a - bQ$ where a and b are positive constants. Cournot assumed that each firm was only interested in its own profit and preferred higher profit to lower profit (that is, each firm is "selfish and greedy").

The profit function of Firm 1 is given by

$$\pi_1(q_1,q_2) = Pq_1 - cq_1 = [a - b(q_1+q_2)]\,q_1 - cq_1 = (a-c)q_1 - b(q_1)^2 - bq_1q_2.$$

Similarly, the profit function of Firm 2 is given by

$$\pi_2(q_1,q_2) = (a-c)q_2 - b(q_2)^2 - bq_1q_2$$

Cournot defined an equilibrium as a pair $(\overline{q}_1,\overline{q}_2)$ that satisfies the following two inequalities:

$$\pi_1(\overline{q}_1,\overline{q}_2) \geq \pi_1(q_1,\overline{q}_2), \quad \text{for every} \quad q_1 \geq 0 \qquad (\clubsuit)$$

$$\pi_2(\overline{q}_1,\overline{q}_2) \geq \pi_2(\overline{q}_1,q_2) \quad \text{for every} \quad q_2 \geq 0. \qquad (\blacklozenge)$$

Of course, this is the same as saying that $(\overline{q}_1,\overline{q}_2)$ is a Nash equilibrium of the game where the players are the two firms, the strategy sets are $S_1 = S_2 = [0,\infty)$ and the payoff functions are the profit functions. How do we find a Nash equilibrium? First of all, note that the profit functions are differentiable. Secondly note that $(\clubsuit)$ says that, having fixed the value of q_2 at $\overline{q}_2$, the function $\pi_1(q_1,\overline{q}_2)$ – viewed as a function of q_1 alone – is maximized at the point $q_1 = \overline{q}_1$. A necessary condition for this (if $\overline{q}_1 > 0$) is that the derivative of this function be zero at the point $q_1 = \overline{q}_1$, that is, it must be that $\frac{\partial \pi_1}{\partial q_1}(\overline{q}_1,\overline{q}_2) = 0$. This condition is also sufficient since the second derivative of this function is always negative ($\frac{\partial^2 \pi_1}{\partial q_1^2}(q_1,q_2) = -2b$ for every (q_1,q_2)). Similarly, by $(\blacklozenge)$, it must be that $\frac{\partial \pi_2}{\partial q_2}(\overline{q}_1,\overline{q}_2) = 0$. Thus the Nash equilibrium is found by solving the system of two equations

$$\begin{cases} \frac{\partial \pi_1}{\partial q_1}(q_1,q_2) = a - c - 2bq_1 - bq_2 = 0 \\ \frac{\partial \pi_2}{\partial q_2}(q_1,q_2) = a - c - 2bq_2 - bq_1 = 0 \end{cases}$$

The solution is $\overline{q}_1 = \overline{q}_2 = \frac{a-c}{3b}$. The corresponding price is $\overline{P} = a - b\left(2\frac{a-c}{3b}\right) = \frac{a+2c}{3}$ and the corresponding profits are $\pi_1(\frac{a-c}{3b},\frac{a-c}{3b}) = \pi_2(\frac{a-c}{3b},\frac{a-c}{3b}) = \frac{(a-c)^2}{9b}$.
For example, if $a = 25$, $b = 2$, $c = 1$ then the Nash equilibrium is given by (4,4) with corresponding profits of 32 for each firm. The analysis can easily be extended to the case of more than two firms.

The reader who is interested in further exploring the topic of competition among firms can consult any textbook on Industrial Organization.

Test your understanding of the concepts introduced in this section, by going through the exercises in Section 2.9.7 at the end of this chapter.

2.8 Proofs of theorems

Theorem [Vickrey, 1961] In a second-price auction, if Player i is selfish and greedy then it is a weakly dominant strategy for Player i to bid her true value, that is, to choose $b_i = v_i$.

Proof. In order to make the notation simpler and the argument more transparent, we give the proof for the case where $n = 2$. We shall prove that bidding v_1 is a weakly dominant strategy for Player 1 (the proof for Player 2 is similar). Assume that Player 1 is selfish and greedy. Then we can take her payoff function to be as follows:

$$\pi_1(b_1, b_2) = \begin{cases} v_1 - b_2 & \text{if } b_1 \geq b_2 \\ 0 & \text{if } b_1 < b_2 \end{cases}$$

We need to show that, whatever bid Player 2 submits, Player 1 cannot get a higher payoff by submitting a bid different from v_1. Two cases are possible (recall that b_2 denotes the actual bid of Player 2, which is unknown to Player 1).

Case 1: $b_2 \leq v_1$. In this case, bidding v_1 makes Player 1 the winner and his payoff is $v_1 - b_2 \geq 0$. Consider a different bid b_1. If $b_1 \geq b_2$ then Player 1 is still the winner and his payoff is still $v_1 - b_2 \geq 0$; thus such a bid is as good as (hence not better than) v_1. If $b_1 < b_2$ then the winner is Player 2 and Player 1 gets a payoff of 0. Thus such a bid is also not better than v_1.

Case 2: $b_2 > v_1$. In this case, bidding v_1 makes Player 2 the winner and thus Player 1 gets a payoff of 0. Any other bid $b_1 < b_2$ gives the same outcome and payoff. On the other hand, any bid $b_1 \geq b_2$ makes Player 1 the winner, giving him a payoff of $v_1 - b_2 < 0$, thus making Player 1 worse off than with a bid of v_1.

■

Theorem [Clarke, 1971] In the pivotal mechanism (under the assumed preferences) truthful revelation (that is, stating $w_i = v_i$) is a weakly dominant strategy for every Player i.

Proof. Consider an individual i and possible statements w_j for $j \neq i$. Several cases are possible.

Case 1: $\sum_{j\neq i} w_j > \sum_{j\neq i} c_j$ and $v_i + \sum_{j\neq i} w_j > c_i + \sum_{j\neq i} c_j = C$. Then

	decision	i's tax	i's utility
if i states v_i	Yes	0	$\overline{m}_i + v_i - c_i$
if i states w_i such that $w_i + \sum_{j\neq i} w_j > C$	Yes	0	$\overline{m}_i + v_i - c_i$
if i states w_i such that $w_i + \sum_{j\neq i} w_j \leq C$	No	$\sum_{j\neq i} w_j - \sum_{j\neq i} c_j$	$\overline{m}_i - \left(\sum_{j\neq i} w_j - \sum_{j\neq i} c_j\right)$

Individual i cannot gain by lying if and only if

$$\overline{m}_i + v_i - c_i \geq \overline{m}_i - \left(\sum_{j\neq i} w_j - \sum_{j\neq i} c_j\right), \quad \text{i.e. if and only if} \quad v_i + \sum_{j\neq i} w_j \geq C,$$

which is true by our hypothesis.

Case 2: $\sum_{j \neq i} w_j > \sum_{j \neq i} c_j$ and $v_i + \sum_{j \neq i} w_j \leq c_i + \sum_{j \neq i} c_j = C$. Then

	decision	i's tax	i's utility
if i states v_i	No	$\sum_{j \neq i} w_j - \sum_{j \neq i} c_j$	$\overline{m}_i - \left(\sum_{j \neq i} w_j - \sum_{j \neq i} c_j\right)$
if i states w_i such that $w_i + \sum_{j \neq i} w_j \leq C$	No	$\sum_{j \neq i} w_j - \sum_{j \neq i} c_j$	$\overline{m}_i - \left(\sum_{j \neq i} w_j - \sum_{j \neq i} c_j\right)$
if i states w_i such that $w_i + \sum_{j \neq i} w_j > C$	Yes	0	$\overline{m}_i + v_i - c_i$

Individual i cannot gain by lying if and only if $\overline{m}_i - \left(\sum_{j \neq i} w_j - \sum_{j \neq i} c_j\right) \geq \overline{m}_i + v_i - c_i$, i.e. if and only if $v_i + \sum_{j \neq i} w_j \leq C$, which is true by our hypothesis.

Case 3: $\sum_{j \neq i} w_j \leq \sum_{j \neq i} c_j$ and $v_i + \sum_{j \neq i} w_j \leq c_i + \sum_{j \neq i} c_j = C$. Then

	decision	i's tax	i's utility
if i states v_i	No	0	$\overline{m}_i$
if i states w_i such that $w_i + \sum_{j \neq i} w_j \leq C$	No	0	$\overline{m}_i$
if i states w_i such that $w_i + \sum_{j \neq i} w_j > C$	Yes	$\left(\sum_{j \neq i} c_j - \sum_{j \neq i} w_j\right)$ (recall that $\sum_{j \neq i} w_j \leq \sum_{j \neq i} c_j$)	$\overline{m}_i + v_i - c_i - \left(\sum_{j \neq i} c_j - \sum_{j \neq i} w_j\right)$

Individual i cannot gain by lying if and only if $\overline{m}_i \geq \overline{m}_i + v_i - c_i - \left(\sum_{j \neq i} c_j - \sum_{j \neq i} w_j\right)$, i.e. if and only if $v_i + \sum_{j \neq i} w_j \leq C$, which is true by our hypothesis.

Case 4: $\sum_{j \neq i} w_j \leq \sum_{j \neq i} c_j$ and $v_i + \sum_{j \neq i} w_j > c_i + \sum_{j \neq i} c_j = C$. Then

	decision	i's tax	i's utility
if i states v_i	Yes	$\left(\sum_{j \neq i} c_j - \sum_{j \neq i} w_j\right)$ (recall that $\sum_{j \neq i} w_j \leq \sum_{j \neq i} c_j$)	$\overline{m}_i + v_i - c_i - \left(\sum_{j \neq i} w_j - \sum_{j \neq i} c_j\right)$
if i states w_i such that $w_i + \sum_{j \neq i} w_j > C$	Yes	$\left(\sum_{j \neq i} c_j - \sum_{j \neq i} w_j\right)$ (recall that $\sum_{j \neq i} w_j \leq \sum_{j \neq i} c_j$)	$\overline{m}_i + v_i - c_i - \left(\sum_{j \neq i} w_j - \sum_{j \neq i} c_j\right)$
if i states w_i such that $w_i + \sum_{j \neq i} w_j \leq C$	No	0	$\overline{m}_i$

Individual i cannot gain by lying if and only if $\overline{m}_i + v_i - c_i - \left(\sum_{j \neq i} w_j - \sum_{j \neq i} c_j\right) \geq \overline{m}_i$,
i.e. if and only if $v_i + \sum_{j \neq i} w_j \geq C$, which is true by our hypothesis.

Since we have covered all the possible cases, the proof is complete. ∎

2.9 Exercises

2.9.1 Exercises for Section 2.1: Game frames and games

The answers to the following exercises are in Section 2.10 at the end of this chapter.

Exercise 2.1 Antonia and Bob cannot decide where to go to dinner. Antonia proposes the following procedure: she will write on a piece of paper either the number 2 or the number 4 or the number 6, while Bob will write on his piece of paper either the number 1 or 3 or 5. They will write their numbers secretly and independently. They then will show each other what they wrote and choose a restaurant according to the following rule: if the sum of the two numbers is 5 or less, they will go to a Mexican restaurant, if the sum is 7 they will go to an Italian restaurant and if the number is 9 or more they will go to a Japanese restaurant.

(a) Let Antonia be Player 1 and Bob Player 2. Represent this situation as a game frame, first by writing out each element of the quadruple of Definition 2.1.1 and then by using a table (label the rows with Antonia's strategies and the columns with Bob's strategies, so that we can think of Antonia as choosing the row and Bob as choosing the column).

(b) Suppose that Antonia and Bob have the following preferences (where M stands for 'Mexican', I for 'Italian' and J for 'Japanese'):
for Antonia: $M \succ_{Antonia} I \succ_{Antonia} J$; for Bob: $I \succ_{Bob} M \succ_{Bob} J$.
Using utility function with values 1, 2 and 3 represent the corresponding reduced-form game as a table.

■

Exercise 2.2 Consider the following two-player game-frame where each player is given a set of cards and each card has a number on it. The players are Antonia (Player 1) and Bob (Player 2). Antonia's cards have the following numbers (one number on each card): 2, 4 and 6, whereas Bob's cards are marked 0, 1 and 2 (thus different numbers from the previous exercise). Antonia chooses one of her own cards and Bob chooses one of his own cards: this is done without knowing the other player's choice. The outcome depends on the sum of the points of the chosen cards. If the sum of the points on the two chosen cards is greater than or equal to 5, Antonia gets \$10 minus that sum; otherwise (that is, if the sum is less than 5) she gets nothing; furthermore, if the sum of points is an odd number, Bob gets as many dollars as that sum; if the sum of points turns out to be an even number and is less than or equal to 6, Bob gets \$2; otherwise he gets nothing. (The money comes from a third party.)

(a) Represent the game-frame described above by means of a table. As in the previous exercise, assign the rows to Antonia and the columns to Bob.

(b) Using the game-frame of part **(a)** obtain a reduced-form game by adding the information that each player is selfish and greedy. This means that each player only cares about how much money he/she gets and prefers more money to less.

■

Exercise 2.3 Alice (Player 1), Bob (Player 2), and Charlie (Player 3) play the following simultaneous game. They are sitting in different rooms facing a keyboard with only one key and each has to decide whether or not to press the key. Alice wins if the number of people who press the key is odd (that is, all three of them or only Alice or only Bob or only Charlie), Bob wins if exactly two people (he may be one of them) press the key and Charlie wins if nobody presses the key.

(a) Represent this situation as a game-frame. Note that we can represent a three-player game with a *set of tables*: Player 1 chooses the row, Player 2 chooses the column and Player 3 chooses the table (that is, we label the rows with Player 1's strategies, the columns with Player 2's strategies and the tables with Player 3's strategies).

(b) Using the game-frame of part **(a)** obtain a reduced-form game by adding the information that each player prefers winning to not winning and is indifferent between any two outcomes where he/she does not win. For each player use a utility function with values from the set $\{0,1\}$.

(c) Using the game-frame of part **(a)** obtain a reduced-form game by adding the information that (1) each player prefers winning to not winning, (2) Alice is indifferent between any two outcomes where she does not win, (3) conditional on not winning, Bob prefers if Charlie wins rather than Alice, (4) conditional on not winning, Charlie prefers if Bob wins rather than Alice. For each player use a utility function with values from the set $\{0,1,2\}$.

■

2.9.2 Exercises for Section 2.2: Strict/weak dominance

The answers to the following exercises are in Section 2.10 at the end of this chapter.

Exercise 2.4 There are two players. Each player is given an unmarked envelope and asked to put in it either nothing or \$300 of his own money or \$600 of his own money. A referee collects the envelopes, opens them, gathers all the money, then adds 50% of that amount (using his own money) and divides the total into two equal parts which he then distributes to the players.

(a) Represent this game frame with two alternative tables: the first table showing in each cell the amount of money distributed to Player 1 and the amount of money distributed to Player 2, the second table showing the change in wealth of each player (money received minus contribution).

(b) Suppose that Player 1 has some animosity towards the referee and ranks the outcomes in terms of how much money the referee loses (the more, the better), while Player 2 is selfish and greedy and ranks the outcomes in terms of her own net gain. Represent the corresponding game using a table.

(c) Is there a strictly dominant-strategy equilibrium?

■

Exercise 2.5 Consider again the game of Part **(b)** of Exercise 2.1.

(a) Determine, for each player, whether the player has *strictly* dominated strategies.

(b) Determine, for each player, whether the player has *weakly* dominated strategies. ■

Exercise 2.6 There are three players. Each player is given an unmarked envelope and asked to put in it either nothing or \$3 of his own money or \$6 of his own money. A referee collects the envelopes, opens them, gathers all the money and then doubles the amount (using his own money) and divides the total into three equal parts which he then distributes to the players.
For example, if Players 1 and 2 put nothing and Player 3 puts \$6, then the referee adds another \$6 so that the total becomes \$12, divides this sum into three equal parts and gives \$4 to each player.
Each player is selfish and greedy, in the sense that he ranks the outcomes exclusively in terms of his net change in wealth (what he gets from the referee minus what he contributed).

(a) Represent this game by means of a set of tables. (Do not treat the referee as a player.)

(b) For each player and each pair of strategies determine if one of the two dominates the other and specify if it is weak or strict dominance.

(c) Is there a strictly dominant-strategy equilibrium? ■

2.9.3 Exercises for Section 2.3: Second price auction

The answers to the following exercises are in Section 2.10 at the end of this chapter.

Exercise 2.7 For the second-price auction partially illustrated in Figure 2.10 – reproduced below (recall that the numbers are the payoffs of Player 1 only) – complete the representation by adding the payoffs of Player 2, assuming that Player 2 assigns a value of \$50M to the field and, like Player 1, ranks the outcomes in terms of the net gain from the oil field (defined as profits minus the price paid, if Player 2 wins, and zero otherwise).

		Player 2				
		\$10*M*	\$20*M*	\$30*M*	\$40*M*	\$50*M*
	\$10*M*	0	0	0	0	0
Player	\$20*M*	20	0	0	0	0
1	\$30*M*	20	10	0	0	0
	\$40*M*	20	10	0	0	0
	\$50*M*	20	10	0	−10	0

■

Exercise 2.8 Consider the following "third-price" auction. There are $n \geq 3$ bidders. A single object is auctioned and Player i values the object $\$v_i$, with $v_i > 0$. The bids are simultaneous and secret.
The utility of Player i is: 0 if she does not win and $(v_i - p)$ if she wins and pays $\$p$.
Every non-negative number is an admissible bid. Let b_i denote the bid of Player i.
The winner is the highest bidder. In case of ties the bidder with the lowest index among those who submitted the highest bid wins (e.g. if the highest bid is \$120 and it is submitted by players 6, 12 and 15, then the winner is Player 6). The losers don't get anything and don't pay anything. The winner gets the object and pays the **third** highest bid, which is defined as follows.
Let i be the winner and fix a Player j such that

$$b_j = \max\left(\{b_1, ..., b_n\} \setminus \{b_i\}\right)$$

[note: if

$$\max\left(\{b_1, ..., b_n\} \setminus \{b_i\}\right)$$

contains more than one element, then we pick any one of them]. Then the third price is defined as

$$\max\left(\{b_1, ..., b_n\} \setminus \{b_i, b_j\}\right).$$

For example, if $n = 3$ and the bids are $b_1 = 30$, $b_2 = 40$ and $b_3 = 40$ then the winner is Player 2 and she pays \$30. If $b_1 = b_2 = b_3 = 50$ then the winner is Player 1 and she pays \$50. For simplicity, let us restrict attention to the case where $n = 3$ and $v_1 > v_2 > v_3 > 0$. Does Player 1 have a weakly dominant strategy in this auction? ■

2.9.4 Exercises for Section 2.4: The pivotal mechanism

The answers to the following exercises are in Section 2.10 at the end of this chapter.

Exercise 2.9 The pivotal mechanism is used to decide whether a new park should be built. There are 5 individuals. According to the proposed project, the cost of the park would be allocated as follows:

Individual	1	2	3	4	5
Share of cost	$c_1 = \$30$	$c_2 = \$25$	$c_3 = \$25$	$c_4 = \$15$	$c_5 = \$5$

For every individual $i = 1, \ldots, 5$, let v_i be the perceived gross benefit (if positive; perceived gross loss, if negative) from having the park built. The v_i's are as follows:

Individual	1	2	3	4	5
Gross benefit	$v_1 = \$60$	$v_2 = \$15$	$v_3 = \$55$	$v_4 = -\$25$	$v_5 = -\$20$

(Thus the net benefit (loss) to individual i is $v_i - c_i$). Individual i has the following utility of wealth function (where m_i denotes the wealth of individual i):

$$U_i(\$m_i) = \begin{cases} m_i & \text{if the project is not carried out} \\ m_i + v_i & \text{if the project is carried out} \end{cases}$$

Let $\overline{m}_i$ be the initial endowment of money of individual i and assume that $\overline{m}_i$ is large enough that it exceeds c_i plus any tax that the individual might have to pay.

(a) What is the Pareto-efficient decision: to build the park or not?

Assume that the pivotal mechanism is used, so that each individual i is asked to state a number w_i which is going to be interpreted as the gross benefit to individual i from carrying out the project. There are no restrictions on the number w_i: it can be positive, negative or zero. Suppose that the individuals make the following announcements:

Individual	1	2	3	4	5
Stated benefit	$w_1 = \$70$	$w_2 = \$10$	$w_3 = \$65$	$w_4 = -\$30$	$w_5 = -\$5$

(b) Would the park be built based on the above announcements?

(c) Using the above announcements and the rules of the pivotal mechanism, fill in the following table:

Individual	1	2	3	4	5
Pivotal?					
Tax					

(d) As you know, in the pivotal mechanism each individual has a dominant strategy. If all the individuals played their dominant strategies, would the park be built?

(e) Assuming that all the individuals play their dominant strategies, find out who is pivotal and what tax (if any) each individual has to pay?

(f) Show that if every other individual reports his/her true benefit, then it is best for Individual 1 to also report his/her true benefit. ■

2.9.5 Exercises for Section 2.5: Iterated deletion procedures

The answers to the following exercises are in Section 2.10 at the end of this chapter.

Exercise 2.10 Consider again the game of Part **(b)** of Exercise 2.1.

(a) Apply the IDSDS procedure (Iterated Deletion of Strictly Dominated Strategies).

(b) Apply the IDWDS procedure (Iterated Deletion of Weakly Dominated Strategies). ■

Exercise 2.11 Apply the IDSDS procedure to the game shown in Figure 2.24. Is there a strict iterated dominant-strategy equilibrium? ■

		Player 2					
		d		e		f	
Player 1	a	8	6	0	9	3	8
	b	3	2	2	1	4	3
	c	2	8	1	5	3	1

Figure 2.24: A strategic-form game with ordinal payoffs

Exercise 2.12 Consider the following game. There is a club with three members: Ann, Bob and Carla. They have to choose which of the three is going to be president next year. Currently, Ann is the president. Each member is both a candidate and a voter. Voting is as follows: each member votes for one candidate (voting for oneself is allowed); if two or more people vote for the same candidate then that person is chosen as the next president; if there is complete disagreement, in the sense that there is exactly one vote for each candidate, then the person for whom Ann voted is selected as the next president.

(a) Represent this voting procedure as a game frame, indicating inside each cell of each table which candidate is elected.

(b) Assume that the players' preferences are as follows: $Ann \succ_{Ann} Carla \succ_{Ann} Bob$, $Carla \succ_{Bob} Bob \succ_{Bob} Ann$, $Bob \succ_{Carla} Ann \succ_{Carla} Carla$. Using utility values 0, 1 and 2, convert the game frame into a game.

(c) Apply the IDWDS to the game of Part **(b)**. Is there an iterated weakly dominant-strategy equilibrium?

(d) Does the extra power given to Ann (in the form of tie-breaking in case of complete disagreement) benefit Ann?

■

Exercise 2.13 Consider the game shown in Figure 2.25.

(a) Apply the IDSDS procedure. Is there an iterated strictly dominant-strategy equilibrium?

(b) Apply the IDWDS procedure. Is there an iterated weakly dominant-strategy equilibrium?

■

		Player 2					
		D		E		F	
	a	2	3	2	2	3	1
Player 1	b	2	0	3	1	1	0
	c	1	4	2	0	0	4

Figure 2.25: A strategic-form game with ordinal payoffs

2.9.6 Exercises for Section 2.6: Nash equilibrium

The answers to the following exercises are in Section 2.10 at the end of this chapter.

Exercise 2.14 Find the Nash equilibria of the game of Exercise 2.2. ■

Exercise 2.15 Find the Nash equilibria of the games of Exercise 2.3 **(b)** and **(c)**. ■

Exercise 2.16 Find the Nash equilibria of the game of Exercise 2.4 **(b)**. ■

Exercise 2.17 Find the Nash equilibria of the game of Exercise 2.6. ■

Exercise 2.18 Find the Nash equilibria of the game of Exercise 2.7. ■

Exercise 2.19 Find a Nash equilibrium of the game of Exercise 2.8 for the case where

$$n = 3 \qquad \text{and} \qquad v_1 > v_2 > v_3 > 0$$

(there are several Nash equilibria: you don't need to find them all). ■

Exercise 2.20 Find the Nash equilibria of the game of Exercise 2.12 **(b)**. ■

Exercise 2.21 Find the Nash equilibria of the game of Exercise 2.13. ■

2.9.7 Exercises for Section 2.7: Games with infinite strategy sets

The answers to the following exercises are in Section 2.10 at the end of this chapter.

Exercise 2.22 Consider a simultaneous n-player game where each Player i chooses an effort level $a_i \in [0,1]$. The payoff to Player i is given by

$$\pi_i(a_1,\ldots,a_n) = 4\min\{a_1,\ldots,a_n\} - 2a_i$$

(interpretation: efforts are complementary and each player's cost per unit of effort is 2).

(a) Find all the Nash equilibria and prove that they are indeed Nash equilibria.
(b) Are any of the Nash equilibria Pareto efficient?
(c) Find a Nash equilibrium where each player gets a payoff of 1.

■

Exercise 2.23 — ★★★ Challenging Question ★★★.

The Mondevil Corporation operates a chemical plant, which is located on the banks of the Sacramento river. Downstream from the chemical plant is a group of fisheries. The Mondevil plant emits by-products that pollute the river, causing harm to the fisheries. The profit Mondevil obtains from operating the chemical plant is $\$\Pi > 0$.

The harm inflicted on the fisheries due to water pollution is equal to $\$L > 0$ of lost profit [without pollution the fisheries' profit is $\$A$, while with pollution it is $\$(A-L)$]. Suppose that the fisheries collectively sue the Mondevil Corporation. It is easily verified in court that Mondevil's plant pollutes the river. However, the values of Π and L cannot be verified by the court, although they are commonly known to the litigants.

Suppose that the court requires the Mondevil attorney (Player 1) and the fisheries' attorney (Player 2) to play the following litigation game. Player 1 is asked to announce a number $x \geq 0$, which the court interprets as a claim about the plant's profits. Player 2 is asked to announce a number $y \geq 0$, which the court interprets as the fisheries' claim about their profit loss. The announcements are made simultaneously and independently.

Then the court uses Posner's nuisance rule to make its decision (R. Posner, *Economic Analysis of Law*, 9th edition, 1997). According to the rule, if $y > x$, then Mondevil must shut down its chemical plant. If $x \geq y$, then the court allows Mondevil to operate the plant, but the court also requires Mondevil to pay the fisheries the amount y. Note that the court cannot force the attorneys to tell the truth: in fact, it would not be able to tell whether or not the lawyers were reporting truthfully. Assume that the attorneys want to maximize the payoff (profits) of their clients.

(a) Represent this situation as a strategic-form game by describing the strategy set of each player and the payoff functions.

(b) Is it a dominant strategy for the Mondevil attorney to make a truthful announcement (i.e. to choose $x = \Pi$)? [Prove your claim.]

(c) Is it a dominant strategy for the fisheries' attorney to make a truthful announcement (i.e. to choose $y = L$)? [Prove your claim.]

(d) For the case where $\Pi > L$ (recall that Π and L denote the true amounts), find **all** the Nash equilibria of the litigation game. [Prove that what you claim to be Nash equilibria are indeed Nash equilibria and that there are no other Nash equilibria.]

(e) For the case where $\Pi < L$ (recall that Π and L denote the true amounts), find **all** the Nash equilibria of the litigation game. [Prove that what you claim to be Nash equilibria are indeed Nash equilibria and that there are no other Nash equilibria.]

(f) Does the court rule give rise to a Pareto efficient outcome? [Assume that the players end up playing a Nash equilibrium.]

■

2.10 Solutions to exercises

Solution to Exercise 2.1.

(a) $I = \{1,2\}$, $S_1 = \{2,4,6\}$, $S_2 = \{1,3,5\}$, $O = \{M,I,J\}$
(where M stands for 'Mexican', I for 'Italian' and J for 'Japanese').
The set of strategy profiles is
$S = \{(2,1), (2,3), (2,5), (4,1), (4,3), (4,5), (6,1), (6,3), (6,5)\}$;
the outcome function is:
$f(2,1) = f(2,3) = f(4,1) = M$,
$f(2,5) = f(4,3) = f(6,1) = I$ and
$f(4,5) = f(6,3) = f(6,5) = J$.
The representation as a table is shown in Figure 2.26.

Player 2 (Bob)

Player 1 (Antonia)	1	3	5
2	M	M	I
4	M	I	J
6	I	J	J

Figure 2.26: The game-frame for Exercise 2.1 **(a)**

(b) Using values 1, 2 and 3, the utility functions are as follows, where U_1 is the utility function of Player 1 (Antonia) and U_2 is the utility function of Player 2 (Bob):

$$\begin{pmatrix} & M & I & J \\ U_1: & 3 & 2 & 1 \\ U_2: & 2 & 3 & 1 \end{pmatrix}$$

The reduced-form game is shown in Figure 2.27. □

Player 2 (Bob)

Player 1 (Antonia)	1	3	5
2	3, 2	3, 2	2, 3
4	3, 2	2, 3	1, 1
6	2, 3	1, 1	1, 1

Figure 2.27: The game for Exercise 2.1 **(b)**

Solution to Exercise 2.2.

(a) The game-frame is shown in Figure 2.28.

(b) When the outcomes are sums of money and Player i is selfish and greedy, we can take the following as Player i's utility function: $U_i(\$x) = x$ (other utility functions would do too: the only requirement is that the utility of a larger sum of money is larger than the utility of a smaller sum of money). Thus the reduced-form game is shown in Figure 2.29. □

		Player 2 (Bob)		
		0	1	2
Player 1 (Antonia)	2	Antonia gets nothing Bob gets $2	Antonia gets nothing Bob gets $3	Antonia gets nothing Bob gets $2
	4	Antonia gets nothing Bob gets $2	Antonia gets $5 Bob gets $5	Antonia gets $4 Bob gets $2
	6	Antonia gets $4 Bob gets $2	Antonia gets $3 Bob gets $7	Antonia gets $2 Bob gets nothing

Figure 2.28: The game-frame for Exercise 2.2 **(a)**

		Player 2 (Bob)					
		0		1		2	
Player 1 (Antonia)	2	0	2	0	3	0	2
	4	0	2	5	5	4	2
	6	4	2	3	7	2	0

Figure 2.29: The game for Exercise 2.2 **(b)**

Solution to Exercise 2.3.

(a) The game-frame is shown in Figure 2.30.

(b) The reduced-form game is shown in Figure 2.31.

(c) The reduced-form game is shown in Figure 2.32. For Alice we chose 1 and 0 as utilities, but one could also use 2 and 1 or 2 and 0. □

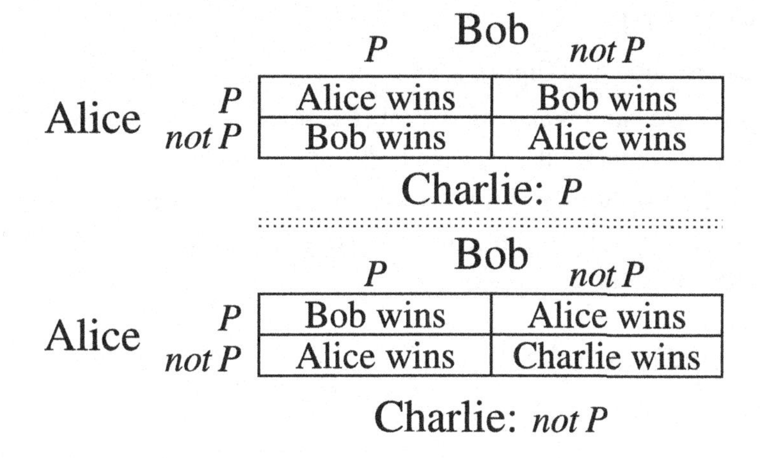

Charlie: *P*

Alice \ Bob	*P*	*not P*
P	Alice wins	Bob wins
not P	Bob wins	Alice wins

Charlie: *not P*

Alice \ Bob	*P*	*not P*
P	Bob wins	Alice wins
not P	Alice wins	Charlie wins

Figure 2.30: **The game-frame for Exercise 2.3 (a)**

Charlie: *P*

Alice \ Bob	*P*			*not P*		
P	1	0	0	0	1	0
not P	0	1	0	1	0	0

Charlie: *not P*

Alice \ Bob	*P*			*not P*		
P	0	1	0	1	0	0
not P	1	0	0	0	0	1

Figure 2.31: **The reduced-form game for Exercise 2.3 (b)**

Charlie: *P*

Alice \ Bob	*P*			*not P*		
P	1	0	0	0	2	1
not P	0	2	1	1	0	0

Charlie: *not P*

Alice \ Bob	*P*			*not P*		
P	0	2	1	1	0	0
not P	1	0	0	0	1	2

Figure 2.32: **The reduced-form game for Exercise 2.3 (c)**

Solution to Exercise 2.4.

(a) The tables are shown in Figure 2.33.

(b) For Player 1 we can take as his payoff the total money lost by the referee and for Player 2 her own net gain as shown in Figure 2.34.

(c) For Player 1 contributing \$600 is a strictly dominant strategy and for Player 2 contributing \$0 is a strictly dominant strategy. Thus (\$600,\$0) is the strictly dominant-strategy equilibrium. □

Distributed money:

		Player 2: 0		Player 2: 300		Player 2: 600	
Player 1	0	0	0	225	225	450	450
	300	225	225	450	450	675	675
	600	450	450	675	675	900	900

Net amounts:

		Player 2: 0		Player 2: 300		Player 2: 600	
Player 1	0	0	0	225	−75	450	−150
	300	−75	225	150	150	375	75
	600	−150	450	75	375	300	300

Figure 2.33: The tables for Exercise 2.4 **(a)**

		Player 2: 0		Player 2: 300		Player 2: 600	
Player 1	0	0	0	150	−75	300	−150
	300	150	225	300	150	450	75
	600	300	450	450	375	600	300

Figure 2.34: The game for Exercise 2.4 **(b)**

Solution to Exercise 2.5. The game under consideration is shown in Figure 2.35.

(a) For Player 1, 6 is strictly dominated by 2. There is no other strategy which is strictly dominated. Player 2 does not have any strictly dominated strategies.

(b) For Player 1, 6 is weakly dominated by 4 (and also by 2, since strict dominance implies weak dominance); 4 is weakly dominated by 2. Player 2 does not have any weakly dominated strategies. □

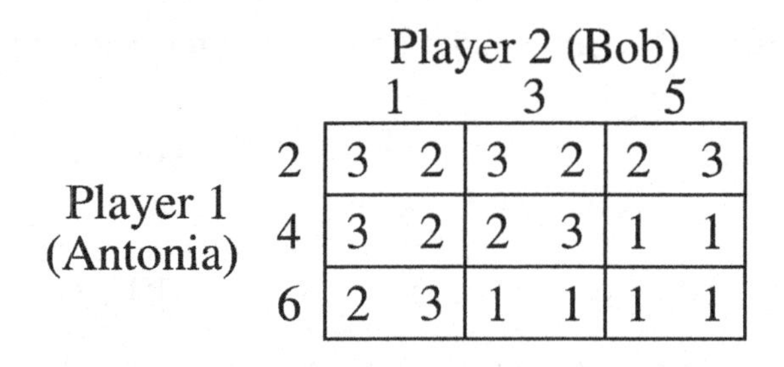

Player 1 (Antonia) \ Player 2 (Bob)	1	3	5
2	3 2	3 2	2 3
4	3 2	2 3	1 1
6	2 3	1 1	1 1

Figure 2.35: The game for Exercise 2.5

Solution to Exercise 2.6.

(a) The game under consideration is shown in Figure 2.36.

(b) For Player 1, 0 strictly dominates 3 and 6, 3 strictly dominates 6 (the same is true for every player). Thus 0 is a strictly dominant strategy.

(c) The strictly dominant-strategy equilibrium is (0,0,0) (everybody contributes 0). □

Player 1 \ Player 2	0	3	6
0	0 0 0	2 −1 2	4 −2 4
3	−1 2 2	1 1 4	3 0 6
6	−2 4 4	0 3 6	2 2 8

Player 3: 0

Player 1 \ Player 2	0	3	6
0	2 2 −1	4 1 1	6 0 3
3	1 4 1	3 3 3	5 2 5
6	0 6 3	2 5 5	4 4 7

Player 3: 3

Player 1 \ Player 2	0	3	6
0	4 4 −2	6 3 0	8 2 2
3	3 6 0	5 5 2	7 4 4
6	2 8 2	4 7 4	6 6 6

Player 3: 6

Figure 2.36: The game for Exercise 2.6

Solution to Exercise 2.7. The game under consideration is shown in Figure 2.37. □

		Player 2 (value \$50M)				
		\$10M	\$20M	\$30M	\$40M	\$50M
	\$10M	0 40	0 40	0 40	0 40	0 40
Player 1	\$20M	20 0	0 30	0 30	0 30	0 30
(value \$30M)	\$30M	20 0	10 0	0 20	0 20	0 20
	\$40M	20 0	10 0	0 0	0 10	0 10
	\$50M	20 0	10 0	0 0	−10 0	0 0

Figure 2.37: The game for Exercise 2.7

Solution to Exercise 2.8. No. Suppose, by contradiction, that $\hat{b}_1$ is a weakly dominant strategy for Player 1. It cannot be that $\hat{b}_1 > v_1$, because when $b_2 = b_3 = \hat{b}_1$ Player 1 wins and pays $\hat{b}_1$, thereby obtaining a payoff of $v_1 - \hat{b}_1 < 0$, whereas bidding 0 would give him a payoff of 0.
It cannot be that $\hat{b}_1 = v_1$ because when $b_2 > \hat{b}_1$ and $b_3 < v_1$ the auction is won by Player 2 and Player 1 gets a payoff of 0, while a bid of Player 1 greater than b_2 would make him the winner with a payoff of $v_1 - b_3 > 0$.
Similarly, it cannot be that $\hat{b}_1 < v_1$ because when $b_2 > v_1$ and $b_3 < v_1$ then with $\hat{b}_1$ the auction is won by Player 2 and Player 1 gets a payoff of 0, while a bid greater than b_2 would make him the winner with a payoff of $v_1 - b_3 > 0$. □

Solution to Exercise 2.9.

(a) Since $\sum_{i=1}^{5} v_i = 85 < \sum_{i=1}^{5} c_i = 100$ the Pareto efficient decision is not to build the park.

(b) Since $\sum_{i=1}^{5} w_i = 110 > \sum_{i=1}^{5} c_i = 100$ the park would be built.

(c) Individuals 1 and 3 are pivotal and each of them has to pay a tax of \$30. The other individuals are not pivotal and thus are not taxed.

(d) For each individual i it is a dominant strategy to report v_i and thus, by Part **(a)**, the decision will be the Pareto efficient one, namely not to build the park.

(e) When every individual reports truthfully, Individuals 4 and 5 are pivotal and Individual 4 has to pay a tax of \$25, while individual 5 has to pay a tax of \$10. The others are not pivotal and do not have to pay a tax.

(f) Assume that all the other individuals report truthfully; then if Individual 1 reports truthfully, he is not pivotal, the project is not carried out and his utility is $\overline{m}_1$. Any other w_1 that leads to the same decision (not to build the park) gives him the same utility.
If, on the other hand, he chooses a w_1 that leads to a decision to build the park, then Individual 1 will become pivotal and will have to pay a tax $t_1 = 45$ with a utility

of $\overline{m}_1 + v_1 - c_1 - t_1 = \overline{m}_1 + 60 - 30 - 45 = \overline{m}_1 - 15$, so that he would be worse off relative to reporting truthfully. □

Solution to Exercise 2.10. The game under consideration is shown in Figure 2.38.

		Player 2 (Bob)		
		1	3	5
Player 1 (Antonia)	2	3, 2	3, 2	2, 3
	4	3, 2	2, 3	1, 1
	6	2, 3	1, 1	1, 1

Figure 2.38: The game for Exercise 2.10

(a) The first step of the procedure eliminates 6 for Player 1. After this step the procedure stops and thus the output is as shown in Figure 2.39.

		Player 2 (Bob)		
		1	3	5
Player 1 (Antonia)	2	3, 2	3, 2	2, 3
	4	3, 2	2, 3	1, 1

Figure 2.39: The output of the IDWDS procedure applied to the game of Figure 2.38

(b) The first step of the procedure eliminates 4 and 6 for Player 1 and nothing for Player 2. The second step of the procedure eliminates 1 and 3 for Player 2. Thus the output is the strategy profile (2,5), which constitutes the iterated weakly dominant-strategy equilibrium of this game. □

Solution to Exercise 2.11. The game under consideration is shown in Figure 2.40.
In this game c is strictly dominated by b;
- after deleting c, d becomes strictly dominated by f;
- after deleting d, a becomes strictly dominated by b;
- after deleting a, e becomes strictly dominated by f;
- deletion of e leads to only one strategy profile, namely (b, f).

Thus (b, f) is the iterated strictly dominant-strategy equilibrium. □

		Player 2: d		Player 2: e		Player 2: f	
Player 1	a	8	6	0	9	3	8
	b	3	2	2	1	4	3
	c	2	8	1	5	3	1

Figure 2.40: The game for Exercise 2.11

Solution to Exercise 2.12.

(a) The game-frame under consideration is shown in Figure 2.41.

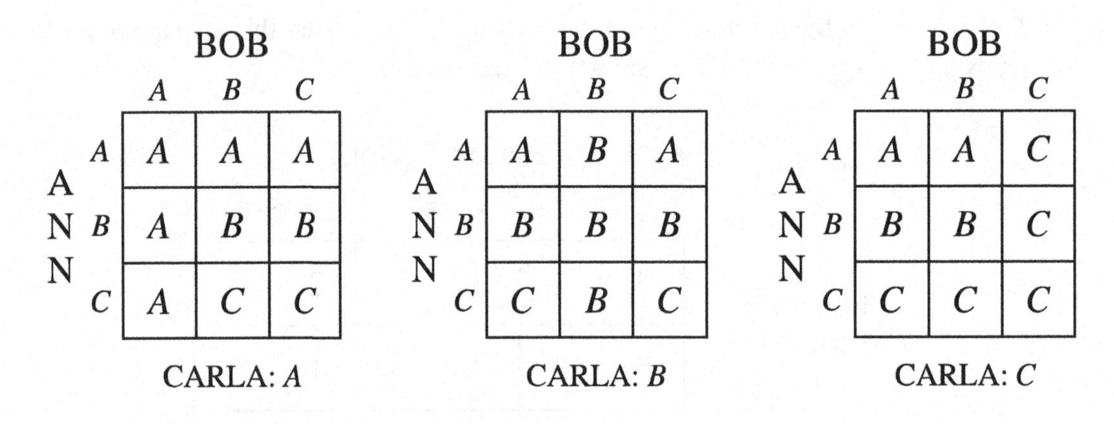

CARLA: A

ANN \ BOB	A	B	C
A	A	A	A
B	A	B	B
C	A	C	C

CARLA: B

ANN \ BOB	A	B	C
A	A	B	A
B	B	B	B
C	C	B	C

CARLA: C

ANN \ BOB	A	B	C
A	A	A	C
B	B	B	C
C	C	C	C

Figure 2.41: The game-frame for Exercise 2.12 **(a)**

(b) The game under consideration is shown in Figure 2.42.

(c) For Ann, both B and C are weakly dominated by A; for Bob, A is weakly dominated by C; for Carla, C is weakly dominated by B.
Thus in the first step of the IDWDS we delete B and C for Ann, A for Bob and C for Carla.
Hence the game reduces to the Figure 2.43. In this game, for Bob, C is weakly dominated by B and for Carla, A is weakly dominated by B.
Thus in the second and final step of the IDWDS we delete C for Bob and A for Carla and we are left with a unique strategy profile, namely (A,B,B), that is, Ann votes for herself and Bob and Carla vote for Bob. This is the iterated weakly dominant-strategy equilibrium.

(d) The elected candidate is Bob, who is Ann's least favorite; thus the extra power given to Ann (tie breaking in case of total disagreement) turns out to be detrimental for Ann! □

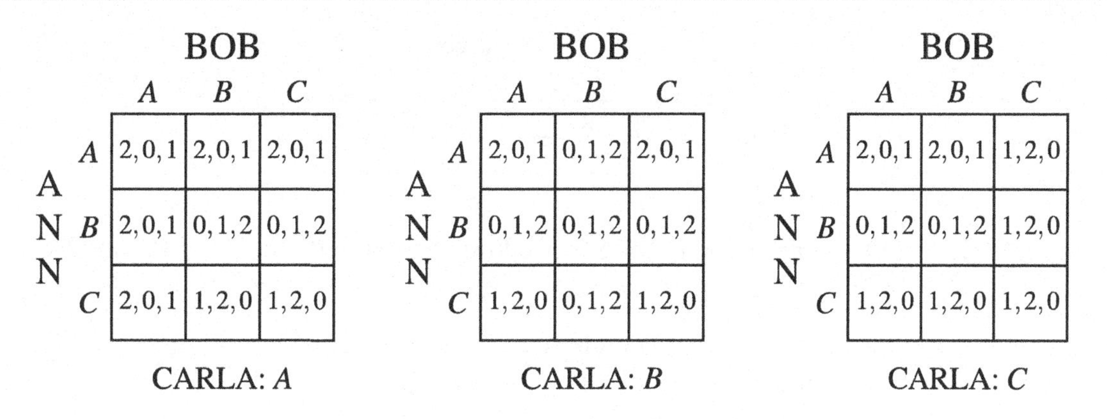

CARLA: *A*

ANN \ BOB	*A*	*B*	*C*
A	2,0,1	2,0,1	2,0,1
B	2,0,1	0,1,2	0,1,2
C	2,0,1	1,2,0	1,2,0

CARLA: *B*

ANN \ BOB	*A*	*B*	*C*
A	2,0,1	0,1,2	2,0,1
B	0,1,2	0,1,2	0,1,2
C	1,2,0	0,1,2	1,2,0

CARLA: *C*

ANN \ BOB	*A*	*B*	*C*
A	2,0,1	2,0,1	1,2,0
B	0,1,2	0,1,2	1,2,0
C	1,2,0	1,2,0	1,2,0

Figure 2.42: The game for Exercise 2.12 **(b)**

		CARLA	
		A	*B*
BOB	*B*	0 1	1 2
	C	0 1	0 1

Figure 2.43: The reduced game for Exercise 2.12 **(c)**

Solution to Exercise 2.13. The game under consideration is shown in Figure 2.44.

		Player 2		
		D	*E*	*F*
Player 1	*a*	2 3	2 2	3 1
	b	2 0	3 1	1 0
	c	1 4	2 0	0 4

Figure 2.44: The game for Exercise 2.13

(a) The output of the IDSDS is shown in Figure 2.45 (first delete c and then F). Thus there is no iterated strictly dominant-strategy equilibrium.

(b) The output of the IDWDS is (b,E) (in the first step delete c and F, the latter because it is weakly dominated by D; in the second step delete a and in the third step delete D). Thus (b,E) is the iterated weakly dominant-strategy equilibrium. □

		Player 2: D		Player 2: E	
Player 1	a	2	3	2	2
	b	2	0	3	1

Figure 2.45: The output of the IDSDS procedure applied to the game of Figure 2.44

Solution to Exercise 2.14. The game under consideration is shown in Figure 2.46. There is only one Nash equilibrium, namely (4,1) with payoffs (5,5). □

		Player 2 (Bob): 0		1		2	
Player 1 (Antonia)	2	0	2	0	3	0	2
	4	0	2	5	5	4	2
	6	4	2	3	7	2	0

Figure 2.46: The game for Exercise 2.14

Solution to Exercise 2.15. The game of Exercise 2.3 **(b)** is shown in Figure 2.47. This game has only one Nash equilibrium, namely *(not P, P, not P)*. The game of Exercise 2.3

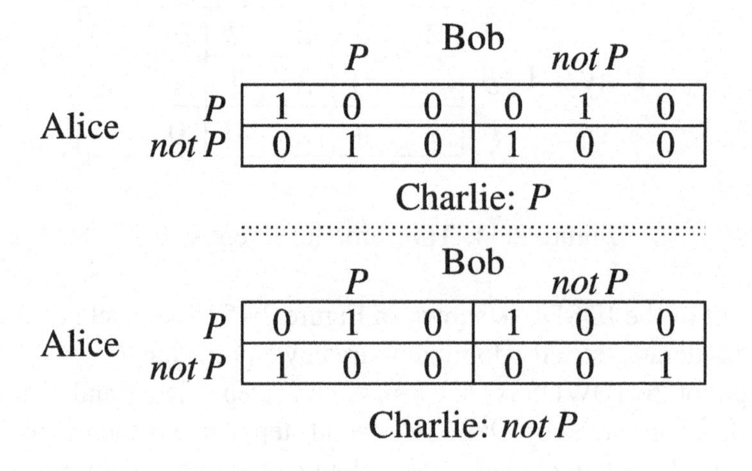

		Bob: *P*			*not P*		
Alice	*P*	1	0	0	0	1	0
	not P	0	1	0	1	0	0

Charlie: *P*

		Bob: *P*			*not P*		
Alice	*P*	0	1	0	1	0	0
	not P	1	0	0	0	0	1

Charlie: *not P*

Figure 2.47: The first game for Exercise 2.15

(c) is shown in Figure 2.48. This game does not have any Nash equilibria. □

		Bob					
		P			*not P*		
Alice	*P*	1	0	0	0	2	1
	not P	0	2	1	1	0	0

Charlie: *P*

		Bob					
		P			*not P*		
Alice	*P*	0	2	1	1	0	0
	not P	1	0	0	0	1	2

Charlie: *not P*

Figure 2.48: The second game for Exercise 2.15

Solution to Exercise 2.16. The game under consideration is shown in Figure 2.49. This game has only one Nash equilibrium, namely (600,0). □

		Player 2					
		0		300		600	
Player 1	0	0	0	150	−75	300	−150
	300	150	225	300	150	450	75
	600	300	450	450	375	600	300

Figure 2.49: The game for Exercise 2.16

Solution to Exercise 2.17. The game under consideration is shown in Figure 2.50. This game has only one Nash equilibrium, namely (0,0,0). □

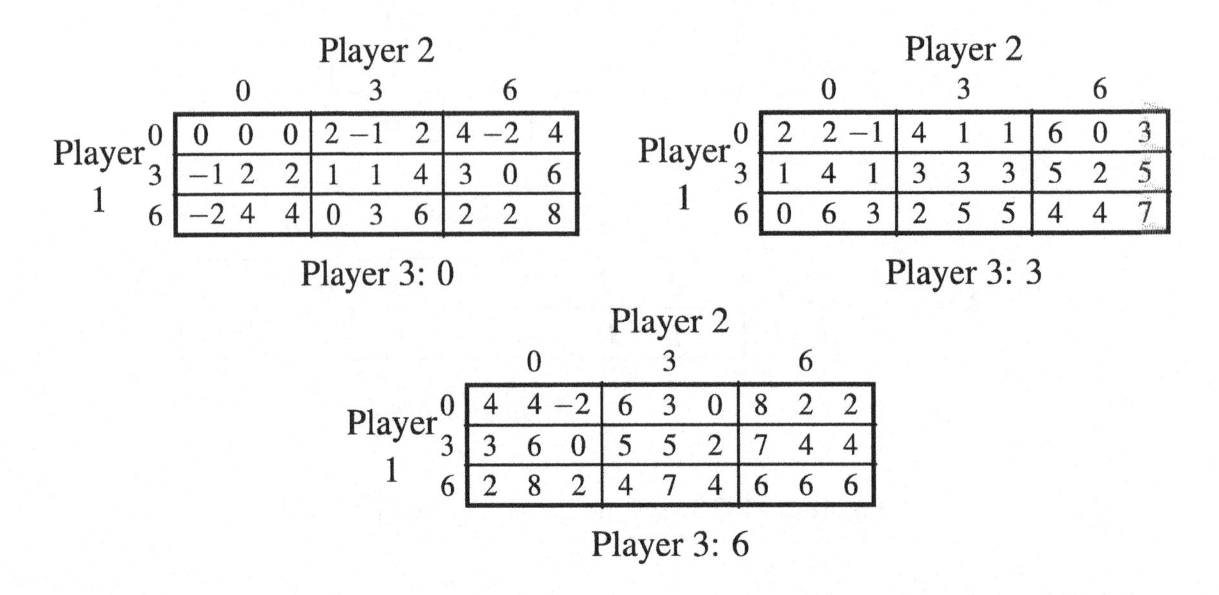

Player 3: 0

Player 1 \ Player 2	0	3	6
0	0 0 0	2 −1 2	4 −2 4
3	−1 2 2	1 1 4	3 0 6
6	−2 4 4	0 3 6	2 2 8

Player 3: 3

Player 1 \ Player 2	0	3	6
0	2 2 −1	4 1 1	6 0 3
3	1 4 1	3 3 3	5 2 5
6	0 6 3	2 5 5	4 4 7

Player 3: 6

Player 1 \ Player 2	0	3	6
0	4 4 −2	6 3 0	8 2 2
3	3 6 0	5 5 2	7 4 4
6	2 8 2	4 7 4	6 6 6

Figure 2.50: The game for Exercise 2.17

Solution to Exercise 2.18. The game under consideration is shown in Figure 2.51. This game has 15 Nash equilibria:
(10,30), (10,40), (10,50), (20,30), (20,40), (20,50), (30,30), (30,40), (30,50), (40,40), (40,50), (50,10), (50,20), (50,30), (50,50). □

Player 1 (value \$30M) \ Player 2 (value \$50M)	\$10M	\$20M	\$30M	\$40M	\$50M
\$10M	0 40	0 40	0 40	0 40	0 40
\$20M	20 0	0 30	0 30	0 30	0 30
\$30M	20 0	10 0	0 20	0 20	0 20
\$40M	20 0	10 0	0 0	0 10	0 10
\$50M	20 0	10 0	0 0	−10 0	0 0

Figure 2.51: The game for Exercise 2.18

Solution to Exercise 2.19. A Nash equilibrium is $b_1 = b_2 = b_3 = v_1$ (with payoffs (0,0,0)). Convince yourself that this is indeed a Nash equilibrium.
There are many more Nash equilibria: for example, any triple (b_1, b_2, b_3) with $b_2 = b_3 = v_1$ and $b_1 > v_1$ is a Nash equilibrium (with payoffs (0,0,0)) and so is any triple (b_1, b_2, b_3) with $b_2 = b_3 = v_2$ and $b_1 \geq v_2$ (with payoffs $(v_1 - v_2, 0, 0)$). □

Solution to Exercise 2.20. The game under consideration is shown in Figure 2.52. There are 5 Nash equilibria: $(A,A,A), (B,B,B), (C,C,C), (A,C,A)$ and (A,B,B). □

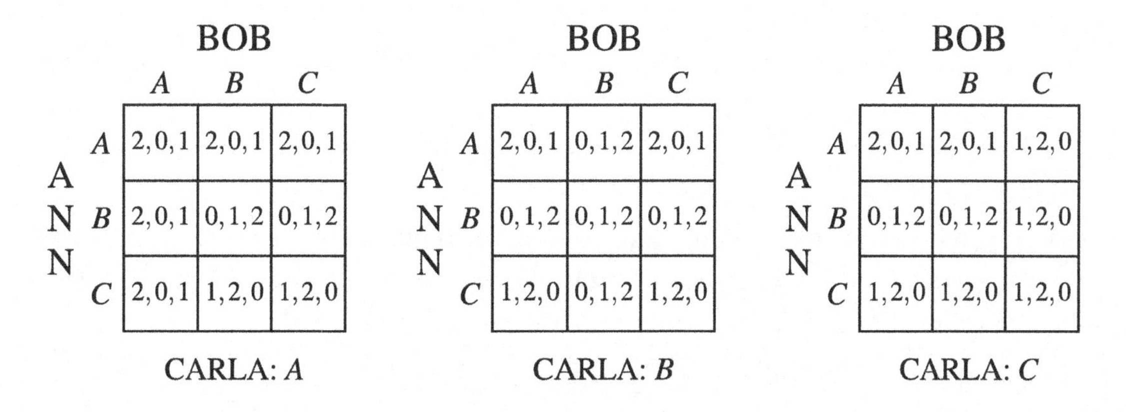

CARLA: *A*

ANN \ BOB	*A*	*B*	*C*
A	2,0,1	2,0,1	2,0,1
B	2,0,1	0,1,2	0,1,2
C	2,0,1	1,2,0	1,2,0

CARLA: *B*

ANN \ BOB	*A*	*B*	*C*
A	2,0,1	0,1,2	2,0,1
B	0,1,2	0,1,2	0,1,2
C	1,2,0	0,1,2	1,2,0

CARLA: *C*

ANN \ BOB	*A*	*B*	*C*
A	2,0,1	2,0,1	1,2,0
B	0,1,2	0,1,2	1,2,0
C	1,2,0	1,2,0	1,2,0

Figure 2.52: The game for Exercise 2.20

Solution to Exercise 2.21. The game under consideration is shown in Figure 2.53. There are 2 Nash equilibria: (a, D) and (b, E). □

Player 1 \ Player 2	*D*		*E*		*F*	
a	2	3	2	2	3	1
b	2	0	3	1	1	0
c	1	4	2	0	0	4

Figure 2.53: The game for Exercise 2.21

Solution to Exercise 2.22.

(a) For every $e \in [0,1]$, $(e,e,...,e)$ is a Nash equilibrium.
The payoff of Player i is $\pi_i(e,e,...,e) = 2e$; if player i increases her effort to $a > e$ (of course, this can only happen if $e < 1$), then her payoff decreases to $4e - 2a$ and if she decreases her effort to $a < e$ (of course, this can only happen if $e > 0$), then her payoff decreases to $2a$.
There is no Nash equilibrium where two players choose different levels of effort.
Proof: suppose there is an equilibrium $(a_1, a_2, \ldots, a_n)$ where $a_i \neq a_j$ for two players i and j.
Let $a_{\min} = \min\{a_1, \ldots, a_n\}$ and let k be a player such that $a_k > a_{\min}$ (such a player exists by our supposition).
Then the payoff to player k is $\pi_k = 4a_{\min} - 2a_k$ and if she reduced her effort to $a_{\min}$ her payoff would increase to $2a_{\min}$.

(b) Any symmetric equilibrium with $e < 1$ is Pareto inefficient, because all the players would be better off if they collectively switched to $(1,1,\ldots,1)$. On the other hand, the symmetric equilibrium $(1,1,\ldots,1)$ is Pareto efficient.

(c) The symmetric equilibrium $\left(\frac{1}{2}, \frac{1}{2}, ..., \frac{1}{2}\right)$. □

Solution to Exercise 2.23.

(a) The strategy sets are $S_1 = S_2 = [0, \infty)$. The payoff functions are as follows:[10]

$$\pi_1(x,y) = \begin{cases} \Pi - y & \text{if } x \geq y \\ 0 & \text{if } y > x \end{cases} \quad \text{and} \quad \pi_2(x,y) = \begin{cases} A - L + y & \text{if } x \geq y \\ A & \text{if } y > x \end{cases}$$

(b) Yes, for player 1 choosing $x = \Pi$ is a weakly dominant strategy.
Proof. Consider an arbitrary y. We must show that $x = \Pi$ gives at least a high a payoff against y as any other x. Three cases are possible.

Case 1: $y < \Pi$. In this case $x = \Pi$ or any other x such that $x \geq y$ yields $\pi_1 = \Pi - y > 0$, while any $x < y$ yields $\pi_1 = 0$.

Case 2: $y = \Pi$. In this case 1's payoff is zero no matter what x he chooses.

Case 3: $y > \Pi$. In this case $x = \Pi$ or any other x such that $x < y$ yields $\pi_1 = 0$, while any $x \geq y$ yields $\pi_1 = \Pi - y < 0$.

(c) No, choosing $y = L$ is not a dominant strategy for Player 2. For example, if $x > L$ then choosing $y = L$ yields $\pi_2 = A$ while choosing a y such that $L < y \leq x$ yields $\pi_2 = A - L + y > A$.

---

[10] We have chosen to use accounting profits as payoffs. Alternatively, one could take as payoffs the changes in profits relative to the initial situation, namely

$$\pi_1(x,y) = \begin{cases} -y & \text{if } x \geq y \\ -\Pi & \text{if } y > x \end{cases} \quad \text{and} \quad \pi_2(x,y) = \begin{cases} y & \text{if } x \geq y \\ L & \text{if } y > x. \end{cases}$$

The answers are the same, whichever choice of payoffs one makes.

(d) **Suppose that** $\Pi > L$. If (x,y) is a Nash equilibrium **with** $x \geq y$ then it must be that $y \leq \Pi$ (otherwise Player 1 could increase its payoff by reducing x below y) and $y \geq L$ (otherwise Player 2 would be better off by increasing y above x).
Thus it must be $L \leq y \leq \Pi$, which is possible, given our assumption.
However, it cannot be that $x > y$, because Player 2 would be getting a higher payoff by increasing y to x.
Thus it must be $x \leq y$, which (together with our hypothesis that $x \geq y$) implies that $x = y$. Thus the following are Nash equilibria:

all the pairs (x,y) with $L \leq y \leq \Pi$ and $x = y$.

Now consider pairs (x,y) **with** $x < y$. Then it cannot be that $y < \Pi$, because Player 1 could increase its payoff by increasing x to y. Thus it must be $y \geq \Pi$ (hence, by our supposition that $\Pi > L$, $y > L$). Furthermore, it must be that $x \leq L$ (otherwise Player 2 could increase its profits by reducing y to (or below) x. Thus

(x,y) with $x < y$ is a Nash equilibrium if and only if $x \leq L$ and $y \geq \Pi$.

(e) **Suppose that** $\Pi < L$. For the same reasons given above, an equilibrium with $x \geq y$ requires $L \leq y \leq \Pi$. However, this is not possible given that $\Pi < L$. Hence,

there is no Nash equilibrium (x,y) with $x \geq y$.

Thus we must restrict attention to pairs (x,y) **with** $x < y$. As explained above, it must be that $y \geq \Pi$ and $x \leq L$. Thus,

(x,y) with $x < y$ is a Nash equilibrium if and only if $\Pi \leq y$ and $x \leq L$.

(f) Pareto efficiency requires that the chemical plant be shut down if $\Pi < L$ and that it remain operational if $\Pi > L$.
Now, when $\Pi < L$ all the equilibria have $x < y$ which leads to shut-down, hence a Pareto efficient outcome.
When $\Pi > L$, there are two types of equilibria: one where $x = y$ and the plant remains operational (a Pareto efficient outcome) and the other where $x < y$ in which case the plant shuts down, yielding a Pareto inefficient outcome. □

3. Perfect-information Games

3.1 Trees, frames and games

Often interactions are not simultaneous but sequential. For example, in the game of Chess the two players, White and Black, take turns moving pieces on the board, having full knowledge of the opponent's (and their own) past moves. Games with sequential interaction are called *dynamic games* or *games in extensive form*. This chapter is devoted to the subclass of dynamic games characterized by *perfect information*, namely the property that, whenever it is her turn to move, a player knows all the preceding moves.

Perfect-information games are represented by means of rooted directed trees.

Definition 3.1.1 A *rooted directed tree* consists of a set of nodes and directed edges joining them.

- The root of the tree has no directed edges leading to it (has indegree 0), while every other node has exactly one directed edge leading to it (has indegree 1).
- There is a unique path (that is, a unique sequence of directed edges) leading from the root to any other node. A node that has no directed edges out of it (has outdegree 0) is called a *terminal node*, while every other node is called a *decision node*.
- We shall denote the set of nodes by X, the set of decision nodes by D and the set of terminal nodes by Z. Thus $X = D \cup Z$.

Definition 3.1.2 A *finite extensive form (or frame) with perfect information* consists of the following items.

- A finite rooted directed tree.
- A set of players $I = \{1, \ldots, n\}$ and a function that assigns one player to every decision node.
- A set of actions A and a function that assigns one action to every directed edge, satisfying the restriction that no two edges out of the same node are assigned the same action.
- A set of outcomes O and a function that assigns an outcome to every terminal node.

■ **Example 3.1** Amy (Player 1) and Beth (Player 2) have decided to dissolve a business partnership whose assets have been valued at $100,000. The charter of the partnership prescribes that the senior partner, Amy, make an offer concerning the division of the assets to the junior partner, Beth. The junior partner can *Accept*, in which case the proposed division is implemented, or *Reject*, in which case the case goes to litigation.

- Litigating involves a cost of $20,000 in legal fees for each partner and the typical verdict assigns 60% of the assets to the senior partner and the remaining 40% to the junior partner.
- Suppose, for simplicity, that there is no uncertainty about the verdict (how to model uncertainty will be discussed in a later chapter). Suppose also that there are only two possible offers that Amy can make: a 50-50 split or a 70-30 split.

This situation can be represented as a finite extensive form with perfect information as shown in Figure 3.1. Each outcome is represented as two sums of money: the top one is what Player 1 gets and the bottom one what Player 2 gets. ■

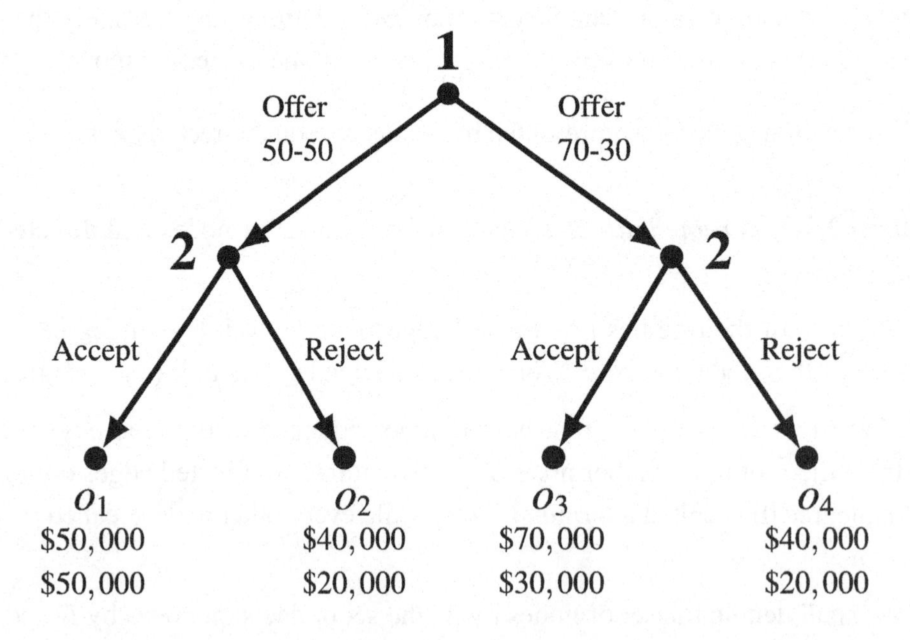

Figure 3.1: A perfect-information extensive form representing the situation described in Example 3.1

What should we expect the players to do in the game of Figure 3.1? Consider the following reasoning, which is called *backward induction* reasoning, because it starts from the end of the game and proceeds backwards towards the root:

- If Player 2 (the junior partner) is offered a 50-50 split then, if she accepts, she will get $50,000, while, if she rejects, she will get $20,000 (the court-assigned 40% minus legal fees in the amount of $20,000); thus, if rational, she will accept.
- Similarly, if Player 2 is offered a 70-30 split then, if she accepts, she will get $30,000, while, if she rejects, she will get $20,000 (the court-assigned 40% minus legal fees in the amount of $20,000); thus, if rational, she will accept.
- Anticipating all of this, Player 1 realizes that, if she offers a 50-50 split then she will end up with $50,000, while if she offers a 70-30 split then she will end up with $70,000; thus, if Player 1 is rational and believes that Player 2 is rational, she will offer a 70-30 split and Player 2, being rational, will accept.

The above reasoning suffers from the same flaw as the reasoning described in Chapter 2: it is not a valid argument because it is based on an implicit assumption about how Player 2 ranks the outcomes, which may or may not be correct. For example, Player 2 may feel that she worked as hard as her senior partner and the only fair division is a 50-50 split; indeed she may feel so strongly about this that – if offered an unfair 70-30 split – she would be willing to sacrifice $10,000 in order to "teach a lesson to Player 1"; in other words, she ranks outcome o_4 above outcome o_3.

Using the terminology introduced in Chapter 2, we say that the situation represented in Figure 3.1 is not a game but a *game-frame*. In order to convert that frame into a game we need to add a ranking of the outcomes for each player.

Definition 3.1.3 A *finite extensive game with perfect information* is a finite extensive form with perfect information together with a ranking $\succsim_i$ of the set of outcomes O, for every player $i \in I$.

As usual, it is convenient to represent the ranking of Player i by means of an ordinal utility function $U_i : O \to \mathbb{R}$. For example, take the extensive form of Figure 3.1 and assume that Player 1 is selfish and greedy, that is, her ranking is:

best o_3
o_1 (or, in the alternative notation, $o_3 \succ_1 o_1 \succ_1 o_2 \sim_1 o_4$).
worst o_2, o_4

while Player 2 is concerned with fairness and her ranking is:

best o_1
o_2, o_4 (or, in the alternative notation, $o_1 \succ_2 o_2 \sim_2 o_4 \succ_2 o_3$)
worst o_3

Then we can represent the players' preferences using the following utility functions:

outcome → utility function ↓	o_1	o_2	o_3	o_4
U_1 (Player 1)	2	1	3	1
U_2 (Player 2)	3	2	1	2

and replace each outcome in Figure 3.1 with a pair of utilities or payoffs, as shown in Figure 3.2, thereby obtaining one of the many possible games based on the frame of Figure 3.1.

Now that we have a game (rather than just a game-frame), we can apply the backward-induction reasoning and conclude that Player 1 will offer a 50-50 split, anticipating that Player 2 would reject the offer of a 70-30 split, and Player 2 will accept Player 1's 50-50 offer. The choices selected by the backward-induction reasoning have been highlighted in Figure 3.2 by doubling the corresponding edges.

Test your understanding of the concepts introduced in this section, by going through the exercises in Section 3.6.1 at the end of this chapter.

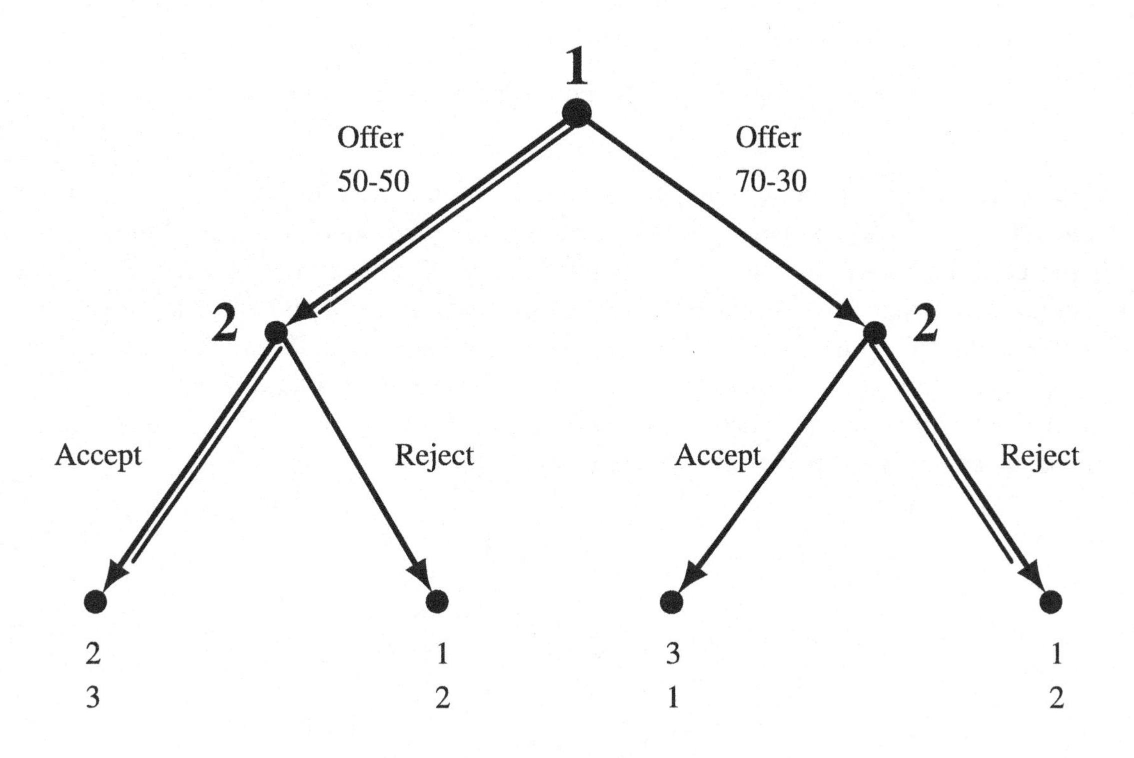

Figure 3.2: A perfect-information game based on the frame of Figure 3.1

3.2 Backward induction

The backward-induction reasoning mentioned above can be formalized as an algorithm for solving any finite perfect-information game. We say that a node is *marked* if a utility vector is associated with it. Initially all and only the terminal nodes are marked; the following procedure provides a way of marking all the nodes.

Definition 3.2.1 The *backward-induction algorithm* is the following procedure for solving a finite perfect-information game:

1. Select a decision node x whose immediate successors are all marked. Let i be the player who moves at x. Select a choice that leads to an immediate successor of x with the highest payoff (or utility) for Player i (highest among the utilities associated with the immediate successors of x). Mark x with the payoff vector associated with the node that follows the selected choice.
2. Repeat the above step until all the nodes have been marked.

Note that, since the game is finite, the above procedure is well defined. In the initial steps one starts at those decision nodes that are followed only by terminal nodes, call them penultimate nodes. After all the penultimate nodes have been marked, there will be unmarked nodes whose immediate successors are all marked and thus the step can be repeated. Note also that, in general, at a decision node there may be several choices that maximize the payoff of the player who moves at that node. If that is the case, then the procedure requires that *one* such choice be selected. This arbitrary selection may lead to the existence of several backward-induction solutions.

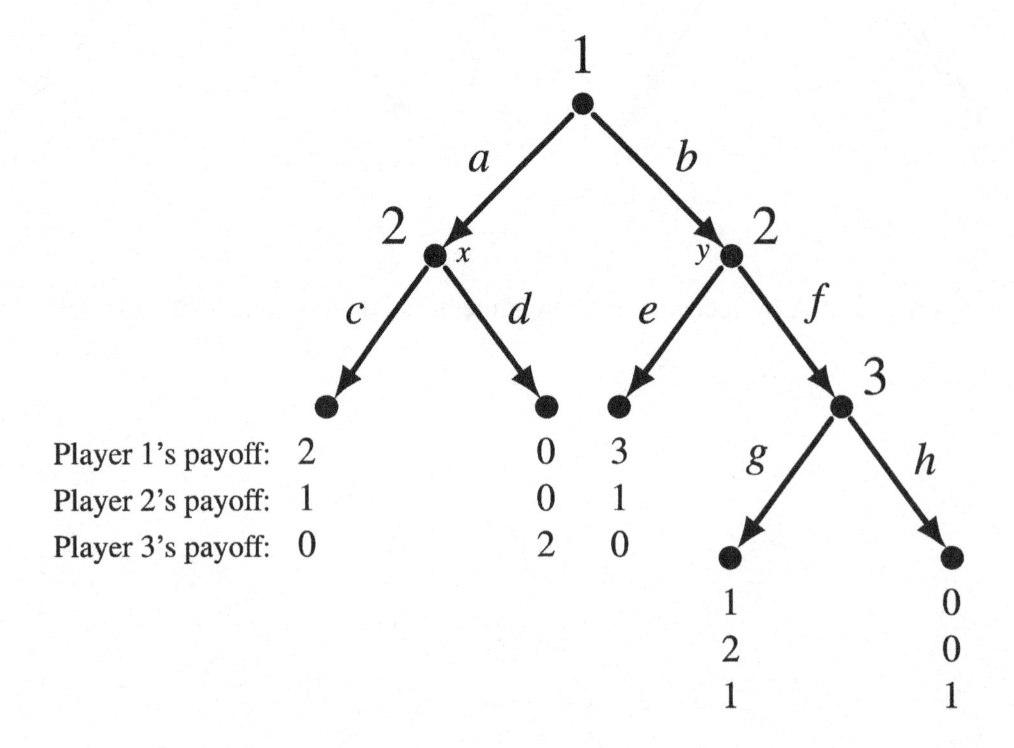

Figure 3.3: A perfect-information game with multiple backward-induction solutions.

For example, in the game of 3.3 starting at node x of Player 2 we select choice c (since it gives Player 2 a higher payoff than d). Then we move on to Player 3's node and we find that both choices there are payoff maximizing for Player 3; thus there are two ways to proceed, as shown in the next two figures.

In Figure 3.4 we show the steps of the backward-induction algorithm with the selection of choice g, while Figure 3.5 shows the steps of the algorithm with the selection of choice h. As before, the selected choices are shown by double edges. In Figures 3.4 and 3.5 the marking of nodes is shown explicitly, but later on we will represent the backward-induction solution more succinctly by merely highlighting the selected choices.

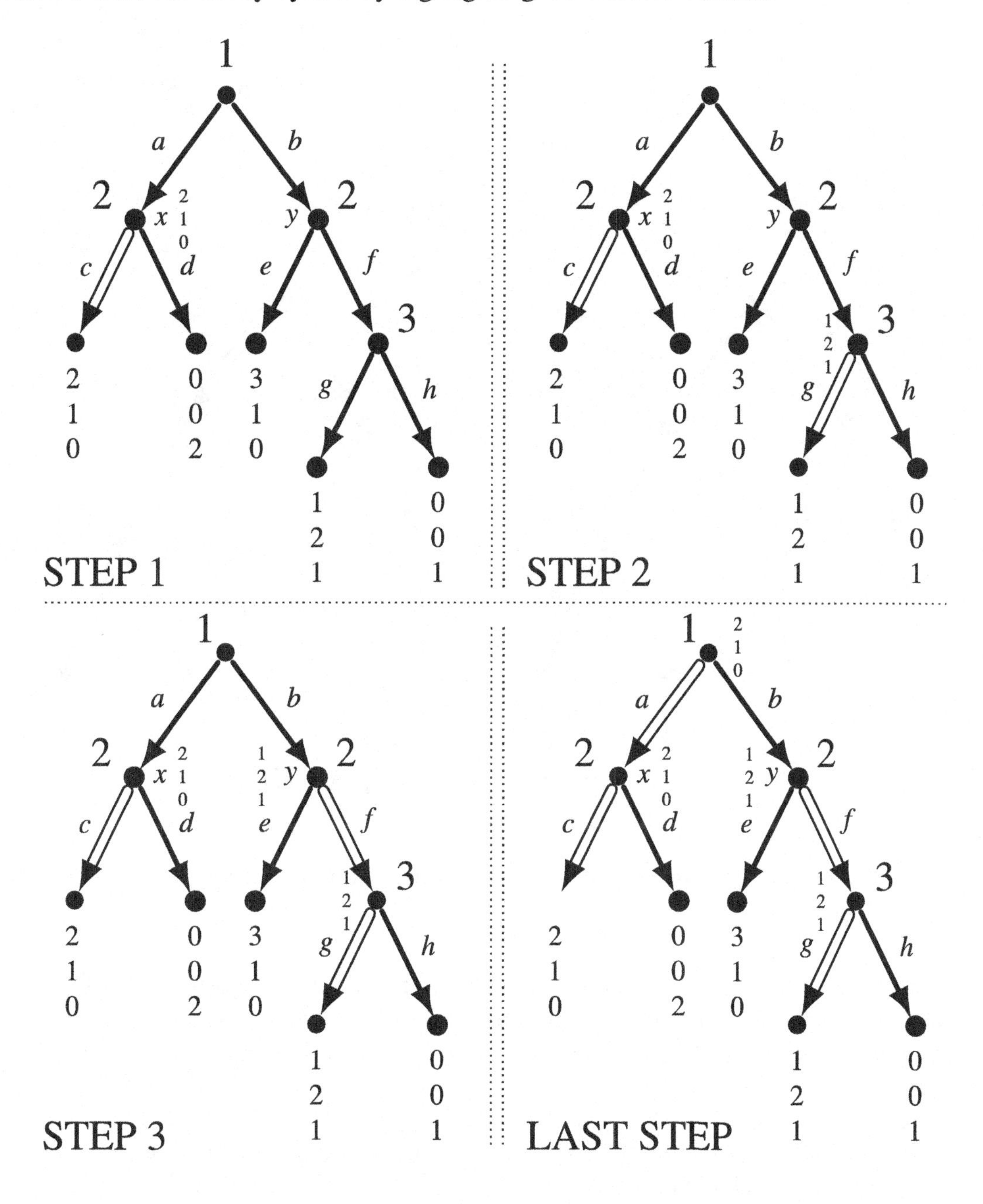

Figure 3.4: One possible output of the backward-induction algorithm applied to the game of Figure 3.3

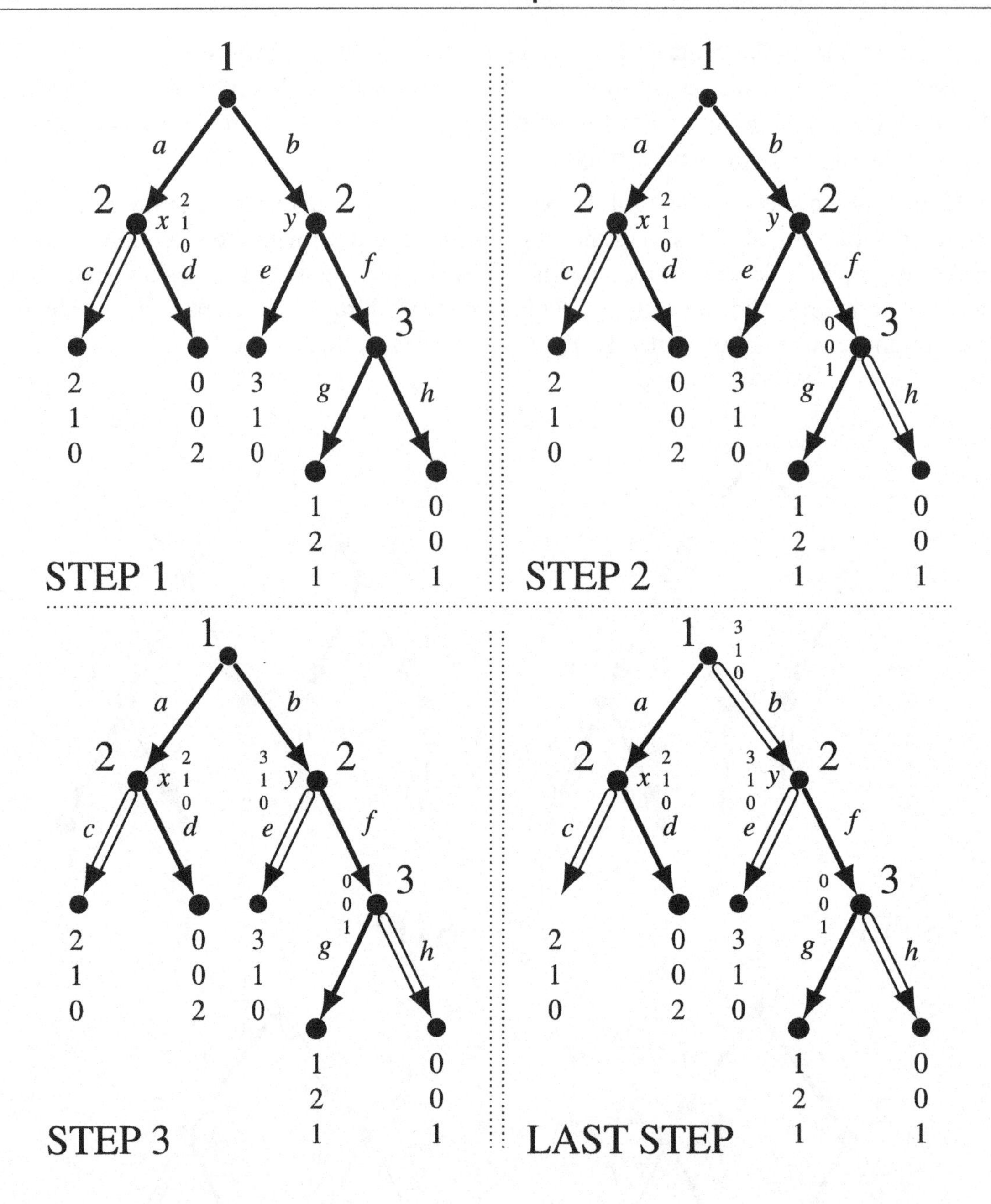

Figure 3.5: Another possible output of the backward-induction algorithm applied to the game of Figure 3.3

How should one define the output of the backward-induction algorithm and the notion of backward-induction solution? What kind of objects are they? Before we answer this question we need to introduce the notion of strategy in a perfect-information game.

Test your understanding of the concepts introduced in this section, by going through the exercises in Section 3.6.2 at the end of this chapter.

3.3 Strategies in perfect-information games

A strategy for a player in a perfect-information game is a complete, contingent plan on how to play the game. Consider, for example, the game shown in Figure 3.6 (which reproduces 3.3) and let us focus on Player 2.

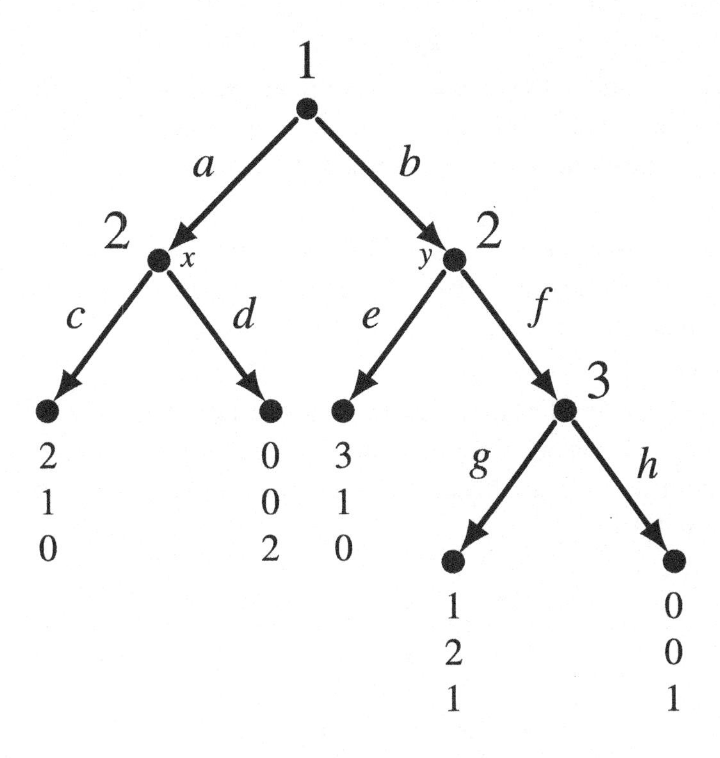

Figure 3.6: Copy of the game of Figure 3.3

Before the game is played, Player 2 does not know what Player 1 will do and thus a complete plan needs to specify what she will do if Player 1 decides to play a and what she will do if Player 1 decides to play b. A possible plan, or strategy, is "if Player 1 chooses a then I will choose c and if Player 1 chooses b then I will choose e", which we can denote more succinctly as (c,e). The other possible plans, or strategies, for Player 2 are $(c,f),(d,e)$ and (d,f). The formal definition of strategy is as follows.

Definition 3.3.1 A *strategy* for a player in a perfect-information game is a list of choices, one for each decision node of that player.

For example, suppose that Player 1 has three decision nodes in a given game: at one node she has three possible choices, a_1, a_2 and a_3, at another node she has two possible choices, b_1 and b_2, and at the third node she has four possible choices, c_1,c_2,c_3 and c_4. Then a strategy for Player 1 in that game can be thought of as a way of filling in three blanks:

$$\Big(\underbrace{\qquad\qquad}_{\text{one of } a_1,a_2,a_3}, \underbrace{\qquad\qquad}_{\text{one of } b_1,b_2}, \underbrace{\qquad\qquad}_{\text{one of } c_1,c_2,c_3,c_4} \Big).$$

Since there are 3 choices for the first blank, 2 for the second and 4 for the third, the total number of possible strategies for Player 1 in this case would be $3\times2\times4 = 24$. One strategy is (a_2,b_1,c_1), another strategy is (a_1,b_2,c_4), etc.

It should be noted that the notion of strategy involves redundancies. To see this, consider the game of Figure 3.7. In this game a possible strategy for Player 1 is (a,g), which means that Player 1 is planning to choose a at the root of the tree and would choose g at her other node. But if Player 1 indeed chooses a, then her other node will *not* be reached and thus why should Player 1 make a plan on what to do there? One could justify this redundancy in the notion of strategy in a number of ways:

1. Player 1 is so cautious that she wants her plan to cover also the possibility that she might make mistakes in the implementation of parts of her plan (in this case, she allows for the possibility that – despite her intention to play a – she might end up playing b), or
2. we can think of a strategy as a set of instructions given to a third party on how to play the game on Player 1's behalf, in which case Player 1 might indeed worry about the possibility of mistakes in the implementation and thus want to cover all contingencies.

An alternative justification relies on a different interpretation of the notion of strategy: not as a plan of Player 1 but as a belief in the mind of Player 2 concerning what Player 1 would do. For the moment we will set this issue aside and simply use the notion of strategy as given in Definition 4.2.1.

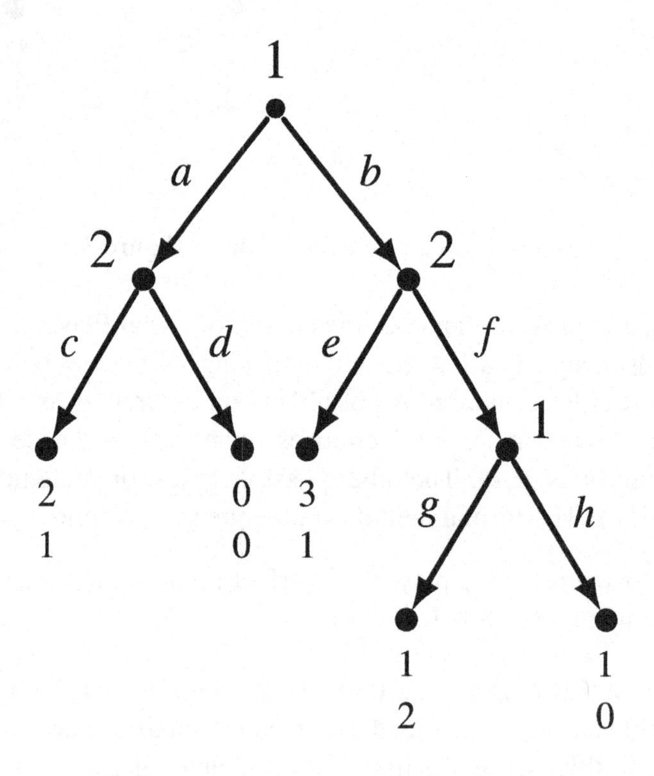

Figure 3.7: A perfect-information game

Using Definition 4.2.1, one can associate with every perfect-information game a strategic-form (or normal-form) game: a strategy profile determines a unique terminal node that is reached if the players act according to that strategy profile and thus a unique vector of payoffs. Figure 3.8 shows the strategic-form associated with the perfect-information game of Figure 3.7, with the Nash equilibria highlighted.

Player 2

Player 1	ce	cf	de	df
ag	2 1	2 1	0 0	0 0
ah	2 1	2 1	0 0	0 0
bg	3 1	1 2	3 1	1 2
bh	3 1	1 0	3 1	1 0

Figure 3.8: The strategic form of the perfect-information game of Figure 3.7 with the Nash equilibria highlighted

Because of the redundancy discussed above, the strategic form also displays redundancies: in this case the top two rows are identical.

Armed with the notion of strategy, we can now revisit the notion of backward-induction solution. Figure 3.9 shows the two backward-induction solutions of the game of Figure 3.7.

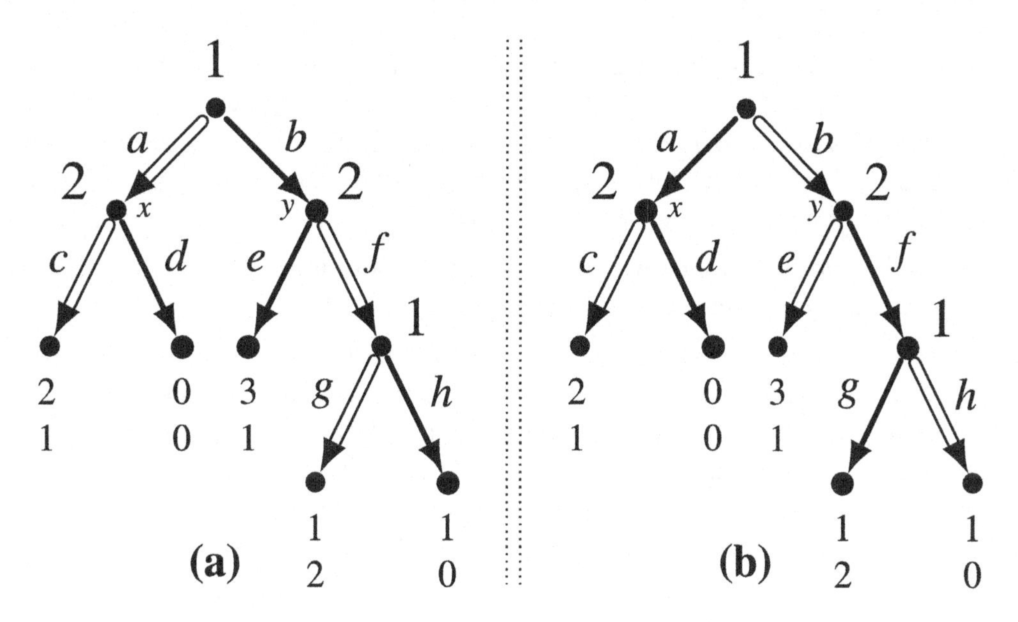

Figure 3.9: The backward-induction solutions of the game of Figure 3.7

It is clear from the definition of backward-induction algorithm (Definition 3.2.1) that the procedure selects a choice at every decision node and thus yields a strategy profile for the entire game: the backward-induction solution shown in Panel (*a*) of Figure 3.9 is the strategy profile $((a,g),(c,f))$, while the backward-induction solution shown in Panel (*b*) is the strategy profile $((b,h),(c,e))$. Both of them are Nash equilibria of the corresponding strategic-form game, but not all the Nash equilibria correspond to backward-induction solutions. The relationship between the two concepts is explained in the next section.

A backward-induction solution is a strategy profile. Since strategies contain a description of what a player actually does and also of what the player would do in circumstances that do not arise, one often draws a distinction between the backward-induction *solution* and the backward-induction *outcome* which is defined as the sequence of actual moves. For example, the backward-induction outcome associated with the solution $((a,g),(c,f))$ is the play *ac* with corresponding payoff $(2,1)$, while the backward-induction outcome associated with the solution $((b,h),(c,e))$ is the play *be* with corresponding payoff $(3,1)$.

Test your understanding of the concepts introduced in this section, by going through the exercises in Section 3.6.3 at the end of this chapter.

3.4 Relationship between backward induction and other solutions

If you have gone through the exercises for the previous three sections, you will have seen that in all those games the backward-induction solutions are also Nash equilibria. This is always true, as stated in the following theorem.

Theorem 3.4.1 Every backward-induction solution of a perfect-information game is a Nash equilibrium of the associated strategic form.

In some games the set of backward-induction solutions coincides with the set of Nash equilibria (see, for example, Exercise 3.9), but typically the set of Nash equilibria is larger than (is a proper superset of) the set of backward-induction solutions (for example the game of Figure 3.7 has two backward-induction solutions – shown in Figure 3.9 – but five Nash equilibria, shown in Figure 3.8).

Nash equilibria that are not backward-induction solutions often involve *incredible threats*. To see this, consider the following game.

An industry is currently a monopoly and the incumbent monopolist is making a profit of \$5 million. A potential entrant is considering whether or not to enter this industry.
- If she does not enter, she will make \$1 million in an alternative investment.
- If she does enter, then the incumbent can either fight entry with a price war whose outcome is that both firms make zero profits, or it can accommodate entry, by sharing the market with the entrant, in which case both firms make a profit of \$2 million.

This situation is illustrated in Figure 3.10 with the associated strategic form. Note that we are assuming that each player is selfish and greedy, that is, cares only about its own profit and prefers more money to less.

The backward-induction solution is *(in, accommodate)* and it is also a Nash equilibrium. However, there is another Nash equilibrium, namely *(out, fight)*. The latter should be discarded as a "rational solution" because it involves an incredible threat on the part of the incumbent, namely that it will fight entry if the potential entrant enters.
- It is true that, if the potential entrant believes the incumbent's threat, then she is better off staying out; however, she should ignore the incumbent's threat because she should realize that – when faced with the *fait accompli* of entry – the incumbent would not want to carry out the threat.

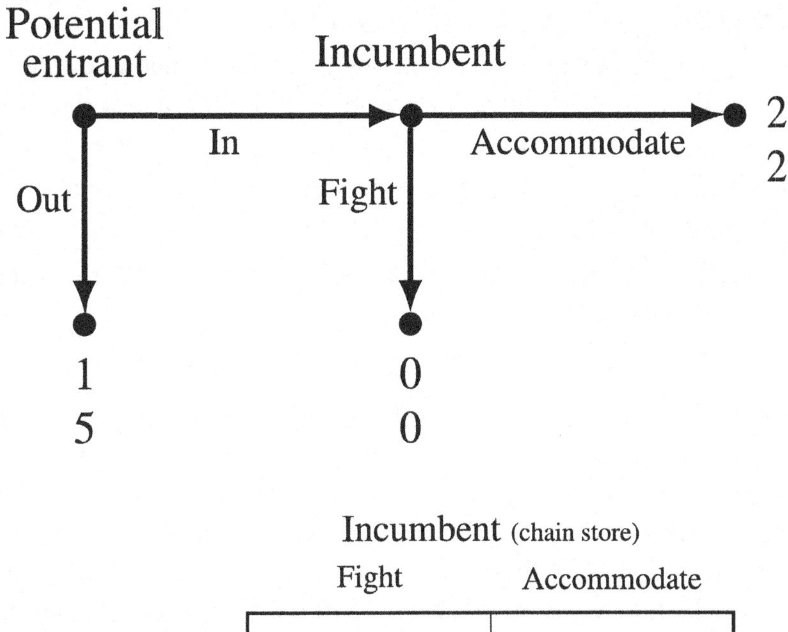

		Incumbent (chain store)			
		Fight		Accommodate	
Potential entrant	In	0	0	2	2
	Out	1	5	1	5

Figure 3.10: The entry game

Reinhard Selten (who shared the 1994 Nobel Memorial prize in economics with two other game theorists, John Harsanyi and John Nash) discussed a repeated version of the above entry game, which has become known as *Selten's Chain Store Game*. The story is as follows:

- A chain store is a monopolist in an industry. It owns stores in m different towns ($m \geq 2$).

- In each town the chain store makes \$5 million if left to enjoy its privileged position undisturbed.

- In each town there is a businesswoman who could enter the industry in that town, but earns \$1 million if she chooses not to enter; if she decides to enter, then the monopolist can either fight the entrant, leading to zero profits for both the chain store and the entrant in that town, or it can accommodate entry and share the market with the entrant, in which case both players make \$2 million in that town.

Thus, in each town the interaction between the incumbent monopolist and the potential entrant is as illustrated in Figure 3.10.

However, decisions are made sequentially, as follows:

At date $t (t = 1, \ldots, m)$ the businesswoman in town t decides whether or not to enter and if she enters then the chain store decides whether or not to fight in that town.

What happens in town t at date t becomes known to everybody. Thus, for example, the businesswoman in town 2 at date 2 knows what happened in town 1 at date 1 (either that there was no entry or that entry was met with a fight or that entry was accommodated).

Intuition suggests that in this game the threat by the incumbent to fight early entrants might be credible, for the following reason. The incumbent could tell Businesswoman 1 the following:

> "It is true that, if you enter and I fight, I will make zero profits, while by accommodating your entry I would make \$2 million and thus it would seem that it cannot be in my interest to fight you. However, somebody else is watching us, namely Businesswoman 2. If she sees that I have fought your entry then she might fear that I would do the same with her and decide to stay out, in which case in town 2, I would make \$5 million, so that my total profits in towns 1 and 2 would be $(0+5) =$ \$5 million. On the other hand, if I accommodate your entry, then she will be encouraged to entry herself and I will make \$2 million in each town, for a total profit of \$4 million. Hence, as you can see, it is indeed in my interest to fight you and thus you should stay out."

Does the notion of backward induction capture this intuition? To check this, let us consider the case where $m = 2$, so that the extensive game is not too large to draw. It is shown in Figure 3.11, where at each terminal node the top number is the profit of the incumbent monopolist (it is the sum of the profits in the two towns), the middle number is the profit of Businesswoman 1 and the bottom number is the profit of Businesswoman 2. All profits are expressed in millions of dollars. We assume that all the players are selfish and greedy, so that we can take the profit of each player to be that player's payoff. The backward-induction solution is unique and is shown by the double directed edges in Figure 3.11. The corresponding outcome is that both businesswomen will enter and the incumbent monopolist accommodates entry in both towns.

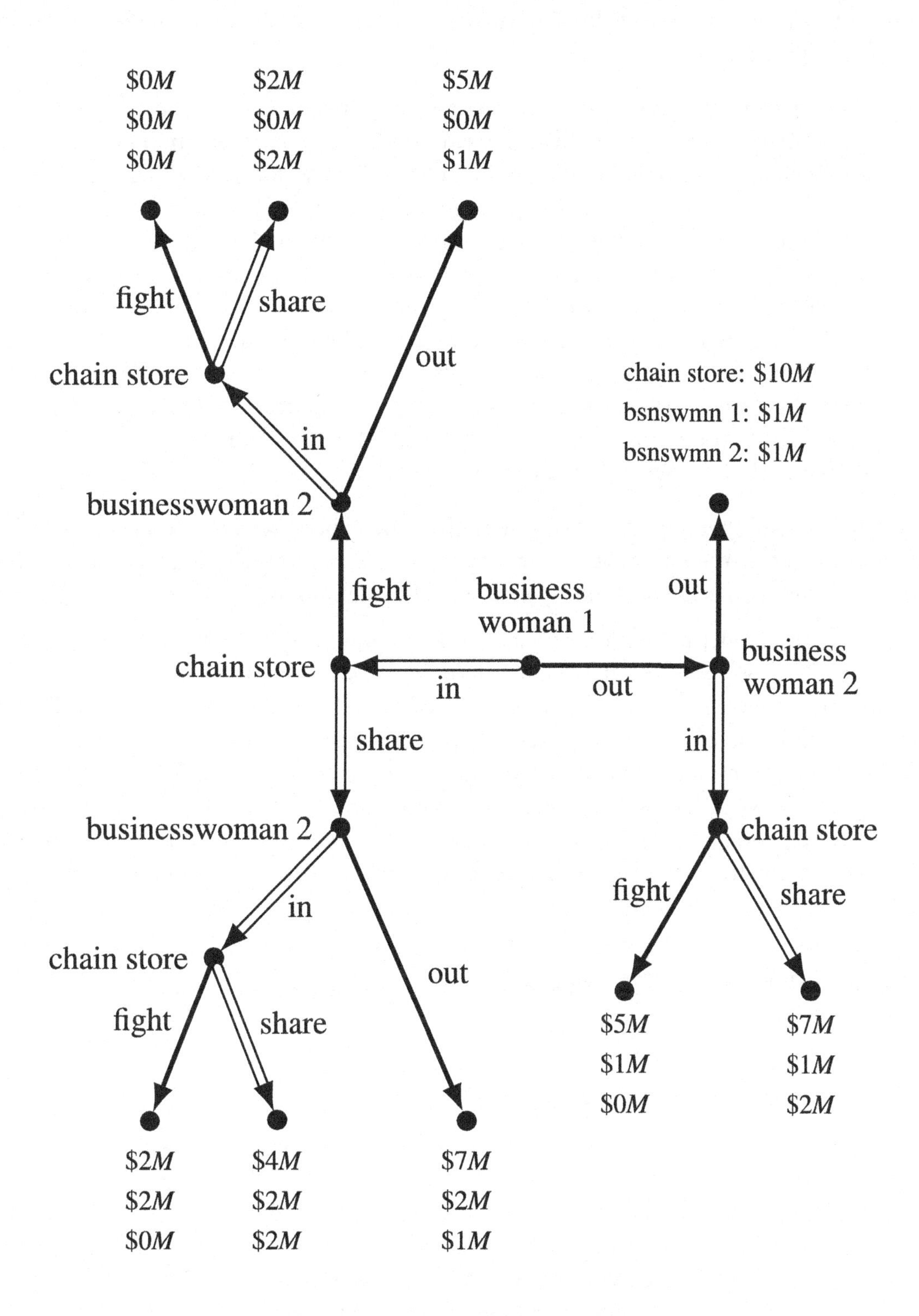

Figure 3.11: Selten's Chain-Store game

Thus the backward-induction solution does not capture the "reputation" argument outlined above. However, the backward-induction solution does seem to capture the notion of rational behavior in this game. Indeed, Businesswoman 1 should reply to the incumbent with the following counter-argument:

> "Your reasoning is not valid. Whatever happens in town 1, it will be common knowledge between you and Businesswoman 2 that your interaction in town 2 will be the last; in particular, nobody else will be watching and thus there won't be an issue of establishing a reputation in the eyes of another player. Hence in town 2 it will be in your interest to accommodate entry, since in essence you will be playing the one-shot entry game of Figure 3.10. Hence a rational Businesswoman 2 will decide to enter in town 2 *whatever happened in town 1*: what you do against me will have no influence on her decision. Thus your "reputation" argument does not apply and it will in fact be in your interest not to fight my entry: your choice will be between a profit of $\$(0+2) = \$2$ million, if you fight me, and a profit of $\$(2+2) = \$4$ million, if you don't fight me. Hence I will enter and you will not fight me."

In order to capture the reputation argument described above we need to allow for some uncertainty in the mind of some of the players, as we will show in a later chapter. In a perfect-information game uncertainty is ruled out by definition.

By Theorem 3.4.1 the notion of backward induction can be seen as a refinement of the notion of Nash equilibrium. Another solution concept that is related to backward induction is the iterated elimination of weakly dominated strategies. Indeed the backward-induction algorithm could be viewed as a step-wise procedure that eliminates dominated choices at decision nodes, and thus strategies that contain those choices. What is the relationship between the two notions? In general this is all that can be said: applying the iterated deletion of weakly dominated strategies to the strategic form associated with a perfect-information game leads to a set of strategy profiles that contains at least one backward-induction solution; however,

(1) it may also contain strategy profiles that are not backward-induction solutions, and

(2) it may fail to contain all the backward-induction solutions, as shown in Exercise 3.8.

3.5 Perfect-information games with two players

We conclude this chapter with a discussion of finite two-player extensive games with perfect information.

We will start with games that have only two outcomes, namely "Player 1 wins" (denoted by W_1) and "Player 2 wins" (denoted by W_2). We assume that Player 1 strictly prefers W_1 to W_2 and Player 2 strictly prefers W_2 to W_1. Thus we can use utility functions with values 0 and 1 and associate with each terminal node either the payoff vector $(1,0)$ (if the outcome is W_1) or the payoff vector $(0,1)$ (if the outcome is W_2). We call these games *win-lose games*. An example of such a game is the following.

Example 3.2 Two players take turns choosing a number from the set $\{1,2,\ldots,10\}$, with Player 1 moving first. The first player who brings the sum of all the chosen numbers to 100 or more wins. ■

The following is one possible play of the game (the bold-face numbers are the ones chosen by Player 1 and the underlined numbers the ones chosen by Player 2):

$$\mathbf{10},\ \underline{9},\ \mathbf{9},\ \underline{10},\ \mathbf{8},\ \underline{7},\ \mathbf{4},\ \underline{10},\ \mathbf{1},\ \underline{8},\ \mathbf{3},\ \underline{3},\ \mathbf{8},\ \underline{10}.$$

In this play Player 2 wins: before her last move the sum is 90 and with her final choice of 10 she brings the total to 100. However, in this game Player 1 has a *winning strategy*, that is, a strategy that guarantees that he will win, no matter what numbers Player 2 chooses. To see this, we can use backward-induction reasoning. Drawing the tree is not a practical option, since the number of nodes is very large: one needs 10,000 nodes just to represent the first 4 moves! But we can imagine drawing the tree, placing ourselves towards the end of the tree and ask what partial sum represents a "losing position", in the sense that whoever is choosing in that position cannot win, while the other player can then win with his subsequent choice. With some thought one can see that 89 is the largest losing position: whoever moves there can take the sum to any number in the set $\{90,91,\ldots,99\}$, thus coming short of 100, while the other player can then take the sum to 100 with an appropriate choice. What is the largest losing position that precedes 89? The answer is 78: whoever moves at 78 must take the sum to a number in the set $\{79,80,\ldots,88\}$ and then from there the other player can make sure to take the sum to 89 and then we know what happens from there! Repeating this reasoning we see that the losing positions are: 89, 78, 67, 56, 45, 34, 23, 12, 1. Since Player 1 moves first he can choose 1 and put Player 2 in the first losing position; then, whatever Player 2 chooses, Player 1 can put her in the next losing position, namely 12, etc. Recall that a strategy for Player 1 must specify what to do in every possible situation in which he might find himself. In his game Player 1's winning strategy is as follows:

> Start with the number 1. Then, at every turn, choose the number $(11-n)$, where n is the number that was chosen by Player 2 in the immediately preceding turn.

Here is an example of a possible play of the game where Player 1 employs the winning strategy and does in fact win:

$$\mathbf{1},\ \underline{9},\ \mathbf{2},\ \underline{6},\ \mathbf{5},\ \underline{7},\ \mathbf{4},\ \underline{10},\ \mathbf{1},\ \underline{8},\ \mathbf{3},\ \underline{3},\ \mathbf{8},\ \underline{9},\ \mathbf{2},\ \underline{5},\ \mathbf{6},\ \underline{1},\ \mathbf{10}$$

We can now state a general result about this class of games.

Theorem 3.5.1 In every finite two-player, win-lose game with perfect information one of the two players has a winning strategy.

Although we will not give a detailed proof, the argument of the proof is rather simple. By applying the backward-induction algorithm we assign to every decision node either the payoff vector $(1,0)$ or the payoff vector $(0,1)$. Imagine applying the algorithm up to the point where the immediate successors of the root have been assigned a payoff vector. Two cases are possible.

Case 1: at least one of the immediate successors of the root has been assigned the payoff vector $(1,0)$. In this case Player 1 is the one who has a winning strategy and his initial choice should be such that a node with payoff vector $(1,0)$ is reached and then his future choices should also be such that only nodes with payoff vector $(1,0)$ are reached.

Case 2: all the immediate successors of the root have been assigned the payoff vector $(0,1)$. In this case it is Player 2 who has a winning strategy. An example of a game where it is Player 2 who has a winning strategy is given in Exercise 3.11.

We now turn to finite two-player games where there are three possible outcomes: "Player 1 wins" (W_1), "Player 2 wins" (W_2) and "Draw" (D). We assume that the rankings of the outcomes are as follows: $W_1 \succ_1 D \succ_1 W_2$ and $W_2 \succ_2 D \succ_2 W_1$.
Examples of such games are Tic-Tac-Toe (`http://en.wikipedia.org/wiki/Tic-tac-toe`), Draughts or Checkers (`http://en.wikipedia.org/wiki/Draughts`) and Chess (although there does not seem to be agreement as to whether the rules of Chess guarantee that every possible play of the game is finite). What can we say about such games? The answer is provided by the following theorem.

Theorem 3.5.2 Every finite two-player, perfect-information game with three outcomes: Player 1 wins (W_1), Player 2 wins (W_2) and Draw (D), and preferences $W_1 \succ_1 D \succ_1 W_2$ and $W_2 \succ_2 D \succ_2 W_1$, falls within one of the following three categories:

1. Player 1 has a strategy that guarantees outcome W_1.
2. Player 2 has a strategy that guarantees outcome W_2.
3. Player 1 has a strategy that guarantees that the outcome will be W_1 or D and Player 2 has a strategy that guarantees that the outcome will be W_2 or D, so that, if both players employ these strategies, the outcome will be D.

The logic of the proof is as follows. By applying the backward-induction algorithm we assign to every decision node either the payoff vector $(2,0)$ (corresponding to outcome W_1) or the payoff vector $(0,2)$ (corresponding to outcome W_2) or the payoff vector $(1,1)$ (corresponding to outcome D). Imagine applying the algorithm up to the point where the immediate successors of the root have been assigned a payoff vector. Three cases are possible.

Case 1: at least one of the immediate successors of the root has been assigned the payoff vector $(2,0)$; in this case Player 1 is the one who has a winning strategy.

Case 2: all the immediate successors of the root have been assigned the payoff vector $(0,2)$; in this case it is Player 2 who has a winning strategy.

Case 3: there is at least one immediate successor of the root to which the payoff vector $(1,1)$ has been assigned and all the other immediate successors of the root have been assigned either $(1,1)$ or $(0,2)$. In this case we fall within the third category of Theorem 3.5.2.

Both Tic-Tac-Toe and Checkers fall within the third category (`http://en.wikipedia.org/wiki/Solved_game#Solved_games`). As of the time of writing this book, it is not known to which category the game of Chess belongs.

Test your understanding of the concepts introduced in this section, by going through the exercises in Section 3.6.4 at the end of this chapter.

3.6 Exercises

3.6.1 Exercises for Section 3.1: Trees, frames and games

The answers to the following exercises are in Section 3.7 at the end of this chapter.

Exercise 3.1 How could they do that! They abducted Speedy, your favorite tortoise! They asked for $1,000 in unmarked bills and threatened to kill Speedy if you don't pay. Call the tortoise-napper Mr. T. Let the possible outcomes be as follows:

o_1 : you don't pay and speedy is released
o_2 : you pay $ 1,000 and speedy is released
o_3 : you don't pay and speedy is killed
o_4 : you pay $ 1,000 and speedy is killed

You are attached to Speedy and would be willing to pay $1,000 to get it back. However, you also like your money and you prefer not to pay, conditional on each of the two separate events "Speedy is released" and "Speedy is killed". Thus your ranking of the outcomes is $o_1 \succ_{you} o_2 \succ_{you} o_3 \succ_{you} o_4$. On the other hand, you are not quite sure of what Mr. T's ranking is.

(a) Suppose first that Mr T has communicated that he wants you to go to Central Park tomorrow at 10:00 a.m. and leave the money in a garbage can; he also said that, two miles to the East and at the exact same time, he will decide whether or not to free Speedy in front of the police station and then go and collect his money in Central Park. What should you do?

(b) Suppose that Mr T is not as dumb as in part **(a)** and instead gives you the following instructions: first you leave the money in a garbage can in Central Park and then he will go there to collect the money. He also told you that if you left the money there then he will free Speedy, otherwise he will kill it. Draw an extensive form or frame to represent this situation.

(c) Now we want to construct a game based on the extensive form of part **(b)**. For this we need Mr T's preferences. There are two types of criminals in Mr T's line of work: the professionals and the one-timers. Professionals are in the business for the long term and thus, besides being greedy, worry about reputation; they want it to be known that (1) every time they were paid they honored their promise to free the hostage and (2) their threats are to be taken seriously: every time they were *not* paid, the hostage was killed. The one-timers hit once and then they disappear; they don't try to establish a reputation and the only thing they worry about, besides money, is not to be caught: whether or not they get paid, they prefer to kill the hostage in order to eliminate any kind of evidence (DNA traces, fingerprints, etc.). Construct two games based on the extensive form of part **(b)** representing the two possible types of Mr T.

■

Exercise 3.2 A three-man board, composed of A, B, and C, has held hearings on a personnel case involving an officer of the company. This officer was scheduled for promotion but, prior to final action on his promotion, he made a decision that cost the company a good deal of money. The question is whether he should be (1) promoted anyway, (2) denied the promotion, or (3) fired. The board has discussed the matter at length and is unable to reach unanimous agreement. In the course of the discussion it has become clear to all three of them that their separate opinions are as follows:

- A considers the officer to have been a victim of bad luck, not bad judgment, and wants to go ahead and promote him but, failing that, would keep him rather than fire him.
- B considers the mistake serious enough to bar promotion altogether; he'd prefer to keep the officer, denying promotion, but would rather fire than promote him.
- C thinks the man ought to be fired but, in terms of personnel policy and morale, believes the man ought not to be kept unless he is promoted, i.e., that keeping an officer who has been declared unfit for promotion is even worse than promoting him.

	PROMOTE	KEEP	FIRE
A:	best	middle	worst
B:	worst	best	middle
C:	middle	worst	best

Assume that everyone's preferences among the three outcomes are fully evident as a result of the discussion. The three must proceed to a vote.
Consider the following voting procedure. First A proposes an action (either promote or keep or fire). Then it is B's turn. If B accepts A's proposal, then this becomes the final decision. If B disagrees with A'a proposal, then C makes the final decision (which may be *any of the three*: promote, keep or fire). Represent this situation as an extensive game with perfect information. (Use utility numbers from the set $\{1,2,3\}$.) ■

3.6.2 Exercises for Section 3.2: Backward induction

The answers to the following exercises are in Section 3.7 at the end of this chapter.

Exercise 3.3 Apply the backward-induction algorithm to the two games of Exercise 3.1 Part **(c)**. ■

Exercise 3.4 Apply the backward-induction algorithm to the game of Exercise 3.2. ■

3.6.3 Exercises for Section 3.3: Strategies in perfect-information games

The answers to the following exercises are in Section 3.7 at the end of this chapter.

Exercise 3.5 Write the strategic form of the game of Figure 3.2, find all the Nash equilibria and verify that the backward-induction solution is a Nash equilibrium. ■

Exercise 3.6 Write the strategic form of the game of Figure 3.3, find all the Nash equilibria and verify that the backward-induction solutions are Nash equilibria. ■

Exercise 3.7 Consider the game of Exercise 3.2.

(a) Write down all the strategies of Player B.

(b) How many strategies does Player C have?

■

Exercise 3.8 Consider the perfect-information game shown in Figure 3.12.

(a) Find the backward-induction solutions.

(b) Write down all the strategies of Player 1.

(c) Write down all the strategies of Player 2.

(d) Write the strategic form associated with this game.

(e) Does Player 1 have a dominant strategy?

(f) Does Player 2 have a dominant strategy?

(g) Is there a dominant-strategy equilibrium?

(h) Does Player 1 have any dominated strategies?

(i) Does Player 2 have any dominated strategies?

(j) What do you get when you apply the iterative elimination of weakly dominated strategies?

(k) What are the Nash equilibria?

■

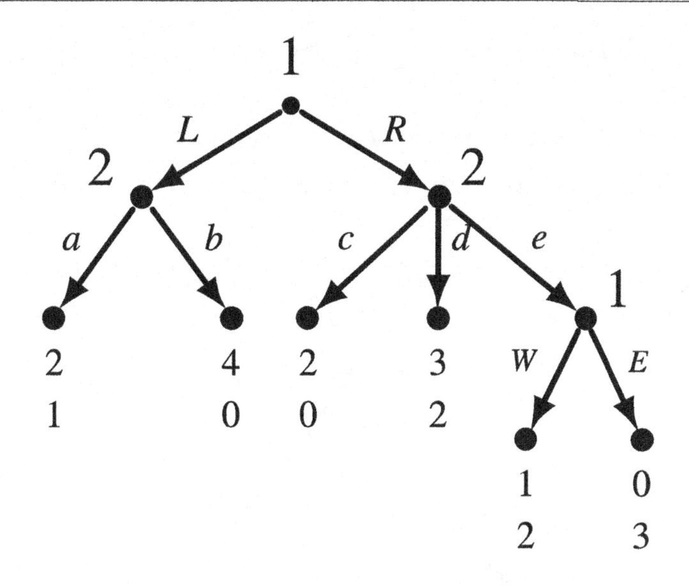

Figure 3.12: The perfect-information game for Exercise 3.8

Exercise 3.9 Consider an industry where there are two firms, a large firm, Firm 1, and a small firm, Firm 2. The two firms produce identical products.
- Let x be the output of Firm 1 and y the output of Firm 2. Industry output is $Q = x + y$.
- The price P at which each unit of output can be sold is determined by the inverse demand function $P = 130 - 10Q$. For example, if Firm 1 produces 4 units and Firm 2 produces 2 units, then industry output is 6 and each unit is sold for $P = 130 - 60 = \$70$.
- For each firm the cost of producing q units of output is $C(q) = 10q + 62.5$.
- Each firm is only interested in its own profits.
- The profit of Firm 1 depends on both x and y and is given by

$$\Pi_1(x,y) = \underbrace{x\,[130 - 10(x+y)]}_{\text{revenue}} - \underbrace{(10x + 62.5)}_{\text{cost}}$$

and similarly the profit function of Firm 2 is given by

$$\Pi_2(x,y) = \underbrace{y\,[130 - 10(x+y)]}_{\text{revenue}} - \underbrace{(10y + 62.5)}_{\text{cost}}.$$

- The two firms play the following sequential game. First Firm 1 chooses its own output x and commits to it; then Firm 2, after having observed Firm 1's output, chooses its own output y; then the price is determined according to the demand function and the two firms collect their own profits. In what follows assume, for simplicity, that x can only be 6 or 6.5 units and y can only be 2.5 or 3 units.

(a) Represent this situation as an extensive game with perfect information.

(b) Solve the game using backward induction.

(c) Write the strategic form associated with the perfect-information game.

(d) Find the Nash equilibria of this game and verify that the backward-induction solutions are Nash equilibria.

■

Exercise 3.10 Consider the perfect-information game shown in Figure 3.13 where x is an integer.

(a) For every value of x find the backward induction solution(s).

(b) Write the corresponding strategic-form and find all the Nash equilibria. ■

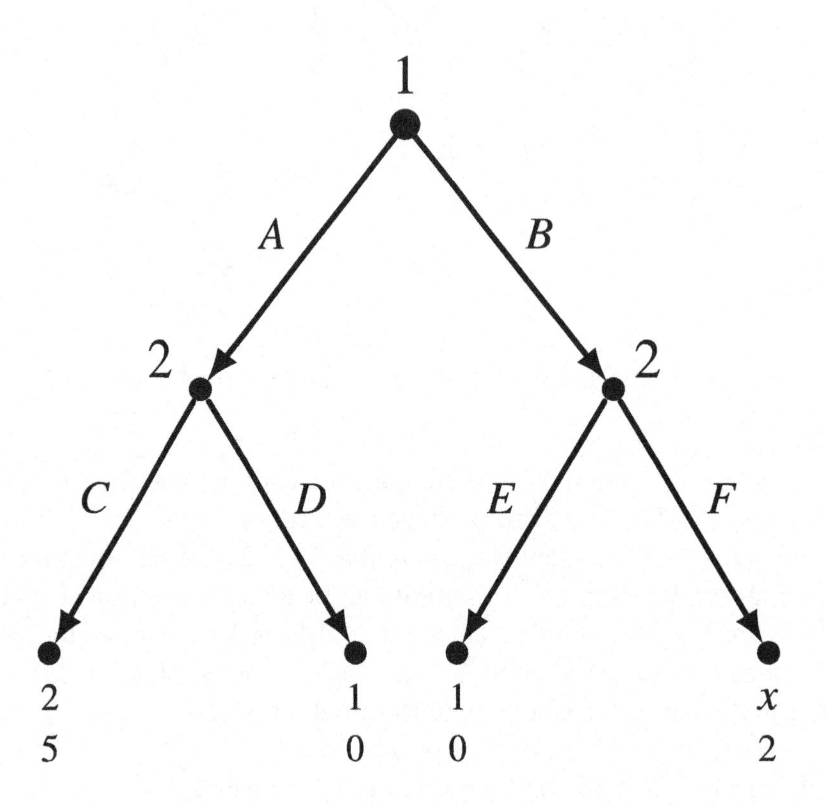

Figure 3.13: A perfect-information game

3.6.4 Exercises for Section 3.5: Two-player games

The answers to the following exercises are in Section 3.7 at the end of this chapter.

Exercise 3.11 Consider the following perfect-information game. Player 1 starts by choosing a number from the set $\{1,2,3,4,5,6,7\}$, then Player 2 chooses a number from this set, then Player 1 again, followed by Player 2, etc. The first player who brings the cumulative sum of all the numbers chosen (up to and including the last one) to 48 or more wins. By Theorem 3.5.1 one of the two players has a winning strategy. Find out who that player is and fully describe the winning strategy. ■

Exercise 3.12 Consider Figure 3.14 and the following two-player, perfect-information game. A coin is placed in the cell marked 'START' (cell A1). Player 1 moves first and can move the coin one cell up (to A2) or one cell to the left (to B1) or one cell diagonally in the left-up direction (to B2). Then Player 2 moves, according to the same rules (e.g. if the coin is in cell B2 then the admissible moves are shown by the directed edges). The players alternate moving the coin. Black cells are not accessible (so that, for example, from A3 the coin can only be moved to A4 or B3 and from F3 it can only be moved to G4, as shown by the directed edge). The player who manages to place the coin in the cell marked 'END' wins.

(a) Represent this game by means of an extensive form with perfect information by drawing the initial part of the tree that covers the first two moves (the first move of Player 1 and the first move of Player 2).

(b) Suppose that the coin is currently in cell G4 and it is Player 1's turn to move. Show that Player 1 has a strategy that allows her to win the game starting from cell G4. Describe the strategy in detail.

(c) Describe a play of the game (from cell A1) where Player 1 wins (describe it by means of the sequence of cells visited by the coin).

(d) Describe a play of the game (from cell A1) where Player 2 wins (describe it by means of the sequence of cells visited by the coin).

(e) Now go back to the beginning of the game. The coin is in cell A1 and player 1 has the first move. By Theorem 3.5.1 one of the two players has a winning strategy. Find out who that player is and fully describe the winning strategy.

■

Figure 3.14: The coin game

Exercise 3.13 — ⋆⋆⋆ **Challenging Question** ⋆⋆⋆.

Two women, Anna and Bess, claim to be the legal owners of a diamond ring that - each claims - has great sentimental value. Neither of them can produce evidence of ownership and nobody else is staking a claim on the ring. Judge Sabio wants the ring to go to the legal owner, but he does not know which of the two women is in fact the legal owner. He decides to proceed as follows. First he announces a fine of $\$F > 0$ and then asks Anna and Bess to play the following game.

Move 1: Anna moves first. Either she gives up her claim to the ring (in which case Bess gets the ring, the game ends and nobody pays the fine) or she asserts her claim, in which case the game proceeds to Move 2.

Move 2: Bess either accepts Anna's claim (in which case Anna gets the ring, the game ends and nobody pays the fine) or challenges her claim. In the latter case, Bess must put in a bid, call it B, and Anna must pay the fine of $\$F$ to Sabio. The game goes on to Move 3.

Move 3: Anna now either matches Bess's bid (in which case Anna gets the ring, Anna pays $\$B$ to Sabio in addition to the fine that she already paid and Bess pays the fine of $\$F$ to Sabio) or chooses not to match (in which case Bess gets the ring and pays her bid of $\$B$ to Sabio and, furthermore, Sabio keeps the fine that Anna already paid).

Denote by C_A the monetary equivalent of getting the ring for Anna (that is, getting the ring is as good, in Anna's mind, as getting $\$C_A$) and C_B the monetary equivalent of getting the ring for Bess. Not getting the ring is considered by both to be as good as getting zero dollars.

(a) Draw an extensive game with perfect information to represent the above situation, assuming that there are only two possible bids: B_1 and B_2. Write the payoffs to Anna and Bess next to each terminal node.

(b) Find the backward-induction solution of the game you drew in part **(a)** for the case where $B_1 > C_A > C_B > B_2 > F > 0$.

Now consider the general case where the bid B can be any non-negative number and assume that both Anna and Bess are very wealthy. Assume also that C_A, C_B and F are positive numbers and that C_A and C_B are common knowledge between Anna and Bess . We want to show that, at the backward-induction solution of the game, the ring always goes to the legal owner. Since we (like Sabio) don't know who the legal owner is, we must consider two cases.

Case 1: the legal owner is Anna. Let us assume that this implies that $C_A > C_B$.

Case 2: the legal owner is Bess. Let us assume that this implies that $C_B > C_A$.

(c) Find the backward-induction solution for Case 1 and show that it implies that the ring goes to Anna.

(d) Find the backward-induction solution for Case 2 and show that it implies that the ring goes to Bess.

(e) How much money does Sabio make at the backward-induction solution? How much money do Ann and Bess end up paying at the backward-induction solution?

■

3.7 Solutions to exercises

Solution to Exercise 3.1.

(a) For you it is a strictly dominant strategy to not pay and thus you should not pay.

(b) The extensive form is shown in Figure 3.15.

(c) For the professional, concern with reputation implies that $o_2 \succ_{MrT} o_4$ and $o_3 \succ_{MrT} o_1$. If we add the reasonable assumption that after all money is what they are after, then we can take the full ranking to be $o_2 \succ_{MrT} o_4 \succ_{MrT} o_3 \succ_{MrT} o_1$.
Representing preferences with ordinal utility functions with values in the set $\{1,2,3,4\}$, we have

outcome → utility function ↓	o_1	o_2	o_3	o_4
U_{you}	4	3	2	1
U_{MrT}	1	4	2	3

The corresponding game is obtained by replacing in Figure 3.15 o_1 with the payoff vector (4,1), o_3 with the payoff vector (2,2), etc.

For the one-timer, the ranking can be taken to be (although this is not the only possibility) $o_4 \succ_{MrT} o_2 \succ_{MrT} o_3 \succ_{MrT} o_1$, with corresponding utility representation:

outcome → utility function ↓	o_1	o_2	o_3	o_4
U_{you}	4	3	2	1
U_{MrT}	1	3	2	4

The corresponding extensive-form game is shown in Figure 3.16. □

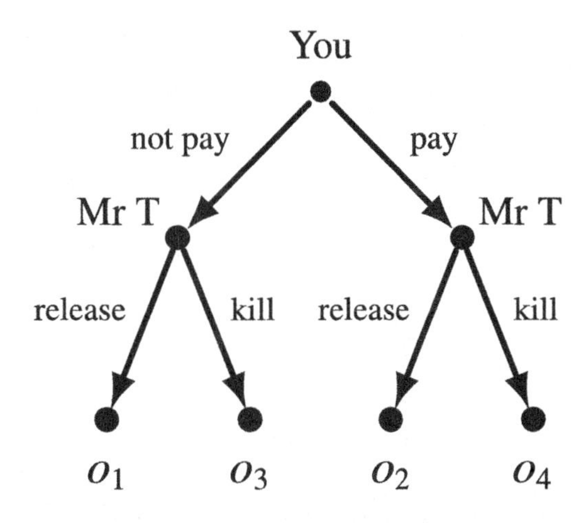

Figure 3.15: The game-frame for Part **(b)** of Exercise 3.1

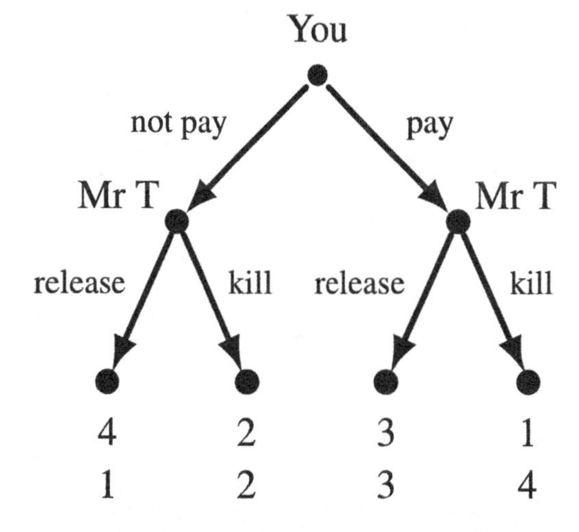

Figure 3.16: The game for Part **(c)** of Exercise 3.1 when Mr T is a one-timer

Solution to Exercise 3.2. The game is shown in Figure 3.17 ('P' stands for promote, 'K' for keep (without promoting), 'F' for fire). □

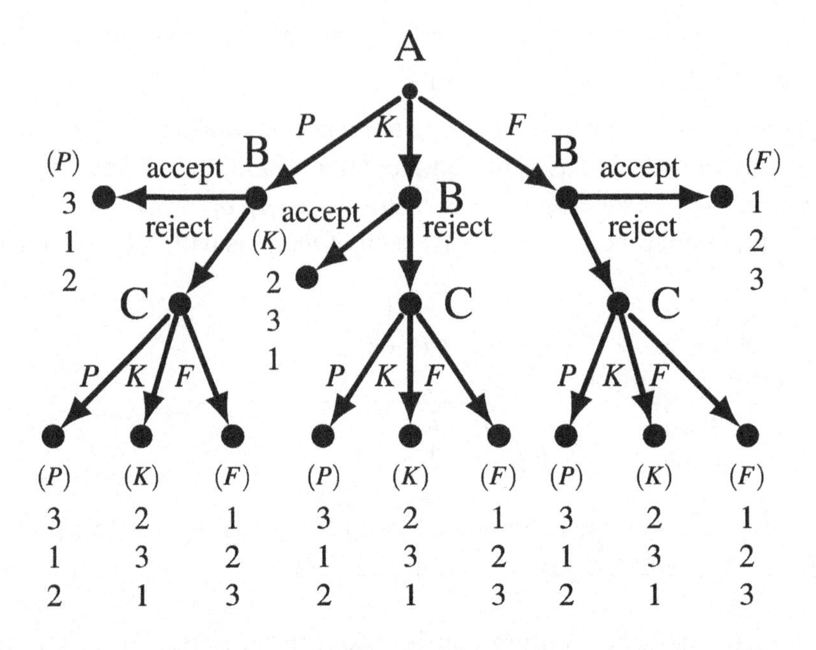

Figure 3.17: The game for Exercise 3.2

Solution to Exercise 3.3. The application of the backward-induction algorithm is shown by double edges in Figure 3.18 for the case of a professional Mr. T and in Figure 3.19 for the case of a one-timer Mr. T. Thus, against a professional you will pay and against a one-timer you would not pay. With the professional you would get Speedy back, with the one-timer you will hold a memorial service for Speedy. □

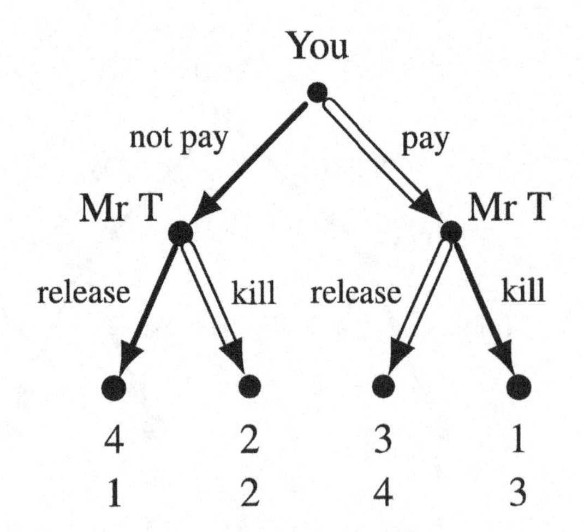

Figure 3.18: The game for Part **(b)** of Exercise 3.3

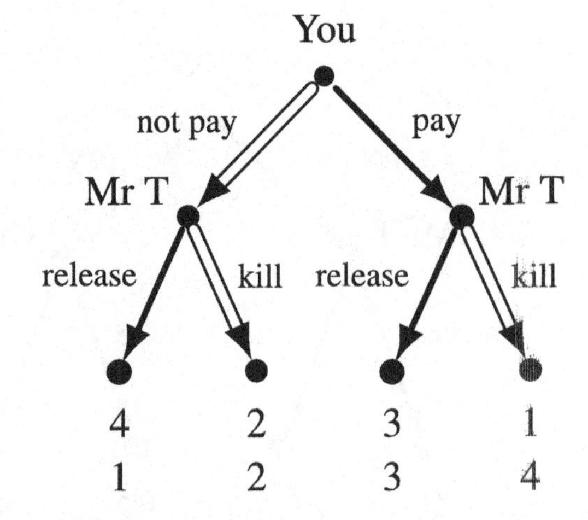

Figure 3.19: The game for Part (**(c)** of Exercise 3.3

Solution to Exercise 3.4. The backward-induction algorithm yields two solutions, shown in Figures 3.20 and 3.21. The difference between the two solutions lies in what Player *B* would do if Player *A* proposed *F*. In both solutions the officer is kept without promotion.□

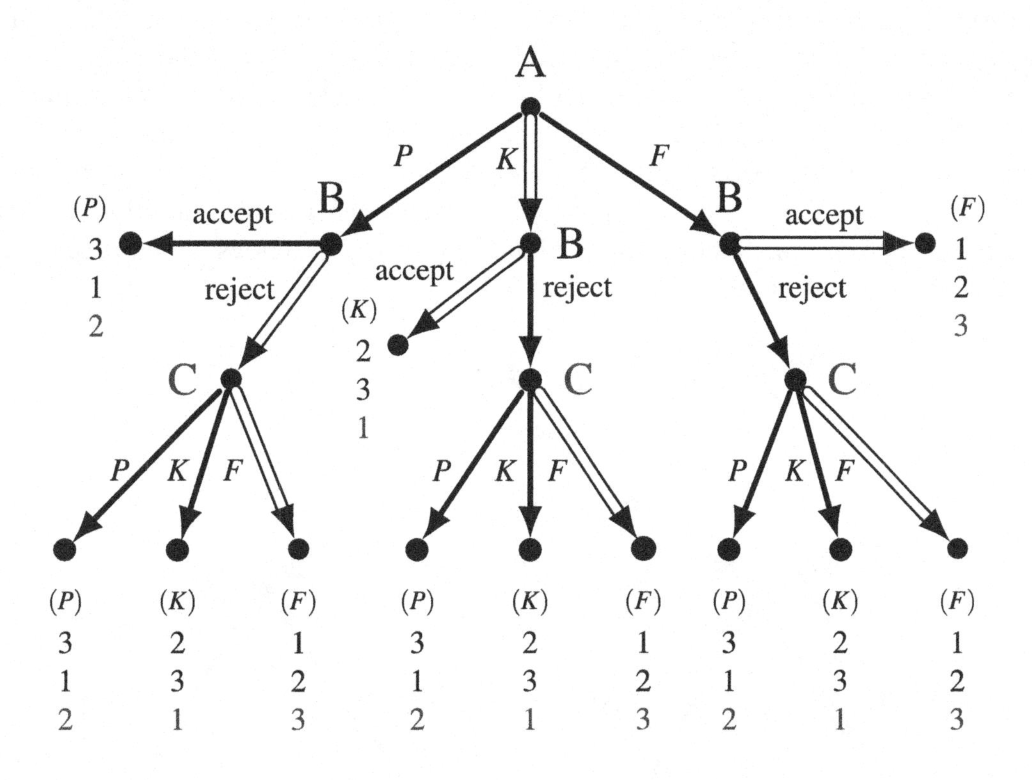

Figure 3.20: The first game for Exercise 3.4

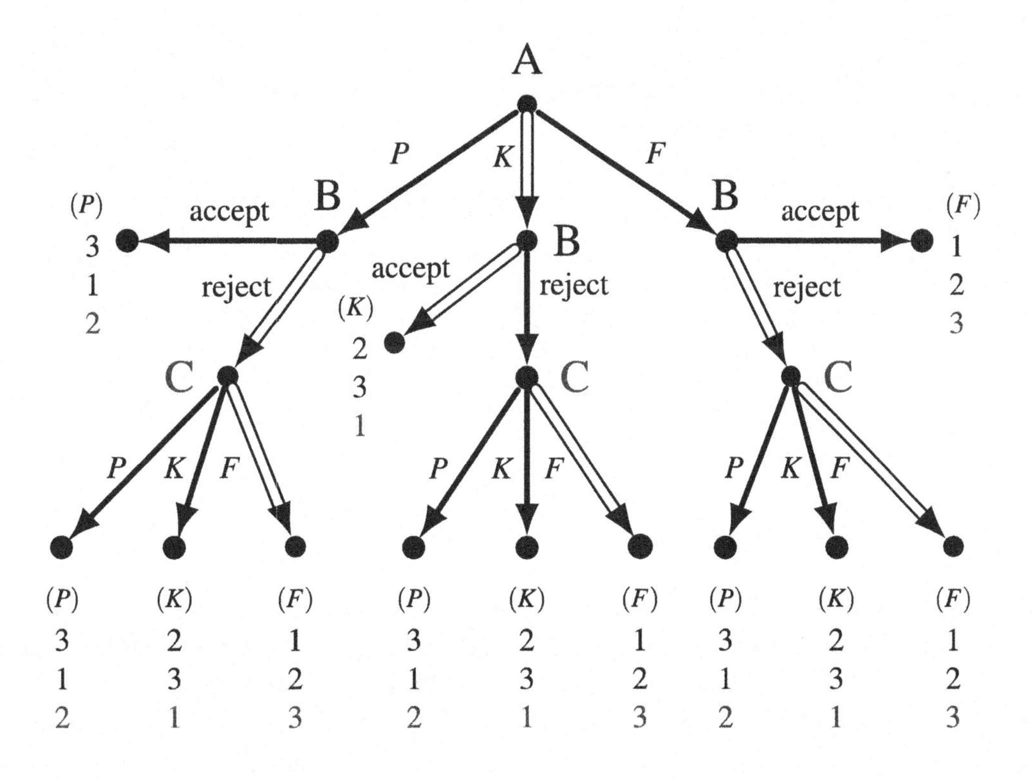

Figure 3.21: The second game for Exercise 3.4

Solution to Exercise 3.5. The game of Figure 3.2 is reproduced in Figure 3.22, with the unique backward-induction solution marked by double edges. The corresponding strategic form is shown In Figure 3.23 (for each of Player 2's strategies, the first element in the pair is what Player 2 would do at her left node and the second element what she would do at her right node). The Nash equilibria are highlighted. One Nash equilibrium, namely *(Offer 50-50,(Accept,Reject))*, corresponds to the backward induction solution, while the other Nash equilibrium, namely *(Offer 70-30,(Reject,Reject))* does not correspond to a backward-induction solution. □

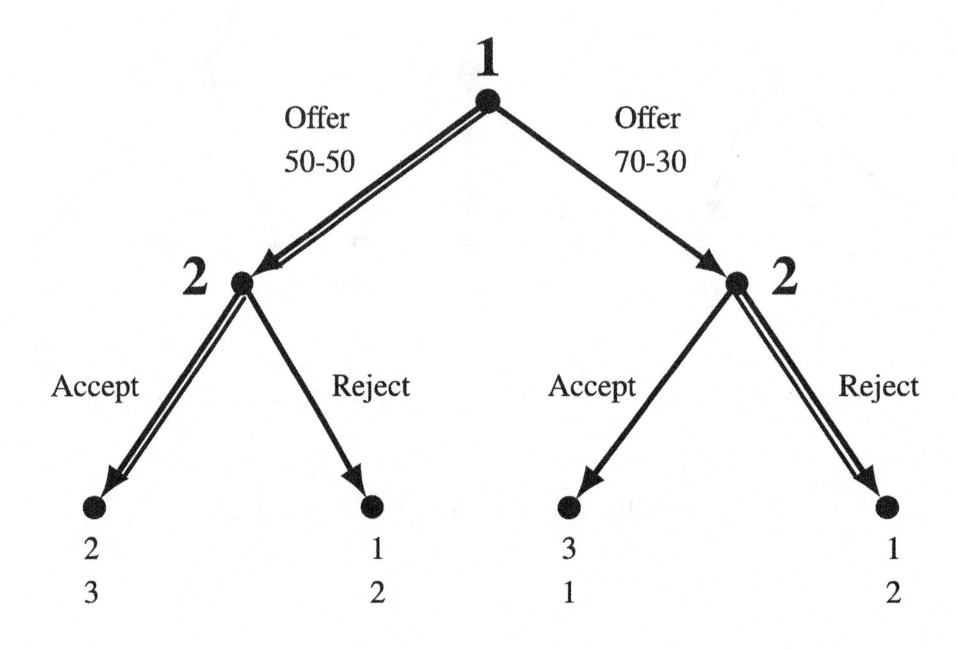

Figure 3.22: The extensive-form game for Exercise 3.5

		Player 2							
		(Accept,Accept)		(Accept,Reject)		(Reject,Accept)		(Reject,Reject)	
Player 1	Offer 50-50	2	3	2	3	1	2	1	2
	Offer 70-30	3	1	1	2	3	1	1	2

Figure 3.23: The strategic-form game for Exercise 3.5

Solution to Exercise 3.6. The game of Figure 3.3 is reproduced in Figure 3.24 with the two backward-induction solutions marked by double edges. The corresponding strategic form is shown in Figure 3.25. The Nash equilibria are highlighted. The backward-induction solutions are $(a,(c,f),g)$ and $(b,(c,e),h)$ and both of them are Nash equilibria. There are three more Nash equilibria which are not backward-induction solutions, namely $(b,(d,f),g),(a,(c,f),h)$ and $(b,(d,e),h)$. □

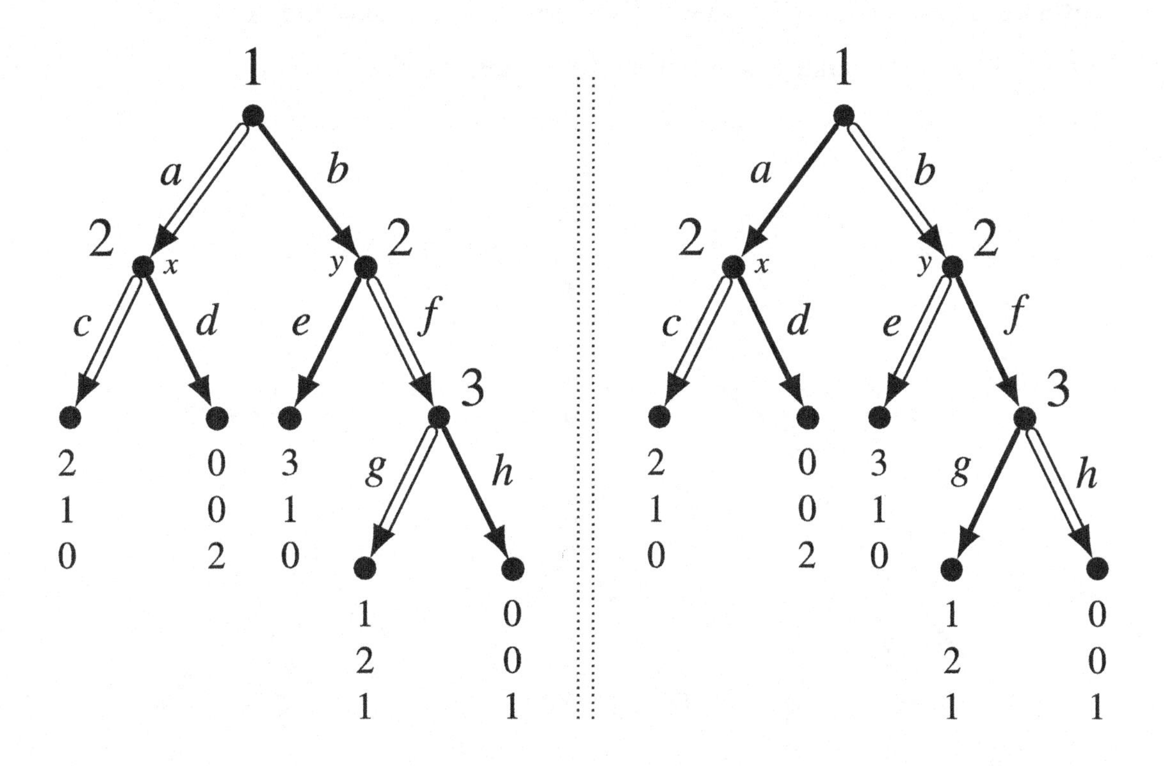

Figure 3.24: The two backward-induction solutions for the game of Exercise 3.6

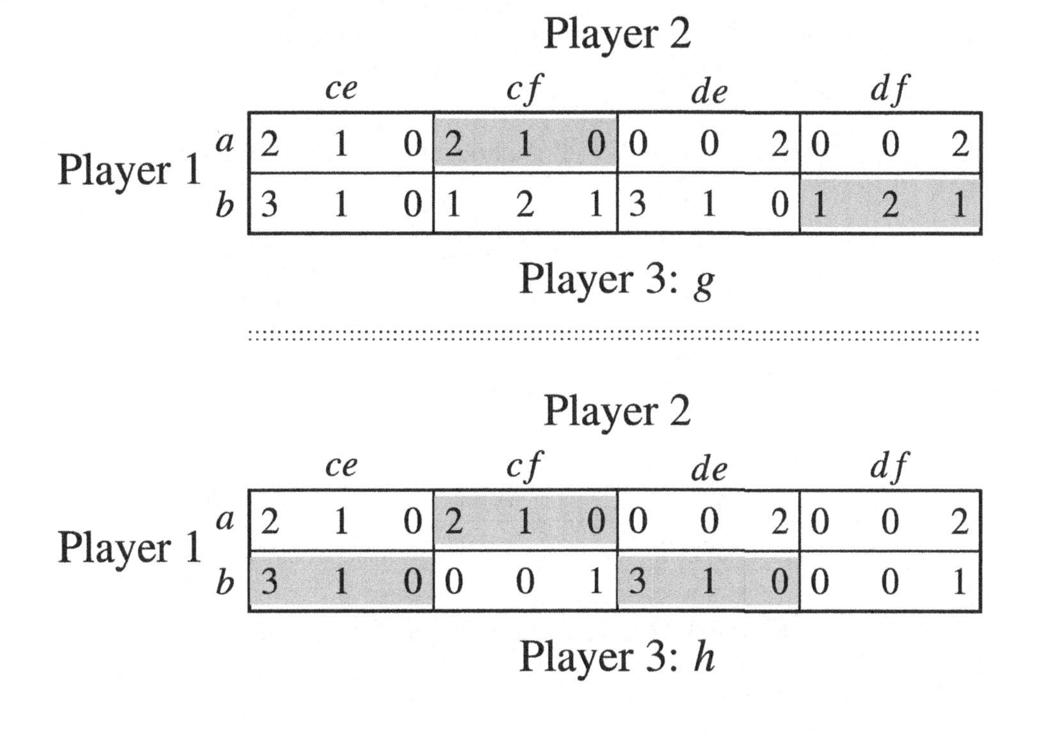

Player 2

Player 1	ce	cf	de	df
a	2 1 0	2 1 0	0 0 2	0 0 2
b	3 1 0	1 2 1	3 1 0	1 2 1

Player 3: g

Player 2

Player 1	ce	cf	de	df
a	2 1 0	2 1 0	0 0 2	0 0 2
b	3 1 0	0 0 1	3 1 0	0 0 1

Player 3: h

Figure 3.25: The strategic-form game for Exercise 3.6

Solution to Exercise 3.7. The game of Exercise 3.2 is reproduced in Figure 3.26.

(a) All the possible strategies of Player B are shown in Figure 3.27.

(b) Player C has three decision nodes and three choices at each of her nodes. Thus she has $3 \times 3 \times 3 = 27$ strategies. □

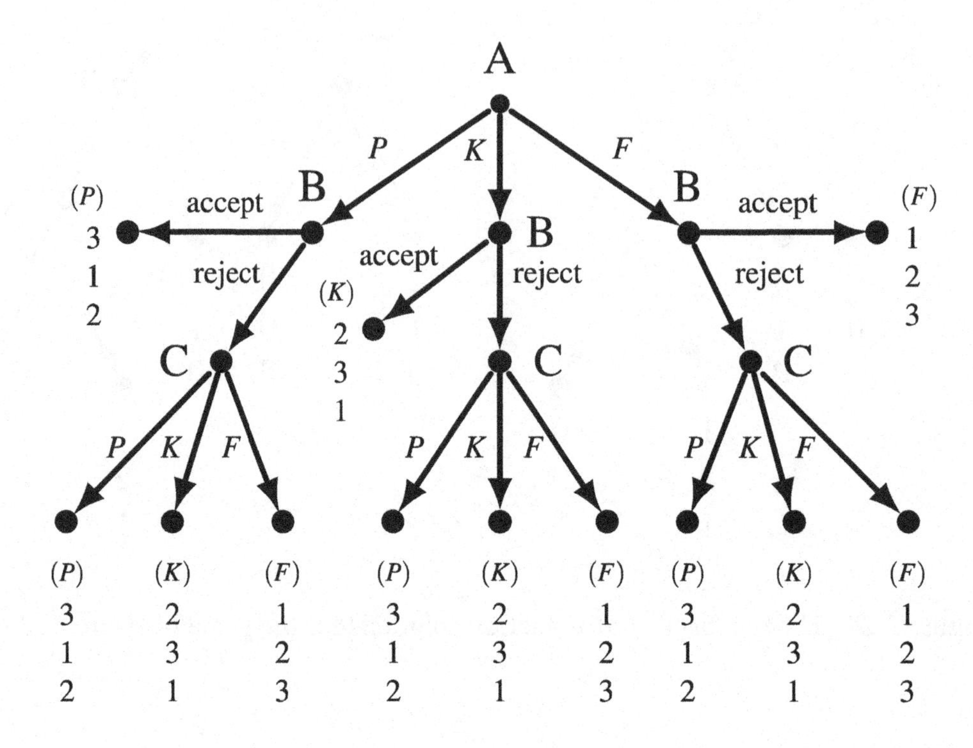

Figure 3.26: The extensive-form game for Exercise 3.7

	If A chooses P	If A chooses K	If A chooses F
1.	accept	accept	accept
2.	accept	accept	reject
3.	accept	reject	accept
4.	accept	reject	reject
5.	reject	accept	accept
6.	reject	accept	reject
7.	reject	reject	accept
8.	reject	reject	reject

Figure 3.27: The eight strategies of Player B

Solution to Exercise 3.8.

(a) One backward-induction solution is the strategy profile $((L,W),(a,e))$ shown by double edges in Figure 3.28. The corresponding backward-induction outcome is the play *La* with associated payoff vector $(2,1)$. The other backward-induction solution is the strategy profile $((R,W),(a,d))$ shown in Figure 3.29. The corresponding backward-induction outcome is the play *Rd* with associated payoff vector $(3,2)$.

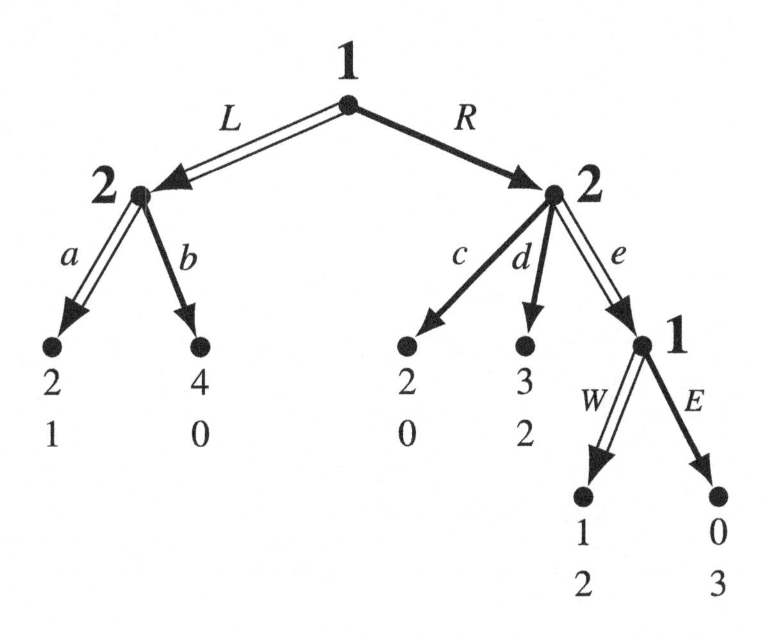

Figure 3.28: One backward-induction solution of the game of Part **(a)** of Exercise 3.8

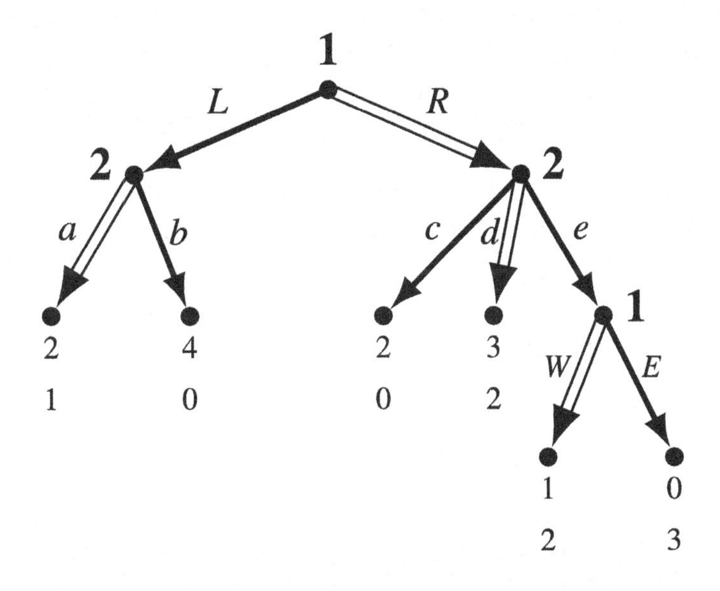

Figure 3.29: A second backward-induction solution of the game of Part **(a)** of Exercise 3.8

(b) Player 1 has four strategies: LW, LE, RW and RE.

(c) Player 2 has six strategies: ac, ad, ae, bc, bd and be.

(d) The strategic form is shown in Figure 3.30.

		Player 2					
		ac	*ad*	*ae*	*bc*	*bd*	*be*
Player 1	*LW*	2 1	2 1	2 1	4 0	4 0	4 0
	LE	2 1	2 1	2 1	4 0	4 0	4 0
	RW	2 0	3 2	1 2	2 0	3 2	1 2
	RE	2 0	3 2	0 3	2 0	3 2	0 3

Figure 3.30: The strategic-form game for Part (d) of Exercise 3.8

(e) Player 1 does not have a dominant strategy.

(f) For Player 2 *ae* is a weakly dominant strategy.

(g) There is no dominant strategy equilibrium.

(h) For Player 1 *RE* is weakly dominated by *RW* (and *LW* and *LE* are equivalent).

(i) For Player 2 *ac* is weakly dominated by *ad* (and *ae*), *ad* is weakly dominated by *ae*, *bc* is (strictly or weakly) dominated by every other strategy, *bd* is weakly dominated by *be* (and by *ae* and *ad*), *be* is weakly dominated by *ae*.
Thus the dominated strategies are: ac, ad, bc, bd and *be*.

(j) The iterative elimination of weakly dominated strategies yields the following reduced game (in Step 1 eliminate *RE* for Player 1 and ac, ad, bc, bd and *be* for Player 2; in Step 2 eliminate *RW* for Player 1):

		Player 2
		ae
Player 1	*LW*	2 , 1
	LE	2 , 1

Thus we are left with one of the two backward-induction solutions, namely $((L,W),(a,e))$ but also with $((L,E),(a,e))$ which is not a backward-induction solution.

(k) The Nash equilibria are highlighted in Figure 3.31.
There are five Nash equilibria: $(LW, ac), (LE, ac), (RW, ad), (LW, ae)$ and (LE, ae).
□

		Player 2					
		ac	*ad*	*ae*	*bc*	*bd*	*be*
Player 1	*LW*	2 1	2 1	2 1	4 0	4 0	4 0
	LE	2 1	2 1	2 1	4 0	4 0	4 0
	RW	2 0	3 2	1 2	2 0	3 2	1 2
	RE	2 0	3 2	0 3	2 0	3 2	0 3

Figure 3.31: The highlighted cells are the Nash equilibria (for Part **(k)** of Exercise 3.8)

Solution to Exercise 3.9.

(a) The extensive game is shown in Figure 3.32.

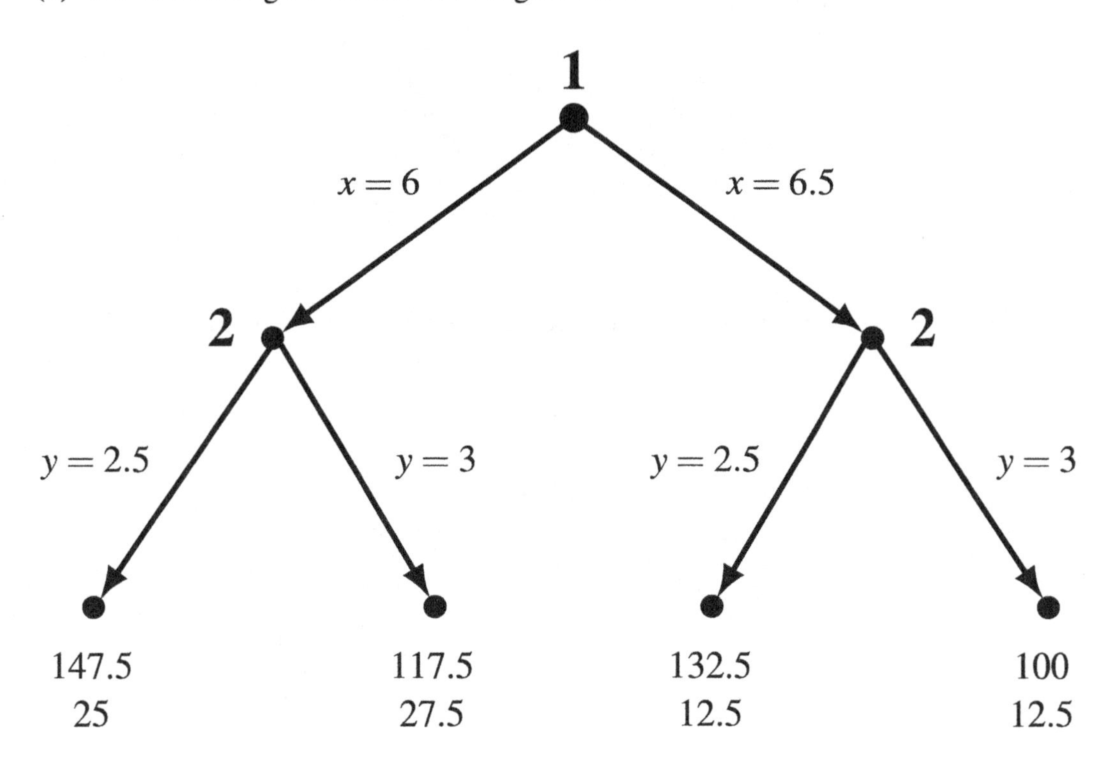

Figure 3.32: The extensive-form game for Exercise 3.9

(b) There are two backward-induction solutions.
The first is the strategy profile shown in Figure 3.33. The corresponding backward-induction outcome is given by Firm 1 producing 6 units and Firm 2 producing 3 units with profits 117.5 for Firm 1 and 27.5 for Firm 2.
The other backward-induction solution is the strategy profile shown in Figure 3.34. The corresponding backward-induction outcome is given by Firm 1 producing 6.5 units and Firm 2 producing 2.5 units with profits 132.5 for Firm 1 and 12.5 for Firm 2.

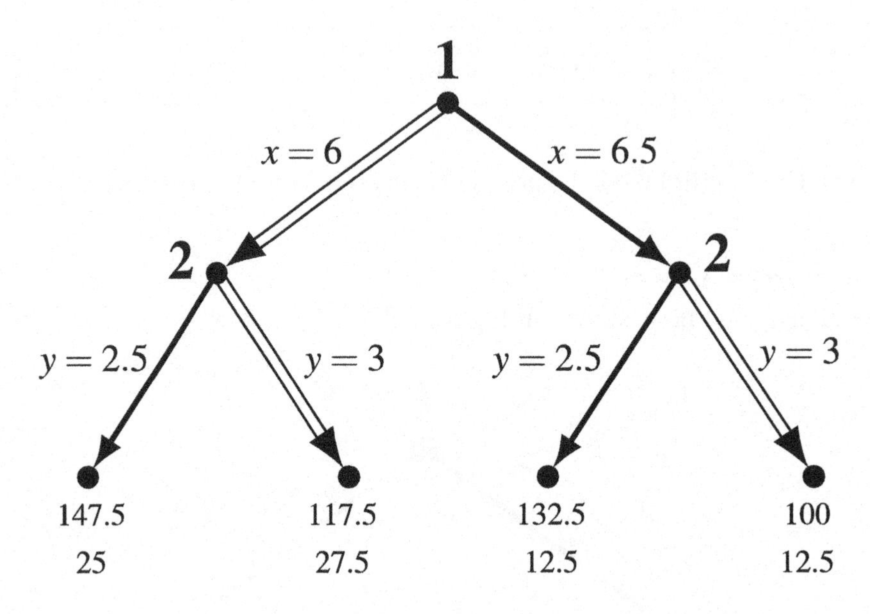

Figure 3.33: One backward-induction solution of the game of Figure 3.32

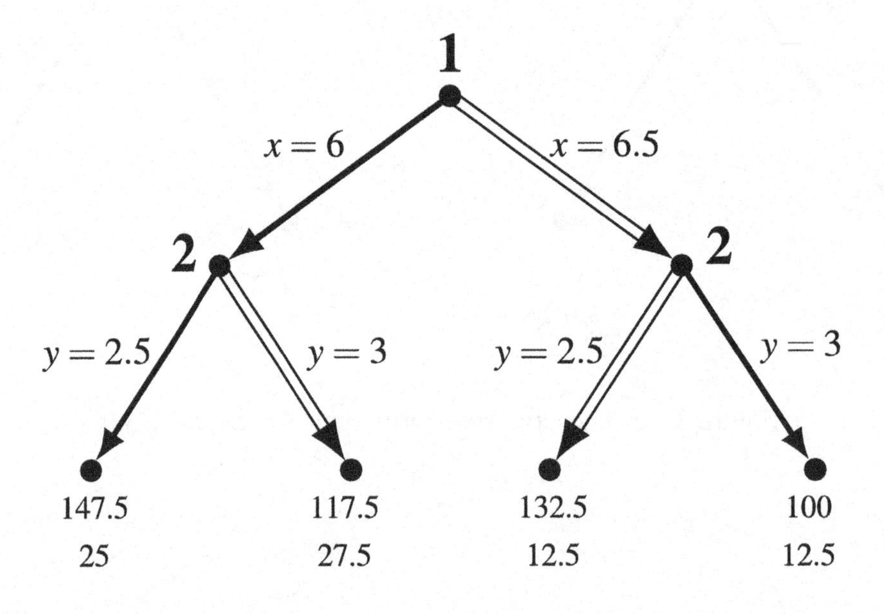

Figure 3.34: A second backward-induction solution of the game of Figure 3.32

(c) The strategic form is shown in Figure 3.35.

		Firm 2			
		(2.5, 2.5)	(2.5, 3)	(3, 2.5)	(3, 3)
Firm 1	6	147.5 25	147.5 25	117.5 27.5	**117.5 27.5**
	6.5	132.5 12.5	100 12.5	**132.5 12.5**	100 12.5

Figure 3.35: The strategic-form game for Part (d) of Exercise 3.9

(d) The Nash equilibria are highlighted in Figure 3.35. In this game the set of Nash equilibria coincides with the set of backward-induction solutions. □

Solution to Exercise 3.10. The game under consideration is shown in Figure 3.36, where x is an integer.

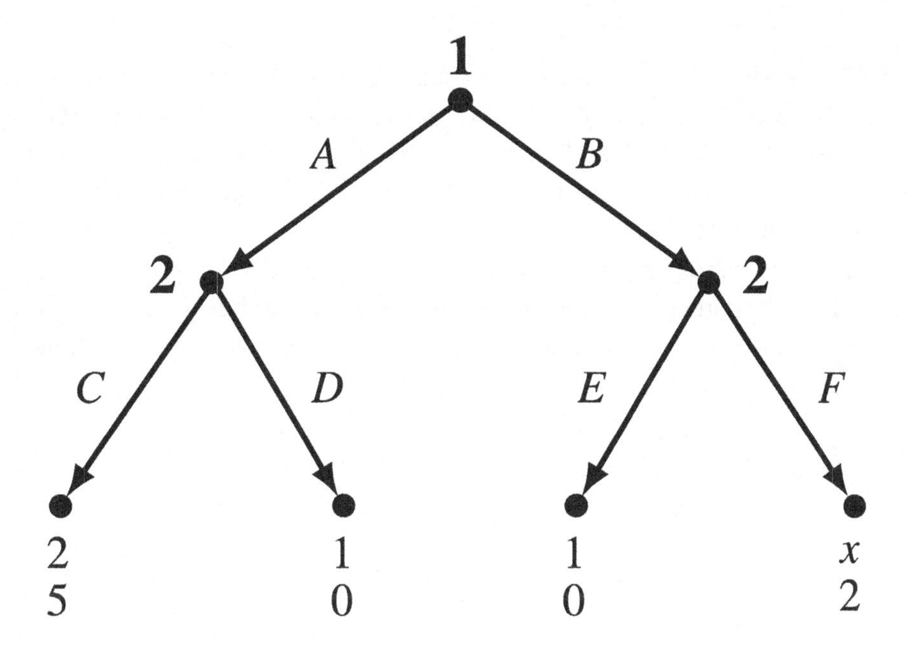

Figure 3.36: The extensive-form game for Part **(a)** of Exercise 3.10

(a) The backward-induction strategy of Player 2 is the same, no matter what x is, namely (C,F). Thus the backward induction solutions are as follows.

- If $x < 2$, there is only one: $(A,(C,F))$.
- If $x = 2$ there are two: $(A,(C,F))$ and $(B,(C,F))$.
- $Ifx > 2$, there is only one: $(B,(C,F))$.

(b) The strategic form is shown in Figure 3.37. First note that $(\mathbf{A},(\mathbf{C},\mathbf{E}))$ **is a Nash equilibrium for every value of** x. Now, depending on the value of x the other Nash equilibria are as follows:

- If $x<1$, $(A,(C,F))$.
- If $1\leq x<2$, $(A,(C,F))$ and $(B,(D,F))$.
- If $x=2$, $(A,(C,F)),(B,(C,F))$ and $(B,(D,F))$.
- If $x>2$, $(B,(C,F))$ and $(B,(D,F))$. □

Player 2

		CE		*CF*		*DE*		*DF*	
Player 1	*a*	2	5	2	5	1	0	1	0
	b	1	0	x	2	1	0	x	2

Figure 3.37: The strategic-form game for Part **(b)** of Exercise 3.10

Solution to Exercise 3.11. Let us find the losing positions. If Player i, with his choice, can bring the sum to **40** then he can win (the other player with her next choice will take the sum to a number between 41 and 47 and then Player i can win with his next choice). Working backwards, the previous losing position is **32** (from here the player who has to move will take the sum to a number between 33 and 39 and after this the opponent can take it to 40). Reasoning backwards, the earlier losing positions are **24**, **16**, **8** and **0**. Thus Player 1 starts from a losing position and therefore it is Player 2 who has a winning strategy. The winning strategy is: at every turn, if Player 1's last choice was n then Player 2 should choose $(\mathbf{8}-\mathbf{n})$. □

Solution to Exercise 3.12.

(a) The initial part of the game is shown in Figure 3.38.

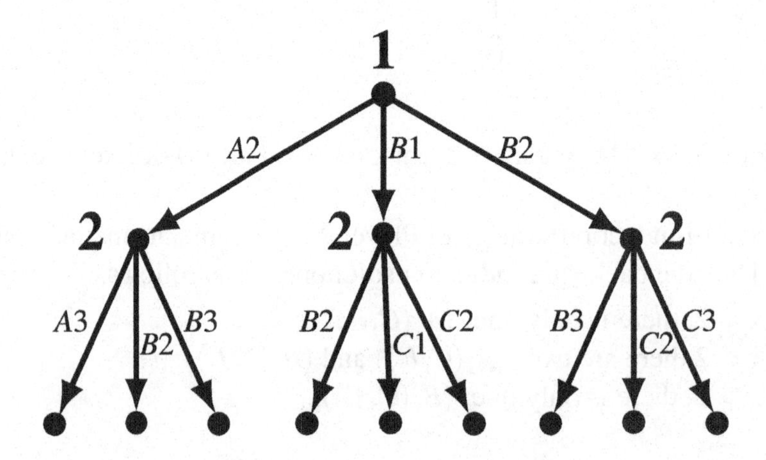

Figure 3.38: The initial part of the game of Part **(a)** of Exercise 3.12

(b) From G4 Player 1 should move the coin to H5. From there Player 2 has to move it to H6 and Player 1 to H7 and Player 2 to H8 and from there Player 1 wins by moving it to H9.

(c) $A1 \xrightarrow{1} B2 \xrightarrow{2} C3 \xrightarrow{1} D4 \xrightarrow{2} E5 \xrightarrow{1} F5 \xrightarrow{2} G6 \xrightarrow{1} H7 \xrightarrow{2} H8 \xrightarrow{1} H9.$

(d) $A1 \xrightarrow{1} B2 \xrightarrow{2} C3 \xrightarrow{1} D4 \xrightarrow{2} E5 \xrightarrow{1} F5 \xrightarrow{2} G6 \xrightarrow{1} G7 \xrightarrow{2} H7 \xrightarrow{1} H8 \xrightarrow{2} H9.$

(e) Using backward induction we can label each cell with a W (meaning that the player who has to move when the coin is there has a winning continuation strategy) or with an L (meaning that the player who has to move when the coin is there can be made to lose).
If all the cells that are accessible from a given cell are marked with a W then that cell must be marked with an L.
If from a cell there is an accessible cell marked with an L then that cell should be marked with a W. See Figure 3.39.
From the picture it is clear that it is Player 1 who has a winning strategy. The winning strategy of Player 1 is: move the coin to cell B1 and from then on, after every move of Player 2, move the coin to a cell marked L. □

	H	G	F	E	D	C	B	A
9	END	W	L	W	L			
8	W	W	W		W	W	L	
7	L	W	L	W	L		W	
6	W	W				W	L	W
5	L	W	L	W		L	W	W
4	W	W		W	L	W		L
3	L		L	W	W	W	L	W
2		W	W		L	W	W	W
1		L	W	L	W	W	L	START

Figure 3.39: Solution for the coin game

Solution to Exercise 3.13.

(a) The game is shown in Figure 3.40.

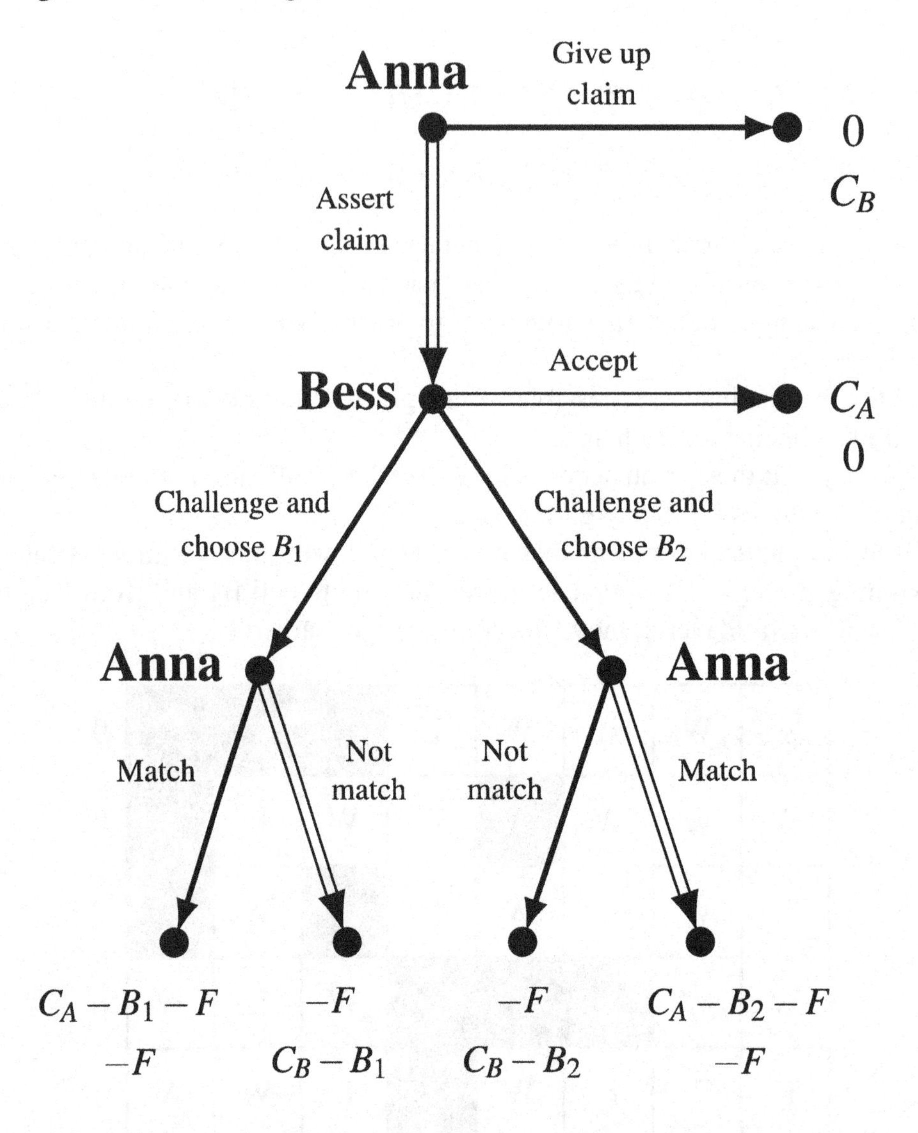

Figure 3.40: The extensive-form game for Part **(a)** of Exercise 3.13

(b) The backward-induction solution is marked by double arrows in Figure 3.40.

(c) The sequence of moves is shown in Figure 3.41.

Suppose that Anna is the legal owner and values the ring more than Bess does: $C_A > C_B$. At the last node Anna will choose "match" if $C_A > B$ and "don't match" if $B > C_A$. In the first case Bess's payoff will be $-F$, while in the second case it will be $C_B - B$, which is negative since $B > C_A$ and $C_A > C_B$. Thus in either case Bess's payoff would be negative. Hence at her decision node Bess will choose "accept" (Bess can get the ring at this stage only if she bids more than the ring is worth to her). Anticipating this, Anna will assert her claim at the first decision node. Thus at the backward-induction solution the ring goes to Anna, the legal owner. The payoffs are C_A for Anna and 0 for Bess. **Note that no money changes hands.**

(d) Suppose that Bess is the legal owner and values the ring more than Anna does: $C_B > C_A$. At the last node Anna will choose "match" if $C_A > B$ and "don't match" if $B > C_A$. In the first case Bess's payoff will be $-F$, while in the second case it will be $C_B - B$, which will be positive as long as $C_B > B$. Hence at her decision node Bess will choose to challenge and bid any amount B such that $C_B > B > C_A$. Anticipating this, at her first decision node Anna will give up (and get a payoff of 0), because if she asserted her claim then her final payoff would be negative. Thus at the backward-induction solution the ring goes to Bess, the legal owner. The payoffs are 0 for Anna and C_B for Bess. **Note that no money changes hands.**

(e) As pointed out above, in both cases no money changes hands at the backward-induction solution. Thus Judge Sabio collects no money at all and both Ann and Bess pay nothing. □

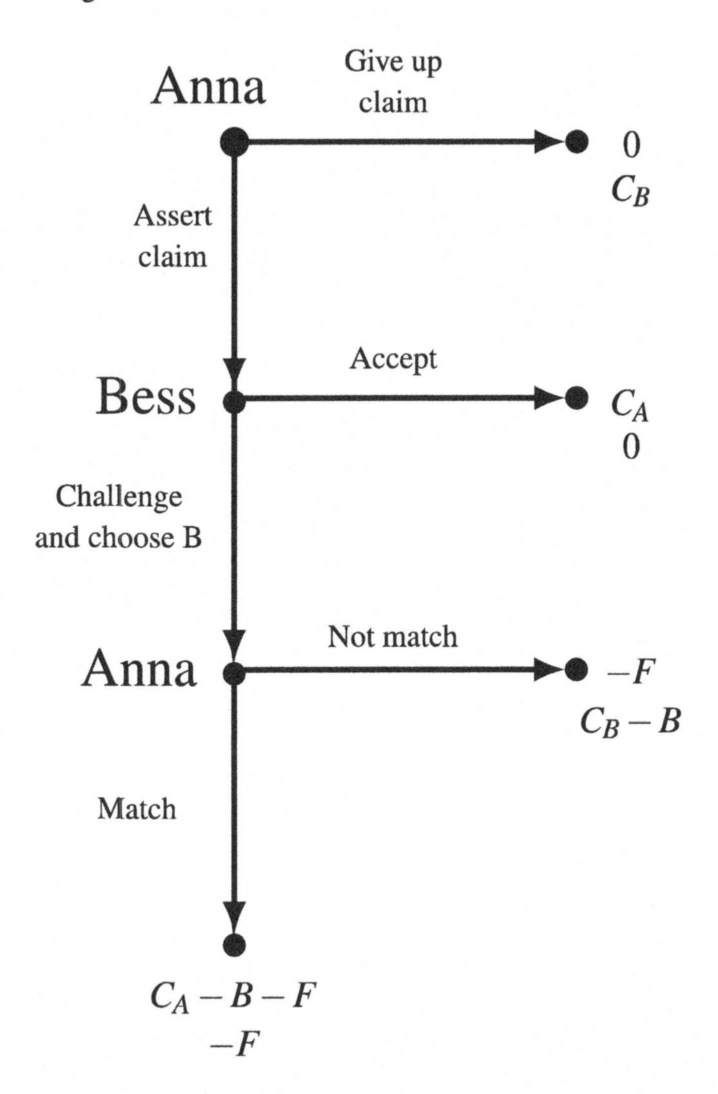

Figure 3.41: The extensive-form game for Part ((**c**) of Exercise 3.13

4. General Dynamic Games

4.1 Imperfect Information

There are many situations where players have to make decisions with only partial information about previous moves by other players. Here is an example from my professional experience: in order to discourage copying and cheating in exams, I prepare two versions of the exam, print one version on white paper and the other on pink paper and distribute the exams in such a way that if a student gets, say, the white version then the students on his left and right have the pink version. For simplicity let us assume that there is only one question in the exam. What matters for my purpose is not that the question is indeed different in the two versions, but rather that the students *believe* that they are different and thus refrain from copying from their neighbors. The students, however, are not naïve and realize that I might be bluffing; indeed, introducing differences between the two versions of the exam involves extra effort on my part. Consider a student who finds himself in the embarrassing situation of not having studied for the final exam and is tempted to copy from his neighbor, whom he knows to be a very good student. Let us assume that, if he does not copy, then he turns in a blank exam; in this case, because of his earlier grades in the quarter, he will get a C; on the other hand, if he copies he will get an A if the two versions are identical but will be caught cheating and get an F if the two versions are slightly different. How can we represent such a situation?

Clearly this is a situation in which decisions are made sequentially: first the Professor decides whether to write identical versions (albeit printed on different-color paper) or different versions and then the Student chooses between copying and leaving the exam blank. We can easily represent this situation using a tree as we did with the case of perfect-information games, but the crucial element here is the fact that the Student *does not know* whether the two versions are identical or different. In order to represent this uncertainty (or lack of information) in the mind of the Student, we use the notion of *information set*. An information set for a player is a collection of decision nodes of that player and the interpretation is that the player does not know at which of these nodes he is making his decision. Graphically, we represent an information set by enclosing the corresponding nodes in a rounded rectangle. Figure 4.1 represents the situation described above.

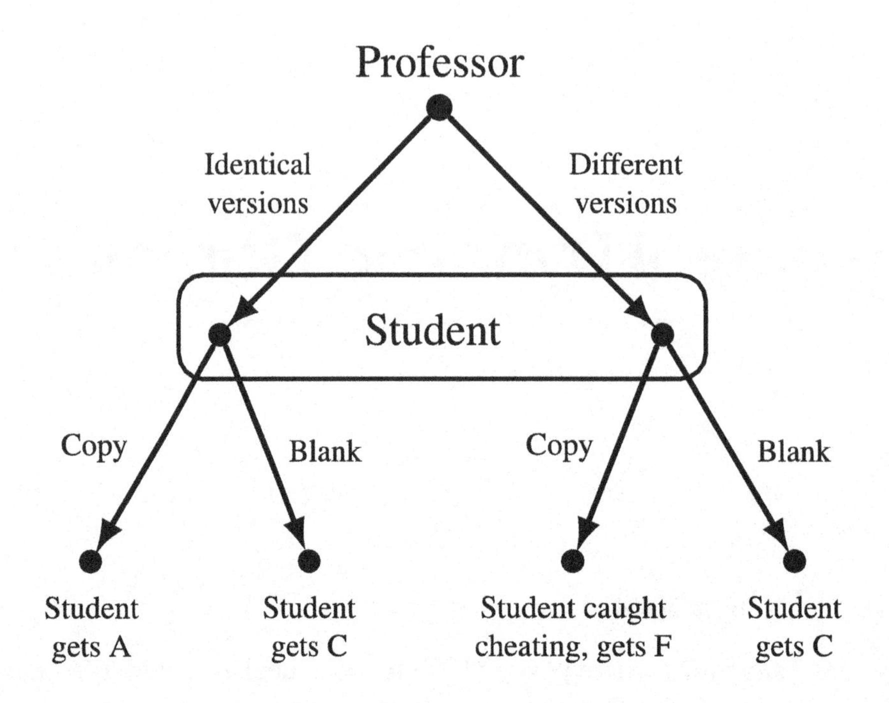

Figure 4.1: An extensive form, or frame, with imperfect information

As usual we need to distinguish between a game-frame and a game. Figure 4.1 depicts a game-frame: in order to obtain a game from it we need to add a ranking of the outcomes for each player. For the moment we shall ignore payoffs and focus on frames. A game-frame such as the one shown in Figure 4.1 is called an *extensive form (or frame) with imperfect information:* in this example it is the Student who has imperfect information, or uncertainty, about the earlier decision of the Professor.

Before we give the definition of extensive form, we need to introduce some additional terminology and notation. Given a directed tree and two nodes x and y we say that *y is a successor of x*, or *x is a predecessor of y*, if there is a sequence of directed edges from x to y.[1]

A *partition* of a set H is a collection $\mathscr{H} = \{H_1, \dots, H_m\}$ $(m \geq 1)$ of non-empty subsets of H such that:
(1) any two elements of $\mathscr{H}$ are disjoint (if $H_j, H_k \in \mathscr{H}$ with $j \neq k$ then $H_j \cap H_k = \emptyset$) and
(2) the elements of $\mathscr{H}$ cover H: $H_1 \cup \ldots \cup H_m = H$.

[1]If the sequence consists of a single directed edge then we say that y is an *immediate successor* of x or x is the *immediate predecessor* of y.

The definition of extensive form given below allows for perfect information as a special case. The first four items of Definition 4.1.1 (marked by the bullet symbol •), coincide with Definition 3.1.1 in Chapter 3 (which covers the case of perfect information); what is new is the additional item marked by the symbol ★.

Definition 4.1.1 A *finite extensive form (or frame) with perfect recall* consists of the following items.

- A finite rooted directed tree.
- A set of players $I = \{1,\ldots,n\}$ and a function that assigns one player to every decision node.
- A set of actions A and a function that assigns one action to every directed edge, satisfying the restriction that no two edges out of the same node are assigned the same action.
- A set of outcomes O and a function that assigns an outcome to every terminal node.

★ For every player $i \in I$, a partition $\mathscr{D}_i$ of the set D_i of decision nodes assigned to player i (thus $\mathscr{D}_i$ is a collection of mutually disjoint subsets of D_i whose union is equal to D_i). Each element of $\mathscr{D}_i$ is called an *information set of player i*.
The elements of $\mathscr{D}_i$ satisfy the following restrictions:
(1) the actions available at any two nodes in the same information set must be the same (that is, for every $D \in \mathscr{D}_i$, if $x, y \in D$ then the outdegree of x is equal to the outdegree of y and the set of actions assigned to the directed edges out of x is equal to the set of actions assigned to the directed edges out of y),
(2) if x and y are two nodes in the same information set then it is not the case that one node is a predecessor of the other,
(3) each player has *perfect recall* in the sense that if node $x \in D \in \mathscr{D}_i$ is a predecessor of node $y \in D' \in \mathscr{D}_i$ (thus, by (2), $D \neq D'$), and a is the action assigned to the directed edge out of x in the sequence of edges leading from x to y, then for every node $z \in D'$ there is a predecessor $w \in D$ such that the action assigned to the directed edge out of w in the sequence of edges leading from w to z is that same action a.

The perfect-recall restriction says that if a player takes action a at an information set and later on has to move again, then at the later time she remembers that she took action a at that earlier information set (because every node she is uncertain about at the later time comes after taking action a at that information set). Perfect recall can be interpreted as requiring that *a player always remember what she knew in the past and what actions she herself took in the past.*

Figure 4.2 shows two examples of violation of perfect recall. In the frame shown in Panel (*i*) Player 1 first chooses between a and b and then chooses between c and d having forgotten his previous choice: he does not remember *what* he chose previously. In the frame shown in Panel (*ii*) when Player 2 has to choose between e and f she is uncertain whether this is the first time she moves (left node) or the second time (right node): she is uncertain *whether* she moved in the past.

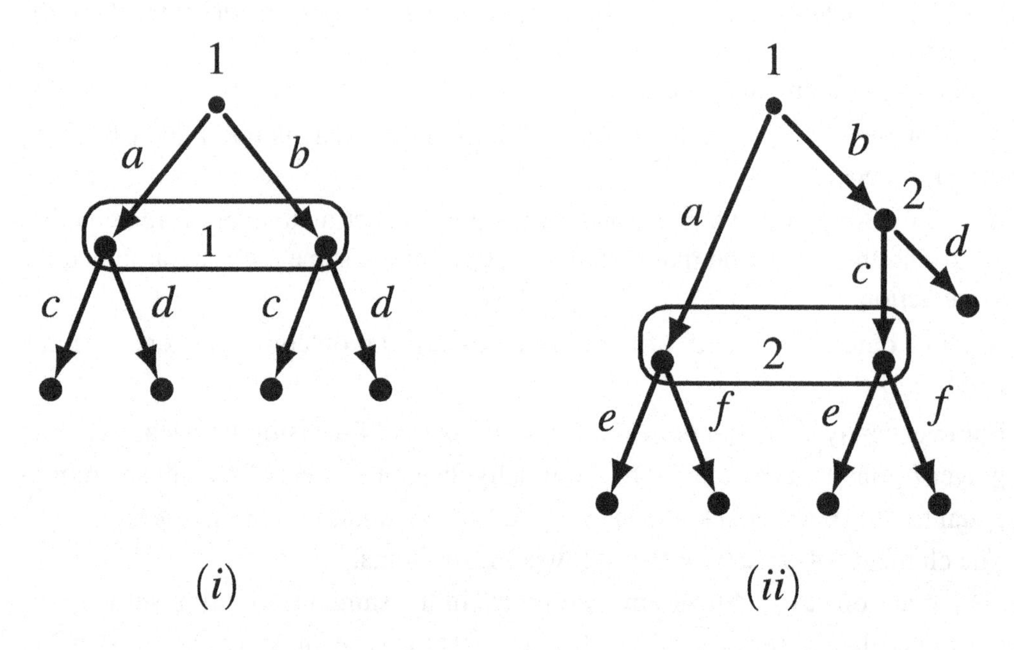

Figure 4.2: Examples of violations of perfect recall

If every information set of every player consists of a single node, then the frame is said to be a *perfect-information frame:* it is easy to verify that, in this case, the last item of Definition 4.1.1 (marked by the symbol ★) is trivially satisfied and thus Definition 4.1.1 coincides with Definition 3.1.1 (Chapter 3). Otherwise (that is, if at least one player has at least one information set that consists of at least two nodes), the frame is said to have *imperfect information.* An example of an extensive frame with imperfect information is the one shown in Figure 4.1. We now give two more examples. In order to simplify the figures, when representing an extensive frame we enclose an information set in a rounded rectangle if and only if that information set contains at least two nodes.

■ **Example 4.1** There are three players, Ann, Bob and Carla. Initially, Ann and Bob are in the same room and Carla is outside the room. Ann moves first, chooses either a red card or a black card from a full deck of cards, shows it to Bob and puts it, face down, on the table. Now Carla enters the room and Bob makes a statement to Carla: he either says "Ann chose a Red card" or he says "Ann chose a Black card"; Bob could be lying or could be telling the truth. After hearing Bob's statement Carla guesses the color of the card that was picked by Ann. The card is then turned and if Carla's guess was correct then Ann and Bob give \$1 each to Carla, otherwise Carla gives \$1 to each of Ann and Bob. When drawing an extensive frame to represent this situation, it is important to be careful about what Carla knows, when she makes her guess, and what she is uncertain about. The extensive frame is shown in Figure 4.3. ■

- Carla's top information set captures the situation she is in after hearing Bob say "Ann chose a black card" and not knowing if he is telling the truth (left node) or he is lying (right node).
- Carla's bottom information set captures the alternative situation where she hears Bob say "Ann chose a black card" and does not know if he is lying (left node) or telling the truth (right node).
- In both situations Carla knows something, namely what Bob tells her, but lacks information about something else, namely what Ann chose.
- The fact that Bob knows the color of the card chosen by Ann is captublack by giving Bob two information sets, each consisting of a single node: Bob's left node represents the situation he is in when he sees that Ann picked a black card, while his right node represents the situation he is in when he sees that Ann picked a black card.

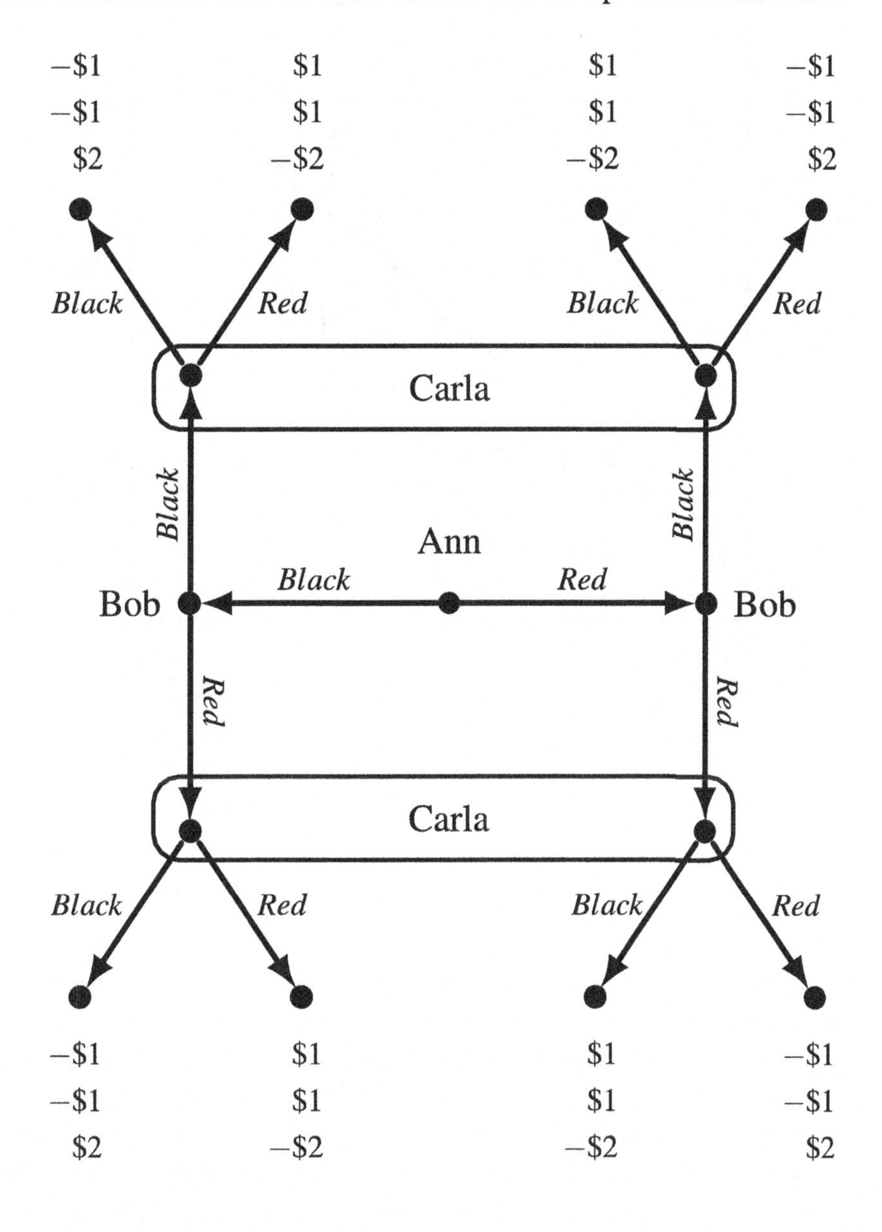

Figure 4.3: The extensive form, or frame, representing Example 4.1

■ **Example 4.2** Yvonne and Fran were both interviewed for the same job, but only one person can be hired. The employer told each candidate: "don't call me, I will call you if I want to offer you the job". He also told them that he desperately needs to fill the position and thus, if turned down by one candidate, he will automatically make the offer to the other candidate, without revealing whether he is making a first offer or a "recycled" offer. This situation is represented in the extensive frame shown in Figure 4.4. ■

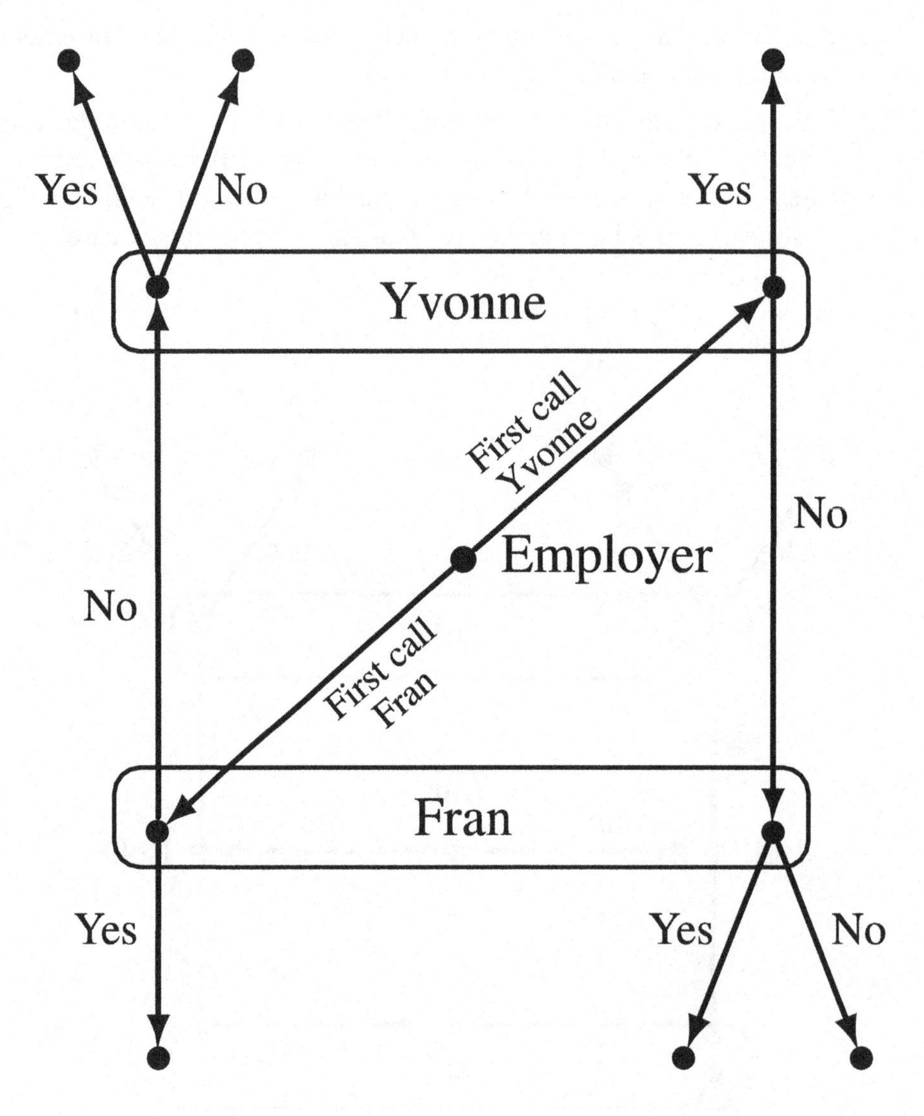

Figure 4.4: The extensive form, or frame, representing Example 4.2

As before, in order to obtain a *game* from an extensive *frame* all we need to do is add a ranking of the outcomes for each player. As usual, the best way to represent such rankings is by means of an ordinal utility function for each player and thus represent an extensive-form game by associating a vector of utilities with each terminal node. For instance, expanding on Example 4.2, suppose that the employer only cares about whether the position is filled or not, prefers filling the position to not filling it, but is indifferent between filling it with Yvonne or with Fran; thus we can assign a utility of 1 for the employer to every outcome where one of the two candidates accepts the offer and a utility of 0 to every other outcome.

Yvonne's favorite outcome is to be hired if she was the recipient of the first call by the employer; her second best outcome is not to be hired and her worst outcome is to accept a recycled offer (in the latter case Fran would take pleasure telling Yvonne "You took that job?! It was offered to me but I turned it down. Who, in her right mind, would want that job? What's wrong with you?!"). Thus for Yvonne we can use utilities of 2 (if she accepts a first offer), 1 (if she is not hired) and 0 (if she accepts a recycled offer). Finally, suppose that Fran has preferences similar (but symmetric) to Yvonne's. Then the extensive frame of Figure 4.4 gives rise to the extensive game shown in Figure 4.5.

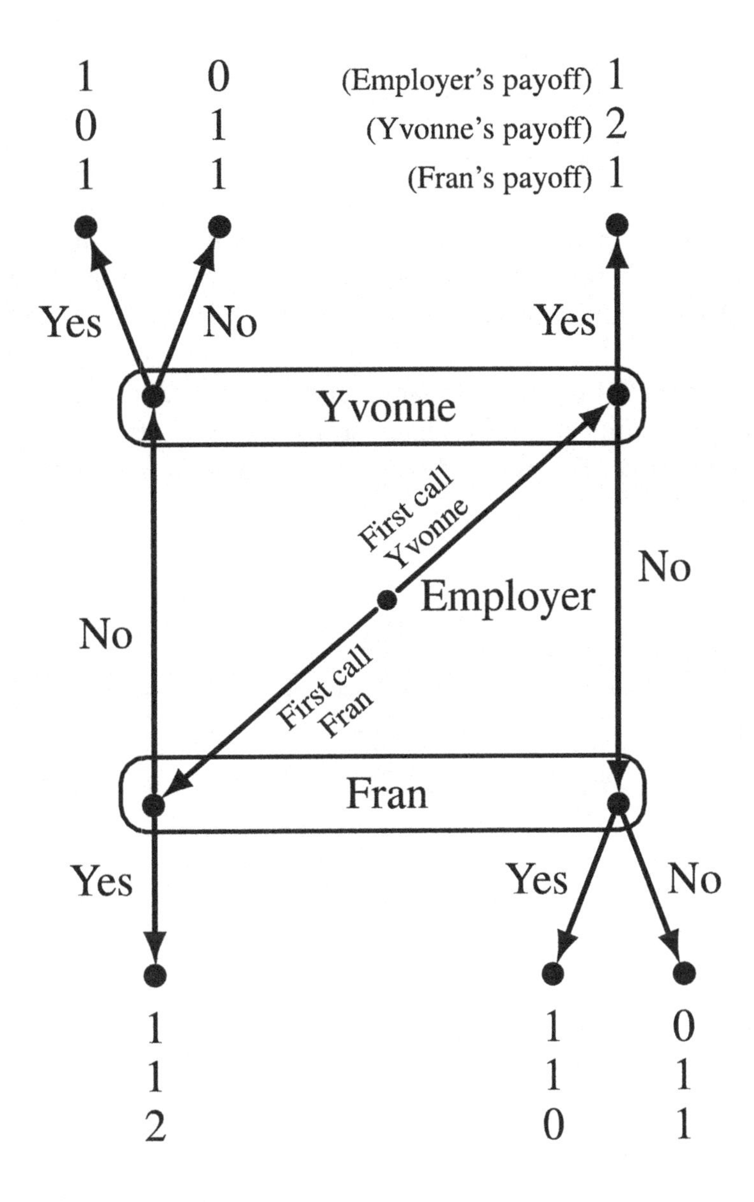

Figure 4.5: A game based on the extensive form of Figure 4.4

Test your understanding of the concepts introduced in this section, by going through the exercises in Section 4.6.1 at the end of this chapter.

4.2 Strategies

The notion of strategy for general extensive games is the same as before: a strategy for Player i is a complete, contingent plan that covers all the possible situations Player i might find herself in. In the case of a perfect-information game a "possible situation" for a player is a decision node of that player; in the general case, where there may be imperfect information, a "possible situation" for a player is an *information set* of that player.

The following definition reduces to Definition 3.3.1 (Chapter 3) if the game is a perfect-information game (where each information set consists of a single node).

Definition 4.2.1 A *strategy* for a player in an extensive-form game is a list of choices, one for every information set of that player.

For example, in the game of Figure 4.5, Yvonne has only one information set and thus a strategy for her is what to do at that information set, namely either say Yes or say No. Yvonne cannot make the plan "if the employer calls me first I will say Yes and if he calls me second I will say No", because when she receives the call she is not told if this is a first call or a recycled call and thus she cannot make her decision dependent on information she does not have.

As in the case of perfect-information games, the notion of strategy allows us to associate with every extensive-form game a strategic-form game. For example, the strategic form associated with the game of Figure 4.5 is shown in Figure 4.6 with the Nash equilibria highlighted.

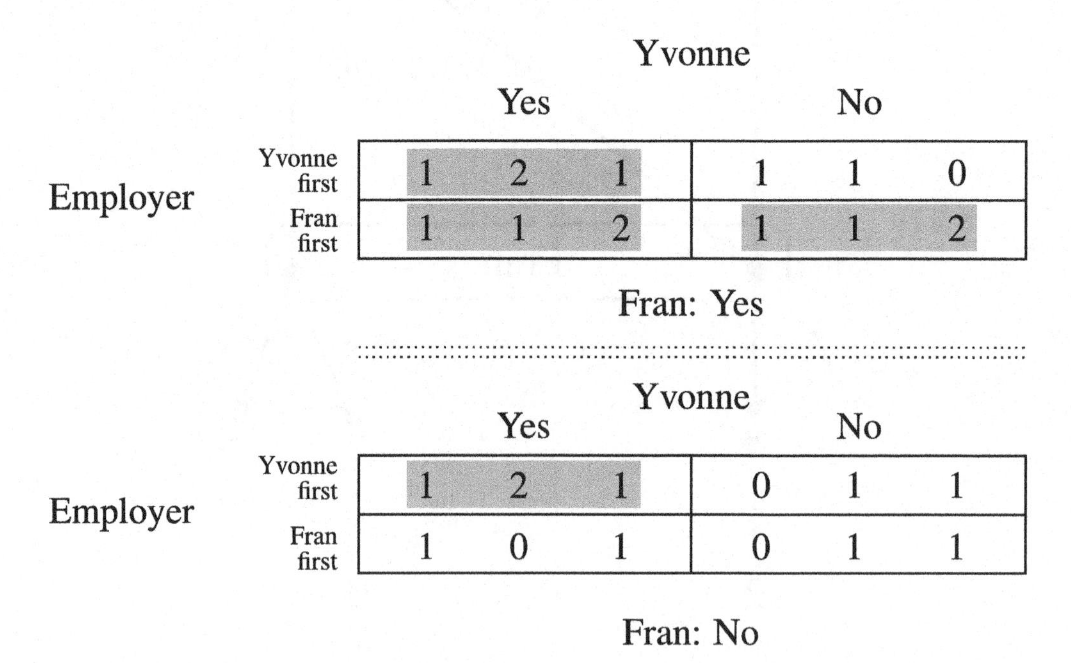

Fran: Yes

Employer	Yvonne: Yes	Yvonne: No
Yvonne first	1 2 1	1 1 0
Fran first	1 1 2	1 1 2

Fran: No

Employer	Yvonne: Yes	Yvonne: No
Yvonne first	1 2 1	0 1 1
Fran first	1 0 1	0 1 1

Figure 4.6: The strategic form of the game of Figure 4.5 with the Nash equilibria highlighted

As another example, consider the extensive form of Figure 4.3 and view it as a game by assuming that each player is selfish and greedy (only cares about how much money he/she gets and prefers more money to less). Then the associated strategic form is shown in Figure 4.7, where Bob's strategy (x, y) means "I say x if Ann chose a black card and I say y if Ann chose a red card". Thus (R, B) means "if Ann chose a black card I say Red and if Ann chose a red card I say Black" (that is, Bob plans to lie in both cases). Similarly, Carla's strategy (x, y) means "I guess x if Bob tells me Black and I guess y if Bob tells me Red". Thus (B, R) means "if Bob tells me Black I guess Black and if Bob tells me Red I guess Red" (that is, Carla plans to repeat what Bob says).

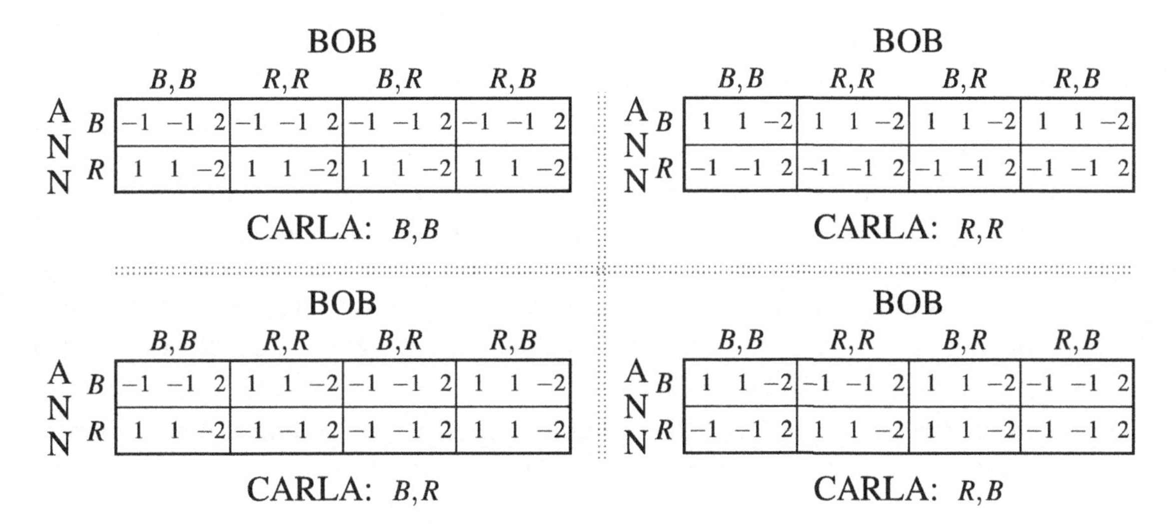

CARLA: B,B

ANN \ BOB	B,B	R,R	B,R	R,B
B	−1 −1 2	−1 −1 2	−1 −1 2	−1 −1 2
R	1 1 −2	1 1 −2	1 1 −2	1 1 −2

CARLA: R,R

ANN \ BOB	B,B	R,R	B,R	R,B
B	1 1 −2	1 1 −2	1 1 −2	1 1 −2
R	−1 −1 2	−1 −1 2	−1 −1 2	−1 −1 2

CARLA: B,R

ANN \ BOB	B,B	R,R	B,R	R,B
B	−1 −1 2	1 1 −2	−1 −1 2	1 1 −2
R	1 1 −2	−1 −1 2	−1 −1 2	1 1 −2

CARLA: R,B

ANN \ BOB	B,B	R,R	B,R	R,B
B	1 1 −2	−1 −1 2	1 1 −2	−1 −1 2
R	−1 −1 2	1 1 −2	1 1 −2	−1 −1 2

Figure 4.7: The strategic form of the game of Figure 4.3

In order to "solve" an extensive-form game we could simply construct the associated strategic-form game and look for the Nash equilibria. However, we saw in Chapter 3 that in the case of perfect-information games not all Nash equilibria of the associated strategic form can be considered "rational solutions" and we introduced the notion of backward induction to select the "reasonable" Nash equilibria. What we now need is a generalization of the notion of backward induction that can be applied to general extensive-form games. This generalization is called *subgame-perfect equilibrium*. First we need to define the notion of subgame.

Test your understanding of the concepts introduced in this section, by going through the exercises in Section 4.6.2 at the end of this chapter.

4.3 Subgames

Roughly speaking, a subgame of an extensive-form game is a portion of the game that could be a game in itself. What we need to be precise about is the meaning of "portion of the game".

Definition 4.3.1 A *proper subgame* of an extensive-form game is obtained as follows:

1. Start from a decision node x, different from the root, whose information set consists of node x only and enclose in an oval node x itself and all its successors.
2. If the oval does not "cut" any information sets (that is, there is no information set S and two nodes $y, z \in S$ such that y is a successor of x while z is not) then what is included in the oval is a proper subgame, otherwise it is not.

The reason why we use the qualifier 'proper' is that one could start from the root, in which case one would end up taking the entire game and consider this as a (trivial) subgame (just like any set is a subset of itself; a proper subgame is analogous to a proper subset).

Consider, for example, the extensive-form game of Figure 4.8. There are three possible starting points for identifying a proper subgame: nodes x, y and z (the other nodes fail to satisfy condition (1) of Definition 4.3.1).

1. Starting from node x and including all of its successors, we do indeed obtain a proper subgame, which is the portion included in the oval on the left.
2. Starting from node y and including all of its successors we obtain the portion of the game that is included in the oval on the right; in this case, condition **(2)** of Definition 4.3.1 is violated, since we are cutting the top information set of Player 3; hence the portion of the game inside this oval is not a proper subgame.
3. Finally, starting from node z and including all of its successors, we do obtain a proper subgame, which is the portion included in the oval at the bottom.

Thus the game of Figure 4.8 has two proper subgames.

Definition 4.3.2 A proper subgame of an extensive-form game is called *minimal* if it does not strictly contain another proper subgame (that is, if there is no other proper subgame which is strictly contained in it).

For example, the game shown in Figure 4.9 on the following page has three proper subgames, one starting at node x, another at node y and the third at node z. The ones starting at nodes x and z are minimal subgames, while the one that starts at node y is not a minimal subgame, since it strictly contains the one that starts at node z.

Test your understanding of the concepts introduced in this section, by going through the exercises in Section 4.6.3 at the end of this chapter.

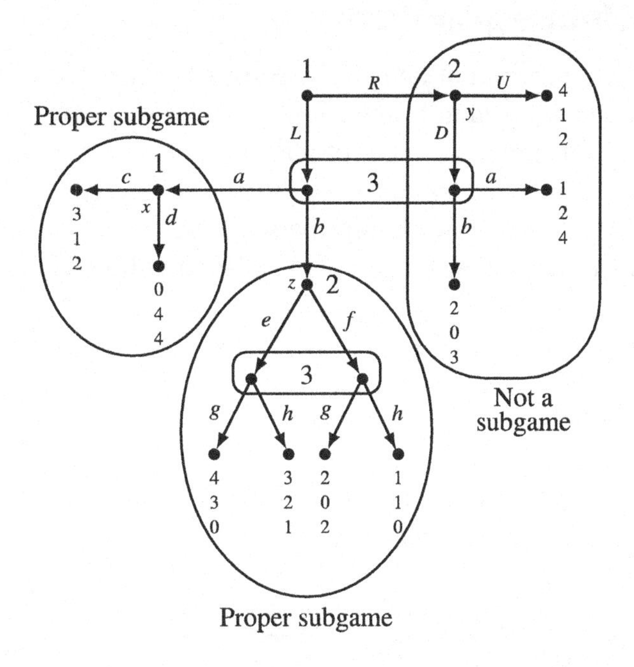

Figure 4.8: An extensive-form game with two proper subgames.

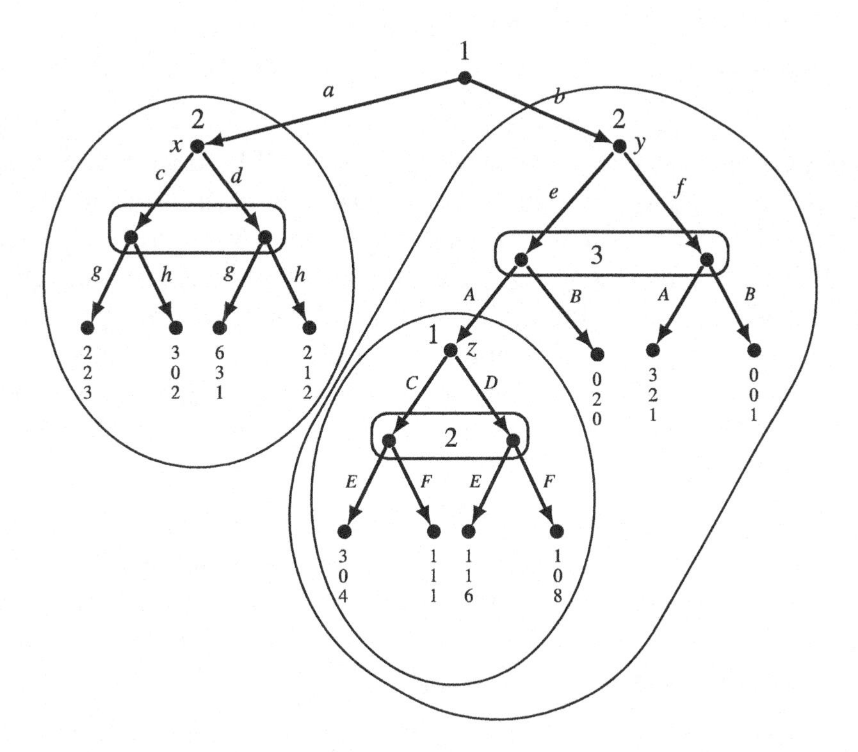

Figure 4.9: An extensive-form game with three proper subgames, two of which are minimal.

4.4 Subgame-perfect equilibrium

A subgame-perfect equilibrium of an extensive-form game is a Nash equilibrium of the entire game which remains an equilibrium in every proper subgame. Consider an extensive-form game and let s be a strategy profile for that game. Let G be a proper subgame. Then the *restriction of s to G*, denoted by $s|_G$, is that part of s which prescribes choices at every information set of G and only at those information sets.

For example, consider the extensive-form game of Figure 4.9 and the strategy profile

$$\left(\underbrace{(a,C)}_{\text{1's strategy}}, \underbrace{(d,f,E)}_{\text{2's strategy}}, \underbrace{(h,B)}_{\text{3's strategy}} \right)$$

Let G be the subgame that starts at node y of Player 2. Then

$$s|_G = \left(\underbrace{C}_{\text{1's strategy in G}}, \underbrace{(f,E)}_{\text{2's strategy in G}}, \underbrace{B}_{\text{3's strategy in G}} \right)$$

Definition 4.4.1 . Given an extensive-form game, let s be a strategy profile for the entire game. Then s is a *subgame-perfect equilibrium* if

1. s is a Nash equilibrium of the entire game and
2. for every proper subgame G, $s|_G$ (the restriction of s to G) is a Nash equilibrium of G.

For example, consider again the extensive-form game of Figure 4.9 and the strategy profile $s = ((a,C),(d,f,E),(h,B))$. Then s is a Nash equilibrium of the entire game: Player 1's payoff is 2 and if he were to switch to any strategy where he plays b his payoff would be 0; Player 2's payoff is 1 and if she were to switch to any strategy where she plays c her payoff would be 0; Player 3's payoff is 2 and if he were to switch to any strategy where he plays g his payoff would be 1. However, s is not a subgame-perfect equilibrium, because the restriction of s to the proper subgame that starts at node z of Player 1, namely (C,E), is not a Nash equilibrium of that subgame: in that subgame, for Player 2 the unique best reply to C is F.

One way of finding the subgame-perfect equilibria of a given game is to first find the Nash equilibria and then, for each of them, check if it satisfies condition **(2)** of Definition 4.4.1. However, this is not a practical way to proceed. A quicker and easier way is to apply the following algorithm, which generalizes the backward-induction algorithm for games with perfect information (Definition 3.2.1, Chapter 3).

Definition 4.4.2 Given an extensive-form game, the *subgame-perfect equilibrium algorithm* is the following procedure.

1. Start with a minimal proper subgame and select a Nash equilibrium of it.
2. Delete the selected proper subgame and replace it with the payoff vector associated with the selected Nash equilibrium, making a note of the strategies that constitute the Nash equilibrium. This yields a smaller extensive-form game.
3. Repeat Steps 1 and 2 in the smaller game so obtained.

For example, let us apply the algorithm to the game of Figure 4.9. Begin with the proper subgame that starts at node x of Player 2, shown in Figure 4.10 with its associated strategic form, where the unique Nash equilibrium (d,h) is highlighted. Note that this is a game only between Players 2 and 3 and thus in Figure 4.10 we only show the payoffs of these two players.

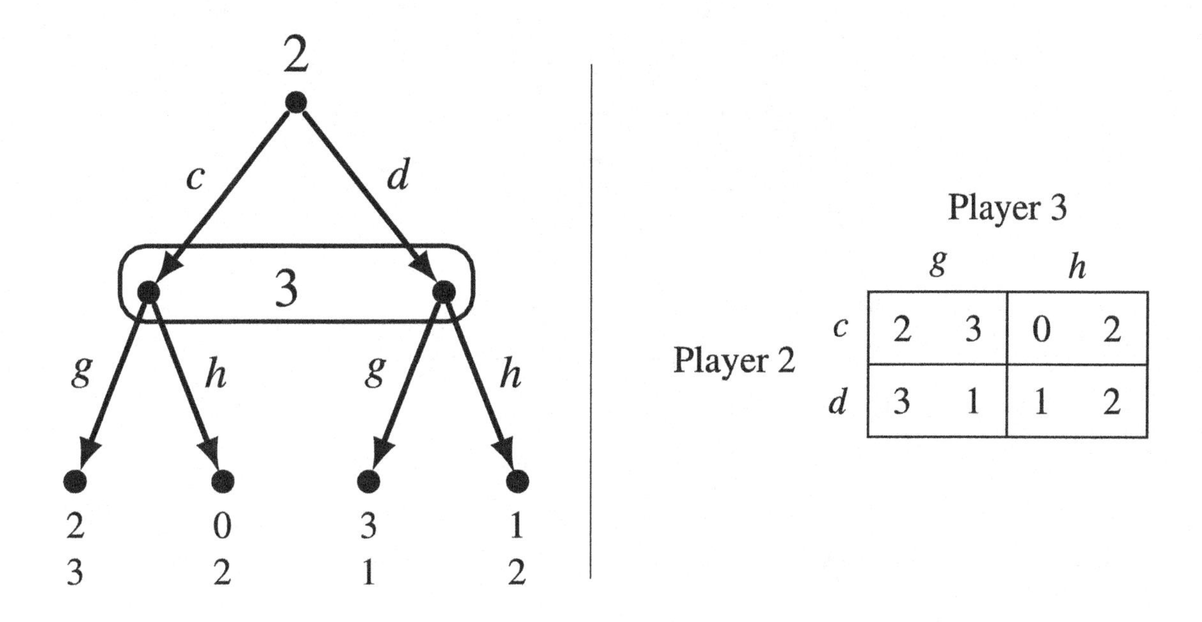

Figure 4.10: A minimal proper subgame of the game of Figure 4.9 and its strategic form

Now we delete the proper subgame, thereby turning node x into a terminal node to which we attach the full payoff vector associated, in the original game, with the terminal node following history *adh*, namely $(2,1,2)$.

Hence we obtain the smaller game shown in Figure 4.11.

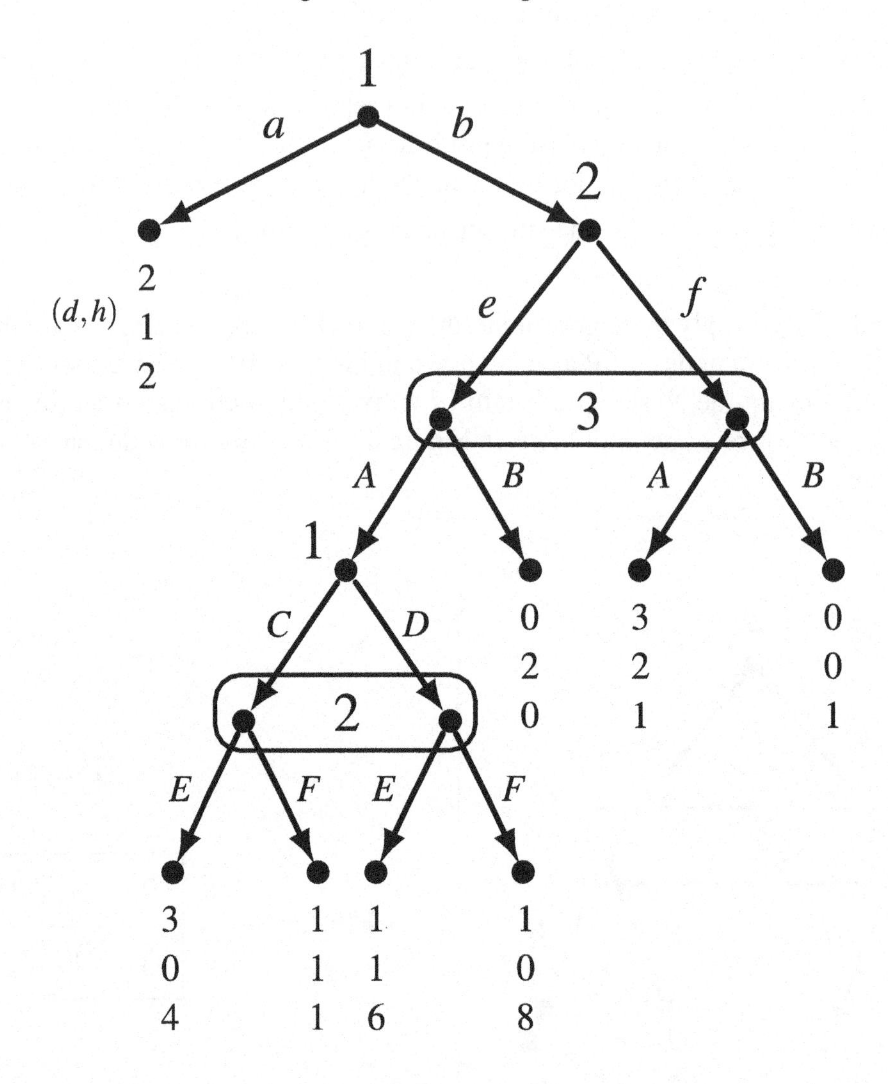

Figure 4.11: The reduced game after replacing a proper minimal subgame in the game of Figure 4.9

Now, in the reduced game of Figure 4.11 we select the only minimal proper subgame, namely the one that starts at the bottom decision node of Player 1. This subgame is shown in Figure 4.12 together with its associated strategic form. The unique Nash equilibrium of this subgame is (C,F).

Then, in the reduced game of Figure 4.11, we replace the selected proper subgame with the payoff vector associated with the history *beACF*, namely $(1,1,1)$, thus obtaining the smaller game shown in Figure 4.13. The game of Figure 4.13 has a unique proper subgame, which has a unique Nash equilibrium, namely (f,A).

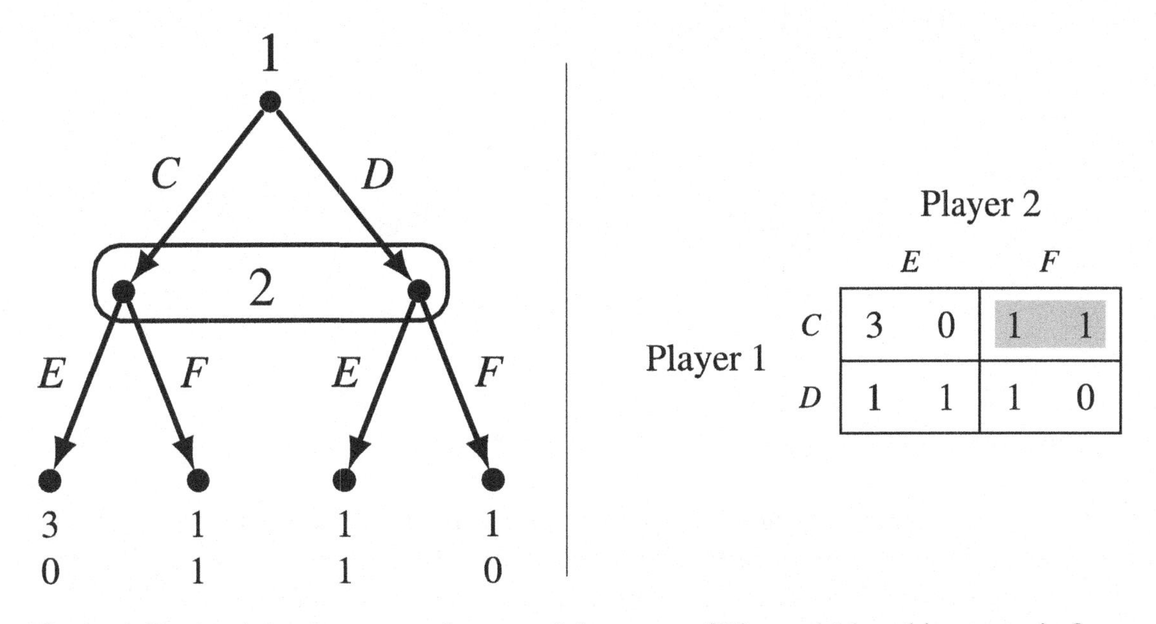

Figure 4.12: A minimal proper subgame of the game of Figure 4.11 and its strategic form

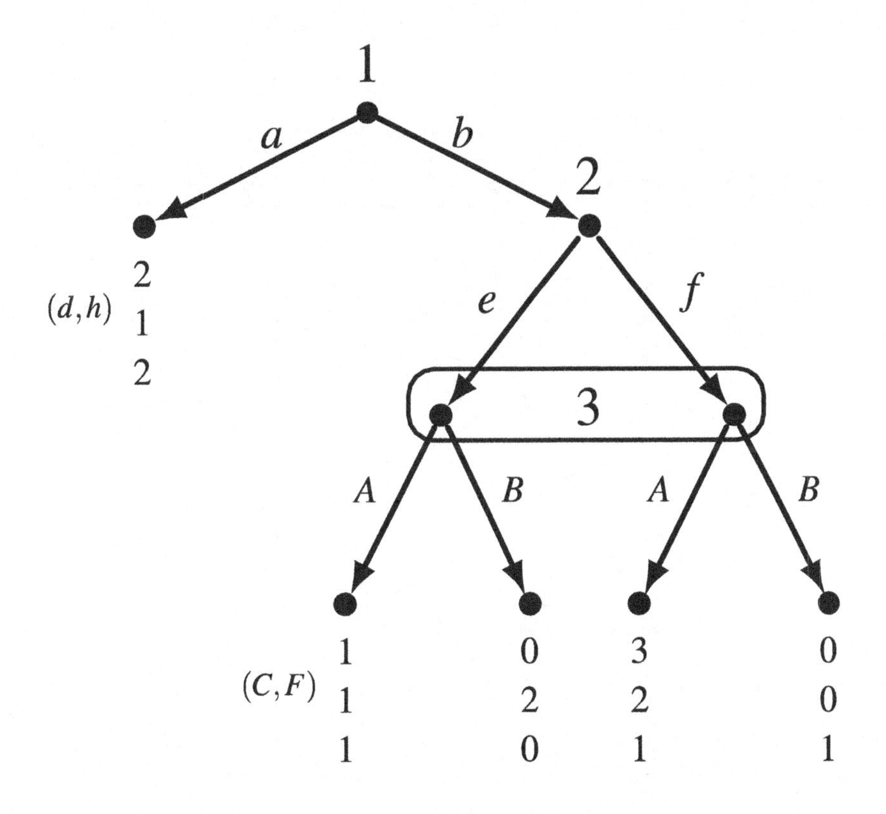

Figure 4.13: The reduced game after replacing a proper minimal subgame in the game of Figure 4.11

Replacing the subgame with the payoff vector associated with the history bfA we get the smaller game shown in Figure 4.14.

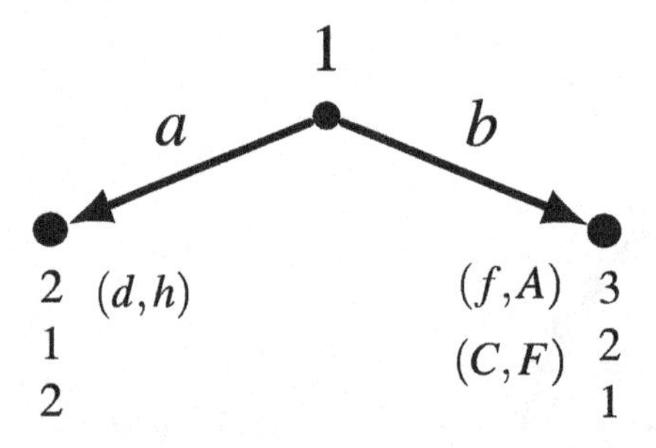

Figure 4.14: The reduced game after replacing the subgame in the game of Figure 4.13

In the reduced game of Figure 4.14 the unique Nash equilibrium is b. Now patching together the choices selected during the application of the algorithm we get the following subgame-perfect equilibrium for the game of Figure 4.9: $((b,C),(d,f,F),(h,A))$.

As a second example, consider the game of Figure 4.15 (which reproduces Figure 4.8).

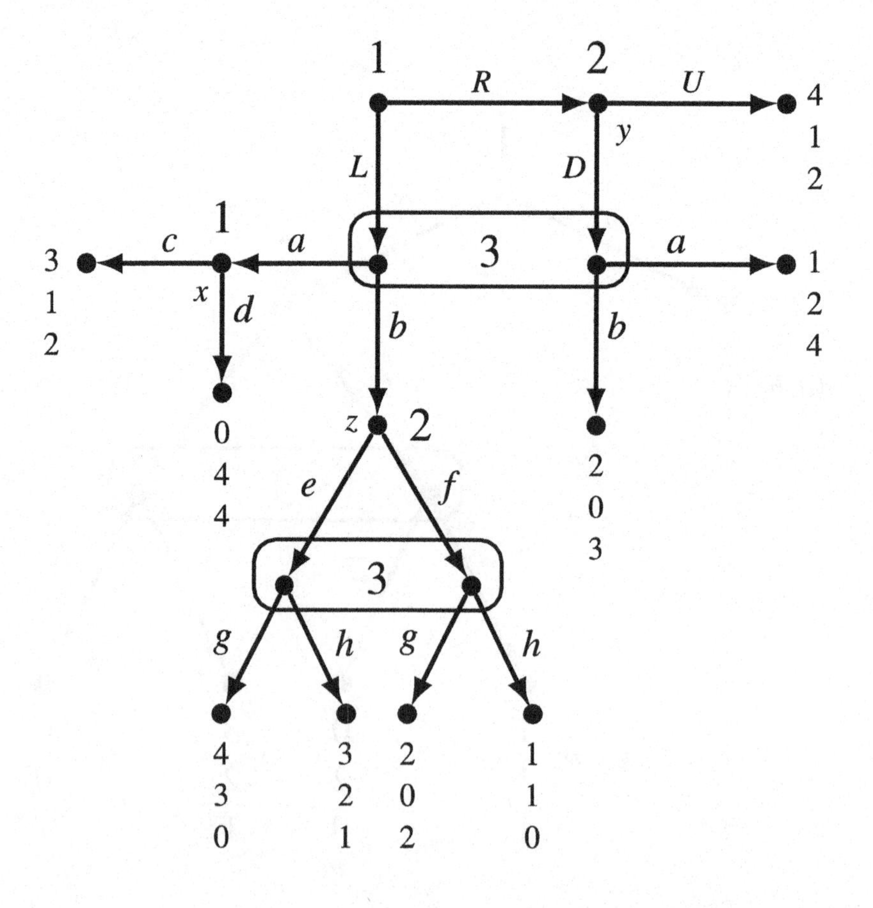

Figure 4.15: Copy of Figure 4.8

Begin with the subgame that starts at node x and replace it with the payoff vector $(3,1,2)$. Next replace the subgame that starts at node z with the payoff vector $(3,2,1)$ which corresponds to the Nash equilibrium (e,h) of that subgame, so that the game is reduced to the one shown in Figure 4.16, together with its strategic form.

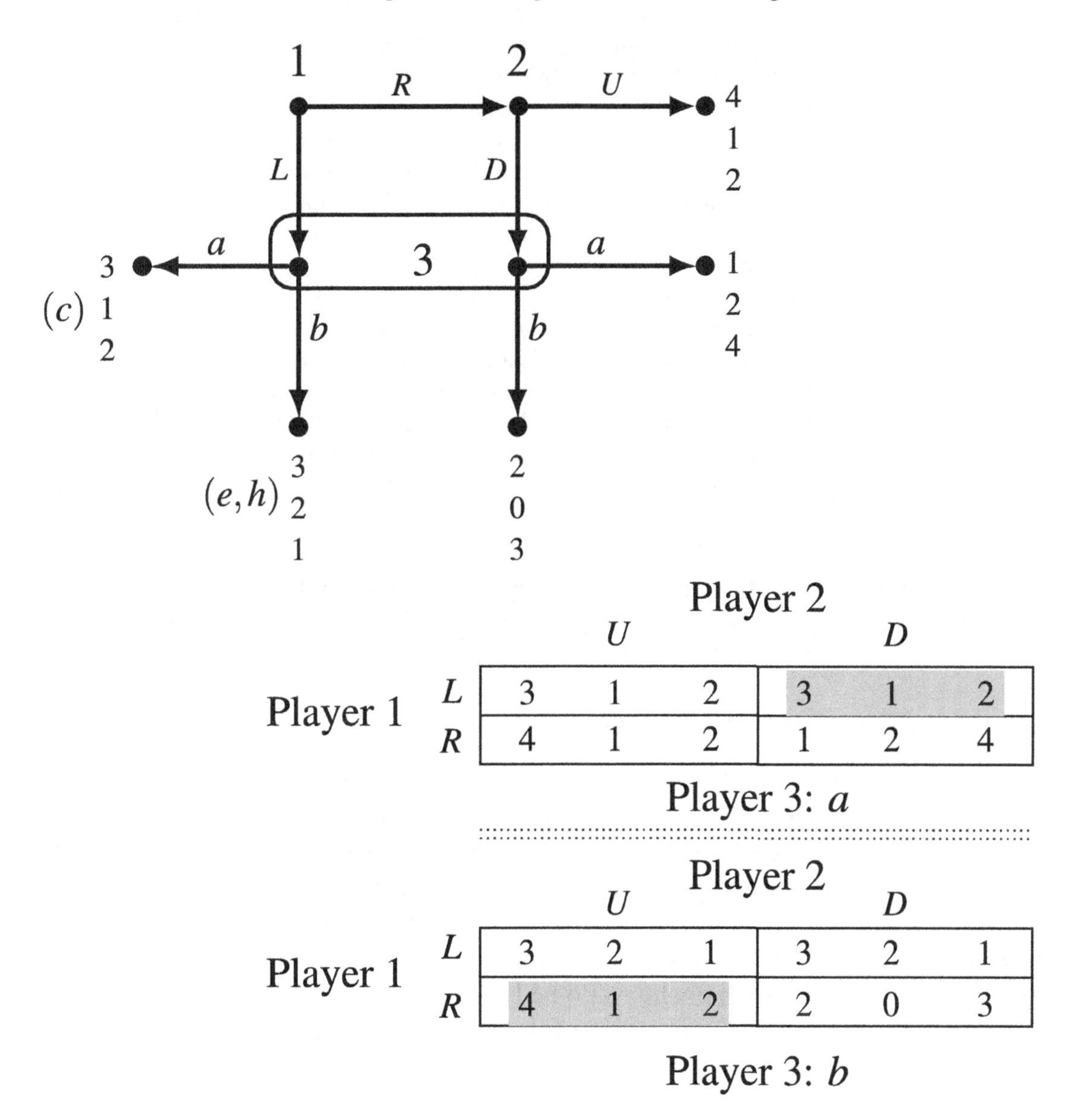

Player 3: a

Player 1 \ Player 2	U	D
L	3 1 2	3 1 2
R	4 1 2	1 2 4

Player 3: b

Player 1 \ Player 2	U	D
L	3 2 1	3 2 1
R	4 1 2	2 0 3

Figure 4.16: The game of Figure 4.15 reduced after solving the proper subgames, together with the associated strategic form with the Nash equilibria highlighted

The reduced game of Figure 4.16 has two Nash equilibria: (L,D,a) and (R,U,b). Thus the game of Figure 4.15 has two subgame-perfect equilibria:

$$\Bigg(\underbrace{(L,c)}_{\text{Player 1}}, \underbrace{(D,e)}_{\text{Player 2}}, \underbrace{(a,h)}_{\text{Player 3}}\Bigg) \quad \text{and} \quad \Bigg(\underbrace{(R,c)}_{\text{Player 1}}, \underbrace{(U,e)}_{\text{Player 2}}, \underbrace{(b,h)}_{\text{Player 3}}\Bigg).$$

- As shown in the last example, it is possible that – when applying the subgame-perfect equilibrium algorithm – one encounters a proper subgame or a reduced game that has several Nash equilibria. In this case one Nash equilibrium must be selected in order to continue the procedure and in the end one obtains one subgame-perfect equilibrium. One then has to repeat the procedure by selecting a different Nash equilibrium and thus obtain a different subgame-perfect equilibrium, and so on. This is similar to what happens with the backward-induction algorithm in perfect-information games.
- It is also possible that – when applying the subgame-perfect equilibrium algorithm –one encounters a proper subgame or a reduced game that has no Nash equilibria.[2] In such a case the game under consideration does not have any subgame-perfect equilibria.
- When applied to perfect-information games, the notion of subgame-perfect equilibrium coincides with the notion of backward-induction solution.
  Thus subgame-perfect equilibrium is a generalization of backward induction.
- For extensive-form games that have no proper subgames (for example, the game of Figure 4.3) the set of Nash equilibria coincides with the set of subgame-perfect equilibria. In general, however, the notion of subgame-perfect equilibrium is a refinement of the notion of Nash equilibrium.

Test your understanding of the concepts introduced in this section, by going through the exercises in Section 4.6.4 at the end of this chapter.

4.5 Games with chance moves

So far we have only considered games where the outcomes do not involve any uncertainty. As a way of introducing the topic discussed in Part II, in this section we consider games where uncertain, probabilistic events are incorporated in the extensive form.

We begin with an example: There are three cards, one black and two red. They are shuffled well and put face down on the table. Adele picks the top card, looks at it without showing it to Ben and then tells Ben either "the top card is black" or "the top card is red": she could be telling the truth or she could be lying. Ben then has to guess the true color of the top card. If he guesses correctly he gets \$9 from Adele, otherwise he gives her \$9. How can we represent this situation?
Whether the top card is black or red is not the outcome of a player's decision, but the outcome of a random event, namely the shuffling of the cards. In order to capture this random event we introduce a fictitious player called *Nature* or *Chance*. We assign a probability distribution to Nature's "choices". In this case, since one card is black and the other two are red, the probability that the top card is black is $\frac{1}{3}$ and the probability that the top card is red is $\frac{2}{3}$. Note that we don't assign payoffs to Nature and thus the only 'real' players are Adele and Ben. The situation can be represented as shown in Figure 4.17, where the numbers associated with the terminal nodes are dollar amounts.

[2]We will see in Part II that, when payoffs are cardinal and one allows for mixed strategies, then every finite game has at least one Nash equilibrium in mixed strategies.

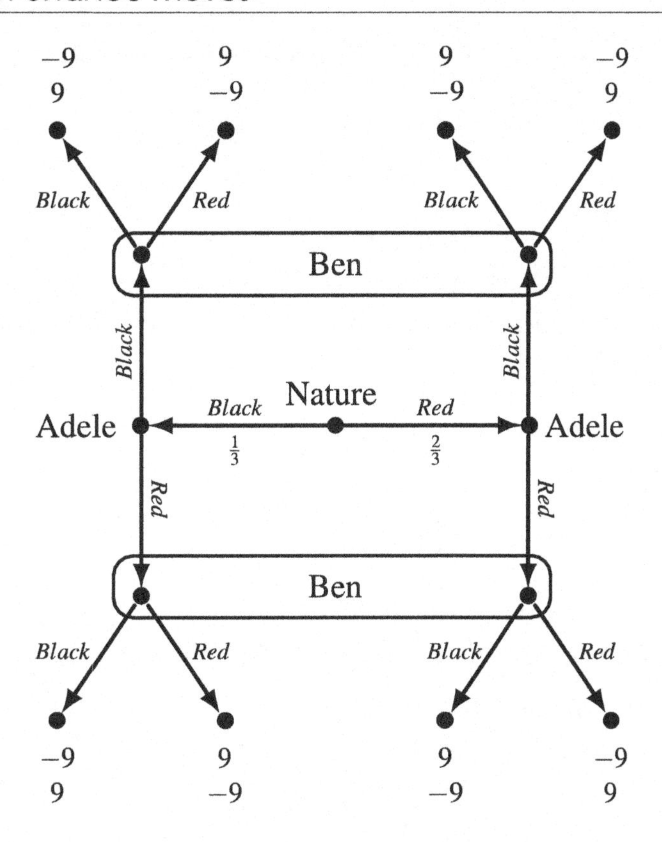

Figure 4.17: An extensive form with a chance move

Clearly, the notion of strategy is not affected by the presence of chance moves. In the game of Figure 4.17 Adele has four strategies and so does Ben. However, we do encounter a difficulty when we try to write the associated strategic form. For example, consider the following strategy profile: $((B,R),(B,B))$ where Adele's strategy is to be truthful (say "Black" if she sees a black card and say "Red" if she sees a red card) and Ben's strategy is to guess Black no matter what Adele says. What is the outcome in this case? It depends on what the true color of the top card is and thus the outcome is a probabilistic one:

$$\begin{pmatrix} \textit{outcome} & \text{Adele gives \$9 to Ben} & \text{Ben gives \$9 to Adele} \\ \textit{probability} & \frac{1}{3} & \frac{2}{3} \end{pmatrix}$$

We call such probabilistic outcomes *lotteries*. In order to convert the game-frame into a game we need to specify how the players rank probabilistic outcomes. Consider the case where Adele is selfish and greedy, in the sense that she only cares about her own wealth and she prefers more money to less. Then, from her point of view, the above probabilistic outcome reduces to the following monetary lottery $\begin{pmatrix} -\$9 & \$9 \\ \frac{1}{3} & \frac{2}{3} \end{pmatrix}$.

If Ben is also selfish and greedy, then he views the same outcome as the lottery $\begin{pmatrix} \$9 & -\$9 \\ \frac{1}{3} & \frac{2}{3} \end{pmatrix}$.

How do we convert a lottery into a payoff or utility? The general answer to this question will be provided in Chapter 5. Here we consider one possibility, which is particularly simple.

Definition 4.5.1 Given a lottery whose outcomes are sums of money

$$\begin{pmatrix} \$x_1 & \$x_2 & \dots & \$x_n \\ p_1 & p_2 & \dots & p_n \end{pmatrix}$$

(with $p_i \geq 0$, for all $i = 1,2,...,n$, and $p_1 + p_2 + \cdots + p_n = 1$) the *expected value* of the lottery is the following sum of money: $(x_1 p_1 + x_2 p_2 + \cdots + x_n p_n)$.
We call lotteries whose outcomes are sums of money, *money lotteries*.

For example, the expected value of the lottery

$$\begin{pmatrix} \$5 & \$15 & \$25 \\ \frac{1}{5} & \frac{2}{5} & \frac{2}{5} \end{pmatrix}$$

is

$$\$\left[5\left(\tfrac{1}{5}\right) + 15\left(\tfrac{2}{5}\right) + 25\left(\tfrac{2}{5}\right)\right] = \$(1+6+10) = \$17$$

and the expected value of the lottery

$$\begin{pmatrix} -\$9 & \$9 \\ \frac{1}{3} & \frac{2}{3} \end{pmatrix}$$

is \$3.

Definition 4.5.2 A player is defined to be *risk neutral* if she considers a money lottery to be just as good as its expected value. Hence a risk neutral person ranks money lotteries according to their expected value.[a]

[a]It is important to stress that our focussing on the case of risk neutrality should *not* be taken to imply that a rational individual ought to be risk neutral nor that risk neutrality is empirically particularly relevant. At this stage we assume risk neutrality only because it yields a very simple type of preference over money lotteries and allows us to introduce the notion of backward induction without the heavy machinery of expected utility theory.

For example, consider the following money lotteries:

$$L_1 = \begin{pmatrix} \$5 & \$15 & \$25 \\ \frac{1}{5} & \frac{2}{5} & \frac{2}{5} \end{pmatrix}, \ L_2 = \begin{pmatrix} \$16 \\ 1 \end{pmatrix} \text{ and } L_3 = \begin{pmatrix} \$0 & \$32 & \$48 \\ \frac{5}{8} & \frac{1}{8} & \frac{1}{4} \end{pmatrix}.$$

The expected value of L_1 is \$17 and the expected value of both L_2 and L_3 is \$16. Thus a risk-neutral player would have the following ranking: $L_1 \succ L_2 \sim L_3$, that is, she would prefer L_1 to L_2 and be indifferent between L_2 and L_3.

For a selfish and greedy player who is risk neutral we can take the expected value of a money lottery as the utility of that lottery. For example, if we make the assumption that, in the extensive form of Figure 4.17, Adele and Ben are selfish, greedy and risk neutral then we can associate a strategic-form game to it as shown in Figure 4.18. Note that inside each cell we have two numbers: the first is the utility (= expected value) of the underlying money lottery as perceived by Adele and the second number is the utility (= expected value) of the underlying money lottery as perceived by Ben.

The first element of Adele's strategy is what she says if she sees a black card and the second element is what she says if she sees a red card. The first element of Ben's strategy is what he guesses if Adele says "Red", the second element is what he guesses if Adele says "Black".

Adele \ Ben	*BB*	*BR*	*RB*	*RR*
BB	3, −3	−3, 3	3, −3	−3, 3
BR	3, −3	9, −9	−9, 9	−3, 3
RB	3, −3	−9, 9	9, −9	−3, 3
RR	3, −3	3, −3	−3, 3	−3, 3

Figure 4.18: The strategic form of the game of Figure 4.17 when the two players are selfish, greedy and risk neutral.

We conclude this section with one more example.

▪ **Example 4.3** There are three unmarked, opaque envelopes. One contains \$100, one contains \$200 and the third contains \$300. They are shuffled well and then one envelope is given to Player 1 and another is given to Player 2 (the third one remains on the table).

- Player 1 opens her envelope and checks its content without showing it to Player 2. Then she either says "pass" – in which case each player gets to keep his/her envelope – or she asks Player 2 to trade his envelope for hers.
- Player 2 is not allowed to see the content of his envelope and has to say either Yes or No. If he says No, then the two players get to keep their envelopes. If Player 2 says Yes, then they trade envelopes and the game ends. Each player is selfish, greedy and risk neutral.

This situation is represented by the extensive-form game shown in Figure 4.19, where $(100,200)$ means that Player 1 gets the envelope with \$100 and Player 2 gets the envelope with \$200, etc.; P stands for "pass" and T for "suggest a trade"; Y for "Yes" and N for "No". ▪

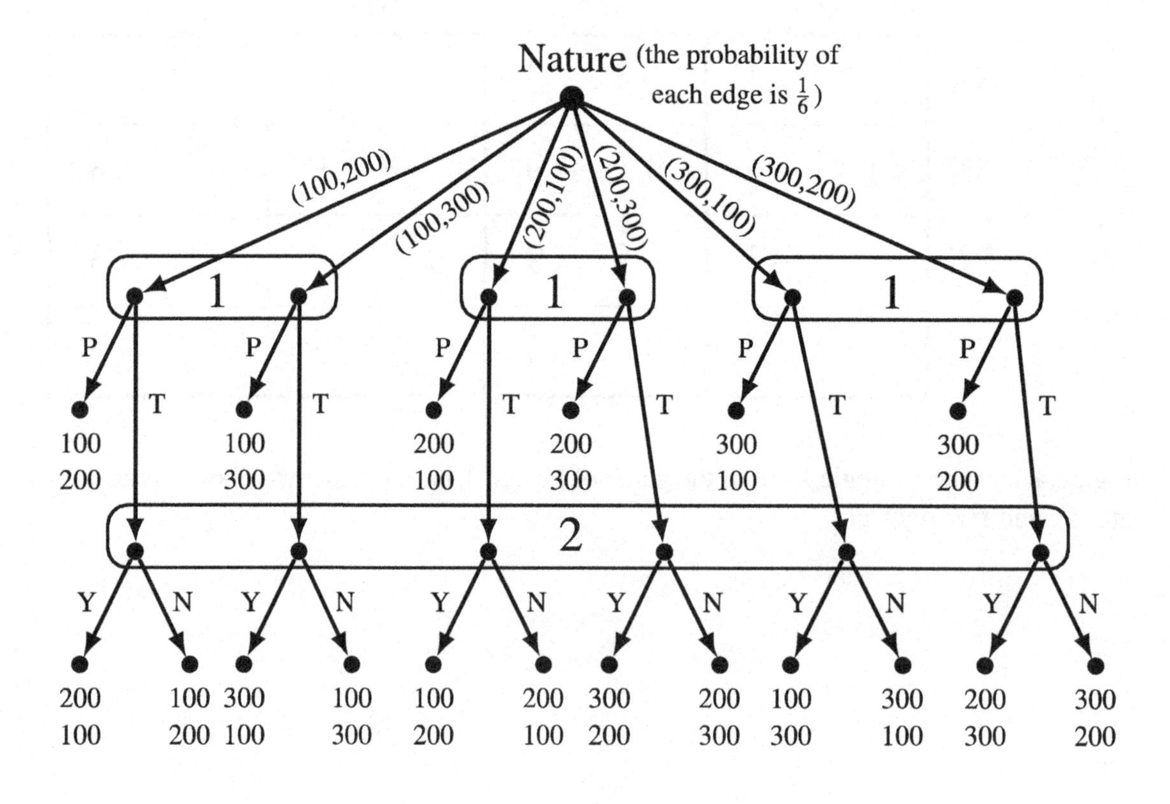

Figure 4.19: The extensive-form game of Example 4.3

In this game Player 1 has eight strategies. One possible strategy is: "if I get \$100 I will pass, if I get \$200 I will propose a trade, if I get \$300 I will pass": we will use the shorthand PTP for this strategy. Similarly for the other strategies. Player 2 has only two strategies: Yes and No.

The strategic form associated with the game of Figure 4.19 is shown in Figure 4.20, where the Nash equilibria are highlighted.

Player 1 \ Player 2	Y		H	
PPP	200	200	200	200
PPT	150	250	200	200
PTP	200	200	200	200
PTT	150	250	200	200
TPP	250	150	200	200
TPT	200	200	200	200
TTP	250	150	200	200
TTT	200	200	200	200

Figure 4.20: The strategic form of the game of Figure 4.19

How did we get those payoffs? Consider, for example, the first cell. Given the strategies PPP and Y, the outcomes are:
$(\$100,\$200)$ with probability $\frac{1}{6}$, $(\$100,\$300)$ with probability $\frac{1}{6}$,
$(\$200,\$100)$ with probability $\frac{1}{6}$, $(\$200,\$300)$ with probability $\frac{1}{6}$,
$(\$300,\$100)$ with probability $\frac{1}{6}$, $(\$300,\$200)$ with probability $\frac{1}{6}$.

Being risk neutral, Player 1 views his corresponding money lottery as equivalent to getting its expected value $\$(100+100+200+200+300+300)(\frac{1}{6}) = \$200$. Similarly for Player 2 and for the other cells.

Since the game of Figure 4.19 has no proper subgames, all the Nash equilibria are also subgame-perfect equilibria. Are some of the Nash equilibria more plausible than others? For Player 1 all the strategies are weakly dominated, except for TPP and TTP. Elimination of the weakly dominated strategies leads to a game where Y is strictly dominated for Player 2. Thus one could argue that (TPP, N) and (TTP, N) are the most plausible equilibria; in both of them Player 2 refuses to trade.

Test your understanding of the concepts introduced in this section, by going through the exercises in Section 4.6.5 at the end of this chapter.

4.6 Exercises

4.6.1 Exercises for Section 4.1: Imperfect information

The answers to the following exercises are in Section 4.7 at the end of this chapter.

Exercise 4.1 Amy and Bill simultaneously write a bid on a piece of paper. The bid can only be either 2 or 3. A referee then looks at the bids, announces the amount of the lowest bid (without revealing who submitted it) and invites Amy to either pass or double her initial bid.

- The outcome is determined by comparing Amy's final bid to Bill's bid: if one is greater than the other then the higher bidder gets the object and pays his/her own bid; if they are equal then Bill gets the object and pays his bid.

Represent this situation by means of two alternative extensive frames.
Note: (1) when there are simultaneous moves we have a choice as to which player we select as moving first: the important thing is that the second player does not know what the first player did;
(2) when representing, by means of information sets, what a player is uncertain about, we typically assume that a player is smart enough to deduce relevant information, even if that information is not explicitly given to him/her. ■

Exercise 4.2 Consider the following situation. An incumbent monopolist decides at date 1 whether to build a small plant or a large plant. At date 2 a potential entrant observes the plant built by the incumbent and decides whether or not to enter.

- If she does not enter then her profit is 0 while the incumbent's profit is \$25 million with a small plant and \$20 million with a large plant.
- If the potential entrant decides to enter, she pays a cost of entry equal to \$K million.
- At date 3 the two firms simultaneously decide whether to produce high output or low output.
- The profits of the firms are as shown in the following table, where 'L' means 'low output' and 'H' means 'high output' (these figure do not include the cost of entry for the entrant; thus you need to subtract that cost for the entrant); in each cell, the first number is the profit of the entrant (in millions of dollars) and the second is the profit of the incumbent.

If Incumbent has small plant:

		Incumbent L	Incumbent H
Entrant	L	10 , 10	7 , 7
	H	7 , 6	4 , 3

If Incumbent has large plant:

		Incumbent L	Incumbent H
Entrant	L	10 , 7	5 , 9
	H	7 , 3	4 , 5

Draw an extensive-form game that represents this situation, assuming that each player is selfish and greedy (that is, cares only about its own profits and prefers more money to less). ■

4.6.2 Exercises for Section 4.2: Strategies

The answers to the following exercises are in Section 4.7 at the end of this chapter.

Exercise 4.3 Write the strategic-form game-frame of the extensive form of Exercise 4.1 (that is, instead of writing payoffs in each cell, you write the outcome). Verify that the strategic forms of the two possible versions of the extensive form are identical. ∎

Exercise 4.4 Consider the extensive-form game of Exercise 4.2.
(a) Write down in words one of the strategies of the potential entrant.
(b) How many strategies does the potential entrant have?
(c) Write the strategic-form game associated with the extensive-form game.
(d) Find the Nash equilibria for the case where $K = 2$.
∎

4.6.3 Exercises for Section 4.3: Subgames

The answers to the following exercises are in Section 4.7 at the end of this chapter.

Exercise 4.5 How many proper subgames does the extensive form of Figure 4.3 have? ∎

Exercise 4.6 How many proper subgames does the extensive form of Figure 4.5 have? ∎

Exercise 4.7 Consider the extensive game Figure 4.21.
(a) How many proper subgames does the game have?
(b) How many of those proper subgames are minimal?
∎

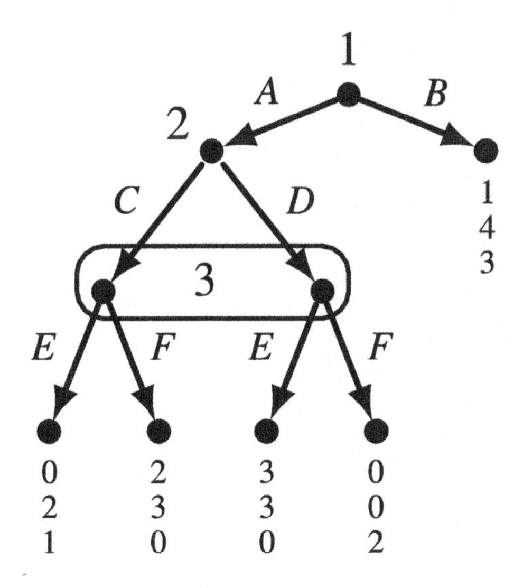

Figure 4.21: The game of Exercise 4.7

Exercise 4.8 Consider the extensive game Figure 4.22.

(a) How many proper subgames does the game have?

(b) How many of those proper subgames are minimal? ■

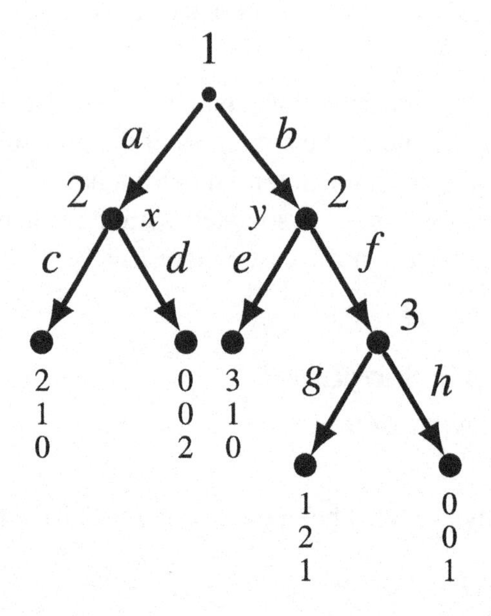

Figure 4.22: The game of Exercise 4.8

4.6.4 Exercises for Section 4.4: Subgame-perfect equilibrium

The answers to the following exercises are in Section 4.7 at the end of this chapter.

Exercise 4.9 Find the Nash equilibria and the subgame-perfect equilibria of the game shown in Figure 4.23. ■

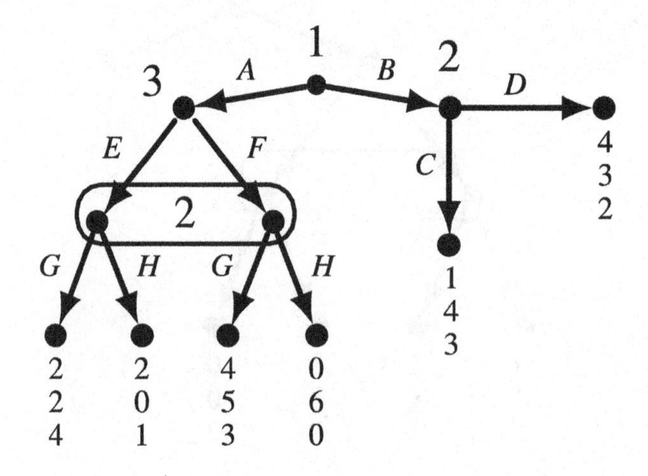

Figure 4.23: The game of Exercise 4.9

Exercise 4.10 Find the subgame-perfect equilibria of the game shown in Figure 4.24. ■

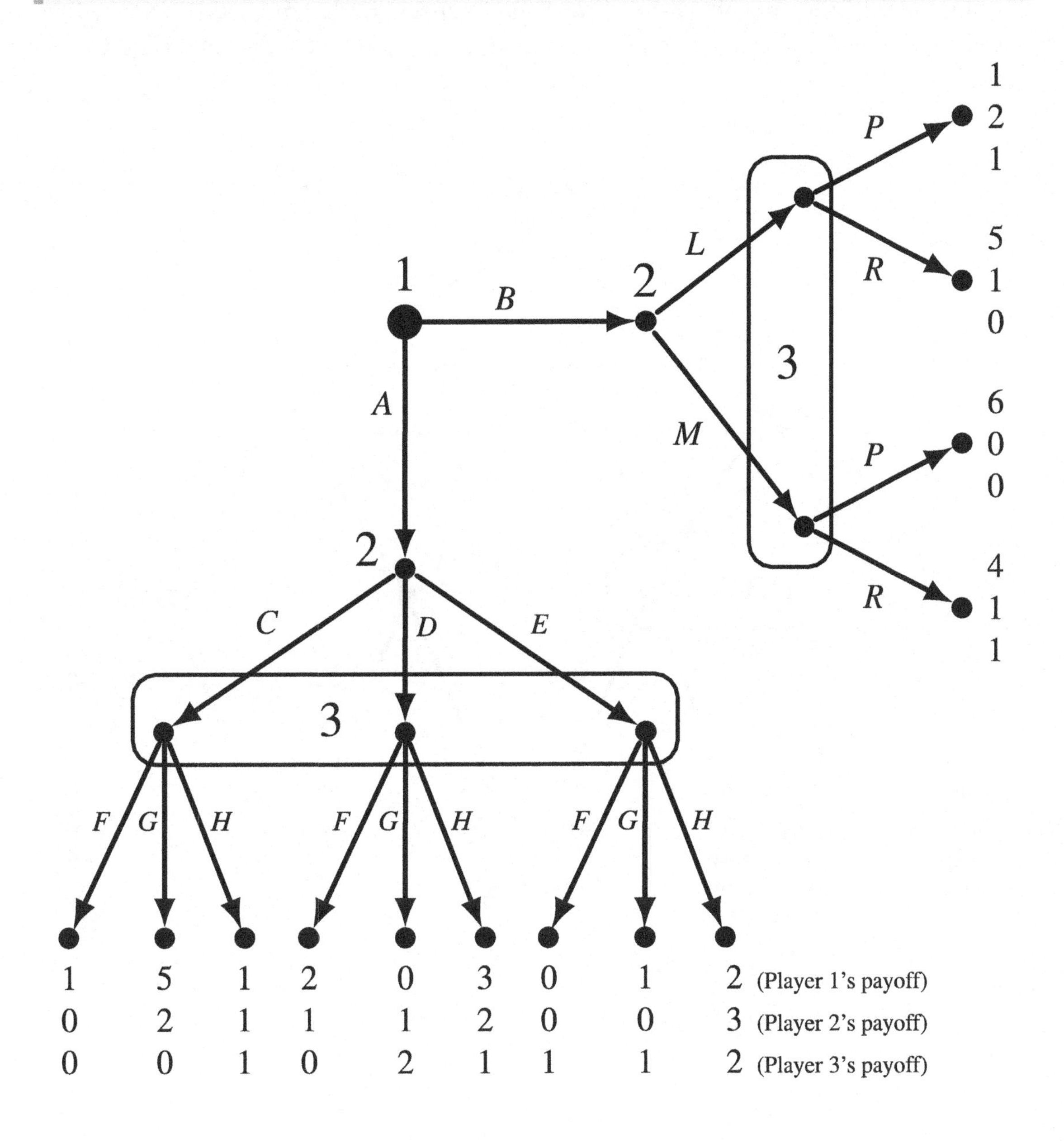

Figure 4.24: The game of Exercise 4.10

Exercise 4.11 Find the subgame-perfect equilibria of the game shown in Figure 4.25, assuming the following about the players' preferences. Both Amy and Bill are selfish and greedy, are interested in their own net gain, Amy values the object at \$5 and Bill at \$4. ■

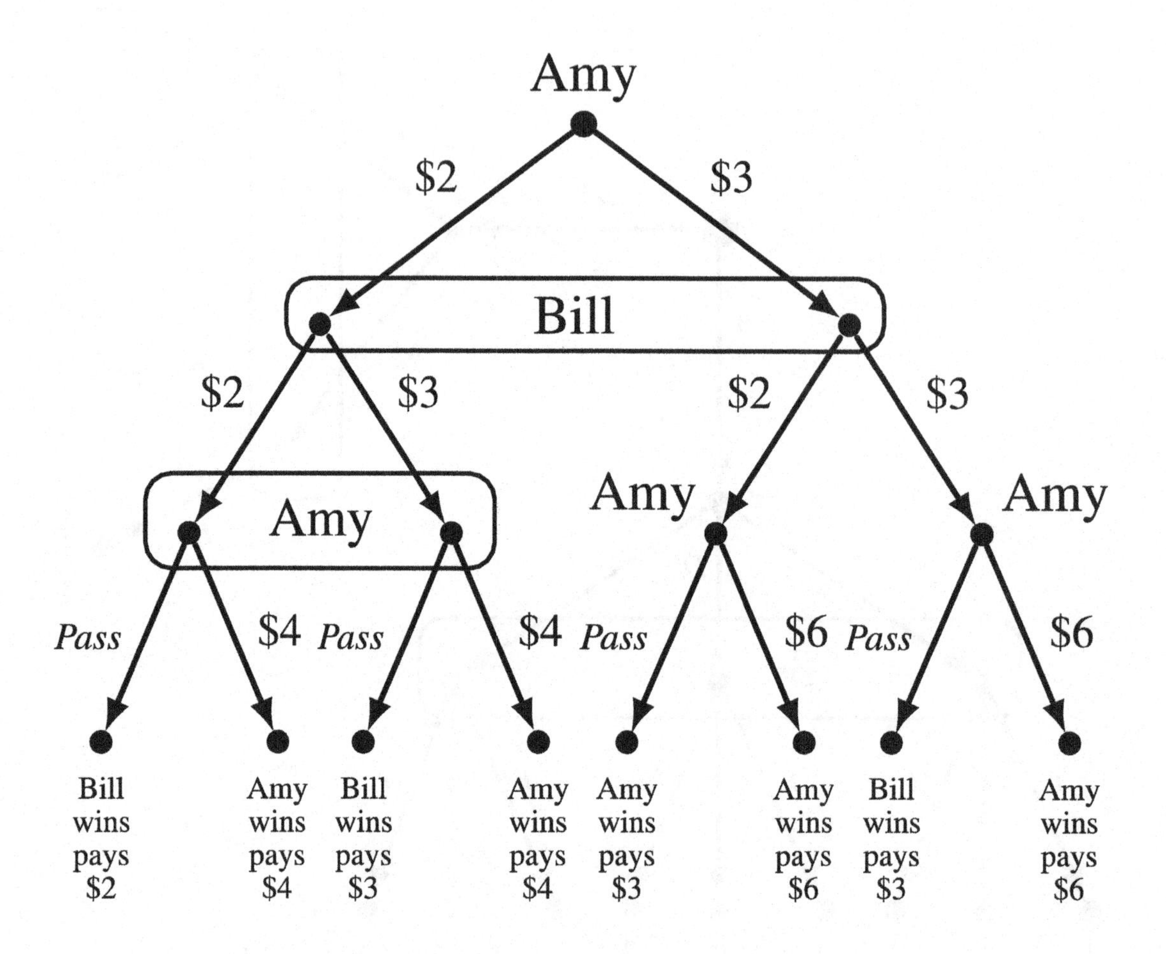

Figure 4.25: The game of Exercise 4.11

4.6.5 Exercises for Section 4.5: Games with chance moves

The answers to the following exercises are in Section 4.7 at the end of this chapter.

Exercise 4.12 Modify the game of Example 4.3 as follows: Player 2 is allowed to privately check the content of his envelope before he decides whether or not to accept Player 1's proposal.

(a) Represent this situation as an extensive-form game game.

(b) List all the strategies of Player 1 and all the strategies of Player 2.

■

Exercise 4.13 Three players, Avinash, Brian and John, play the following game. Two cards, one red and the other black, are shuffled well and put face down on the table. Brian picks the top card, looks at it without showing it to the other players (Avinash and John) and puts it back face down on the table. Then Brian whispers either "Black" or "Red" in Avinash's ear, making sure that John doesn't hear. Avinash then tells John either "Black" or "Red". Note that both players could be lying. Finally John announces either "Black" or "Red" and this exciting game ends.

The payoffs are as follows: if John's final announcement matches the true color of the card Brian looked at, then Brian and Avinash give $2 each to John. In every other case John gives $2 each to Brian and Avinash.

(a) Represent this situation as an extensive-form.

(b) Write the corresponding strategic form assuming that the players are selfish, greedy and risk neutral. [At least try to fill in a few cells in at least one table.]

■

Exercise 4.14 Consider the following highly simplified version of Poker.

- There are three cards, marked A, B and C. A beats B and C, B beats C.
- There are two players, Yvonne and Zoe. Each player contributes $1 to the pot before the game starts. The cards are then shuffled and the top card is given, face down, to Yvonne and the second card (face down) to Zoe. Each player looks at, and only at, her own card: she does not see the card of the other player nor the remaining card.
- Yvonne, the first player, may pass, or bet $1. If she passes, the game ends, the cards are turned and the pot goes to the high-card holder (recall that A beats B and C, B beats C).
- If Yvonne bets, then Zoe can fold, in which case the game ends and the pot goes to Yvonne, or Zoe can see by betting $1, in which case the game ends, the cards are turned and the pot goes to the high-card holder. Both players are selfish, greedy and risk neutral.

(a) Draw the extensive-form game.

(b) How many strategies does Yvonne have?

(c) How many strategies does Zoe have?

(d) Consider the following strategies. For Yvonne: If A pass, if B pass, if C bet.
For Zoe: if Yvonne bets, I will fold no matter which card I get.
Calculate the corresponding payoffs.

(e) Redo the same with the following strategies. For Yvonne: If A pass, if B pass, if C bet. For Zoe: see always (that is, no matter what card she gets).

(f) Now that you have understood how to calculate the payoffs, represent the entire game as a normal form game, assigning the rows to Yvonne and the columns to Zoe. [This might take you the entire night, so make sure you have a lot of coffee!]

(g) What strategies of Yvonne are weakly dominated? What strategies of Zoe are weakly dominated?

(h) What do you get when you apply the procedure of iterative elimination of weakly dominated strategies?

■

Exercise 4.15 — ⋆⋆⋆ **Challenging Question** ⋆⋆⋆.

In an attempt to reduce the deficit, the government of Italy has decided to sell a 14th century palace near Rome. The palace is in disrepair and is not generating any revenue for the government. From now on we will call the government Player G. A Chinese millionaire has offered to purchase the palace for $\$p$. Alternatively, Player G can organize an auction among n interested parties ($n \geq 2$). The participants to the auction (we will call them players) have been randomly assigned labels $1, 2, \ldots, n$. Player i is willing to pay up to $\$p_i$ for the palace, where $\$p_i$ is a positive integer. For the auction assume the following:

1. it is a simultaneous, sealed-bid second-price auction,
2. bids must be non-negative integers,
3. each player only cares about his own wealth,
4. the tie-breaking rule for the auction is that the palace is given to that player who has the lowest index (e.g. if the highest bid was submitted by Players 3, 7 and 12 then the palace is given to Player 3).

All of the above is commonly known among everybody involved, as is the fact that for every $i, j \in \{1, \ldots, n\}$ with $i \neq j$, $p_i \neq p_j$.

We shall consider four different scenarios. In all scenarios you can assume that the p_i's are common knowledge.

Scenario 1. Player G first decides whether to sell the palace to the Chinese millionaire or make a public and irrevocable decision to auction it.

(a) Draw the extensive form of this game for the case where $n = 2$ and the only possible bids are \$1 and \$2. [List payoffs in the following order: first G then 1 then 2; don't forget that this is a *second-price* auction.]

(b) For the general case where $n \geq 2$ and every positive integer is a possible bid, find a pure-strategy subgame-perfect equilibrium of this game. What are the players' payoffs at the equilibrium?

Scenario 2. Here we assume that $n = 2$, and $p_1 > p_2 + 1 > 2$.

- First Player G decides whether to sell the palace to the Chinese or make a public and irrevocable decision to auction it. In the latter case he first asks Player 2 to publicly announce whether or not he is going to participate in the auction.
- If Player 2 says Yes, then he has to pay \$1 to Player G as a participation fee, which is non-refundable. If he says No, then she is out of the game.
- After Player 2 has made his announcement (and paid his fee if he decided to participate), Player 1 is asked to make the same decision (participate and pay a non-refundable fee of \$1 to Player G or stay out); Player 1 knows Player 2's decision when he makes his own decision.

After both players have made their decisions, player G proceeds as follows:

- if both 1 and 2 said Yes, then he makes them play a simultaneous second-price auction,
- if only one player said Yes, then he is asked to put an amount $\$x$ of his choice in an envelope (where x is a positive integer) and give it to Player G in exchange for the palace,
- if both 1 and 2 said No, then G is no longer bound by his commitment to auction the palace and he sells it to the Chinese.

(c) Draw the extensive form of this game for the case where the only possible bids are \$1 and \$2 and also $x \in \{1,2\}$ [List payoffs in the following order: first G then 1 then 2; again, don't forget that this is a *second-price* auction.]

(d) For the general case where all possible bids are allowed (subject to being positive integers) and x can be any positive integer, find a pure-strategy subgame-perfect equilibrium of this game. What are the players' payoffs at the equilibrium?

Scenario 3. Same as Scenario 2; the only difference is that if both Players 1 and 2 decide to participate in the auction then Player G gives to the loser the fraction a (with $0 < a < 1$) of the amount paid by the winner in the auction (note that player G still keeps 100% of the participation fees). This is publicly announced at the beginning and is an irrevocable commitment.

(e) For the general case where all possible bids are allowed (subject to being positive integers) find a subgame-perfect equilibrium of this game. What are the players' payoff at the equilibrium?

Scenario 4. Player G tells the Chinese millionaire the following:

> "First you (= the Chinese) say Yes or No; if you say No I will sell you the palace at the price that you offered me, namely \$100 (that is, we now assume that $p = 100$); if you say Yes then we play the following perfect information game. I start by choosing a number from the set $\{1,2,3\}$, then you (= the Chinese) choose a number from this set, then I choose again, followed by you, etc. The first player who brings the cumulative sum of all the numbers chosen (up to and including the last one) to 40 wins. If you win I will sell you the palace for \$50, while if I win I will sell you the palace for \$200."

Thus there is no auction in this scenario. Assume that the Chinese would actually be willing to pay up to \$300 for the palace.

(f) Find a pure-strategy subgame-perfect equilibrium of this game.

■

4.7 Solutions to exercises

Solution to Exercise 4.1. One possible extensive frame is shown in Figure 4.26, where Amy moves first. Note that we have only one non-trivial information set for Amy, while each of the other three consists of a single node. The reason is as follows: if Amy initially bids \$3 and Bill bids \$2 then the referee announces "the lowest bid was \$2"; this announcement does not directly reveal to Amy that Bill's bid was \$2, but she can figure it out from her knowledge that her own bid was \$3; similarly, if the initial two bids are both \$3 then the referee announces "the lowest bid was \$3", in which case Amy is able to figure out that Bill's bid was also \$3. If we included those two nodes in the same information set for Amy, we would not show much faith in Amy's reasoning ability!

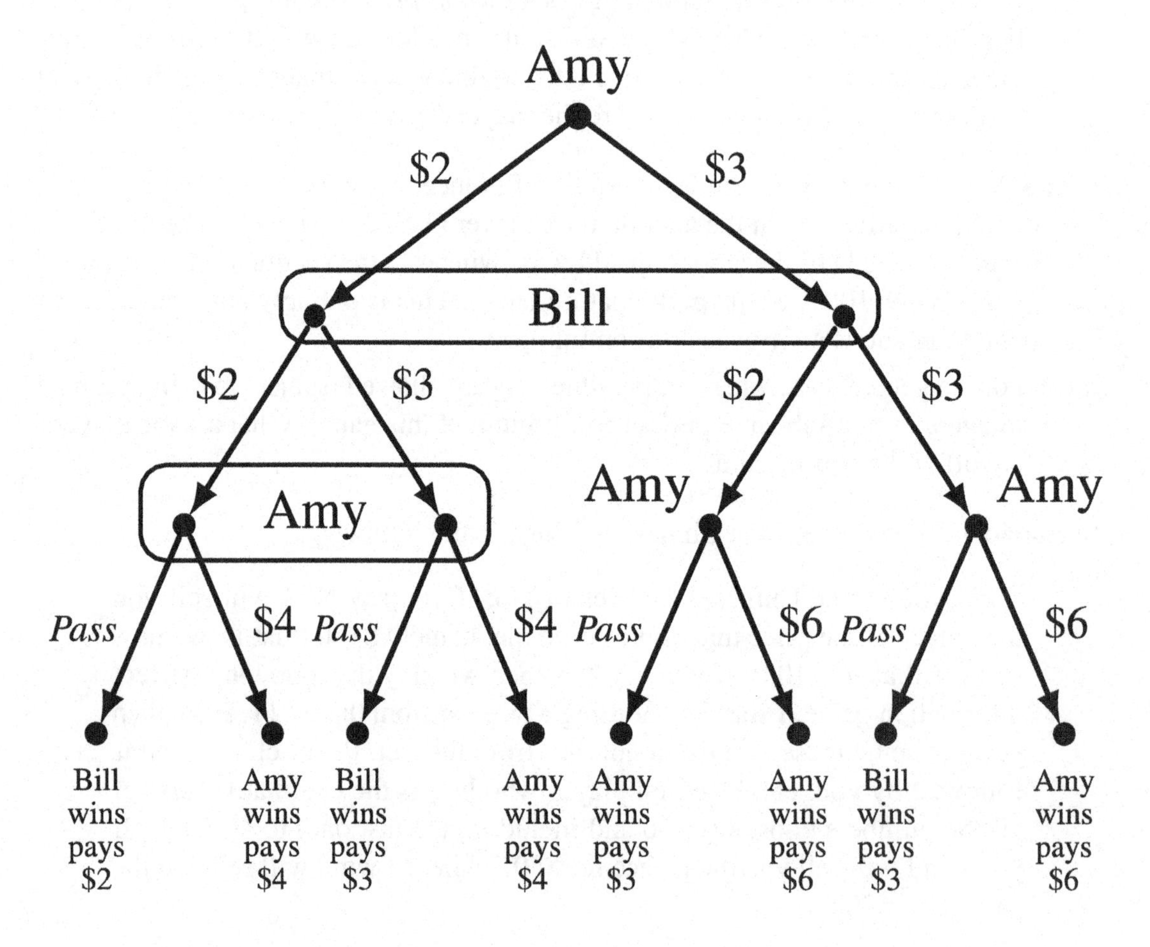

Figure 4.26: One possible game-frame for Exercise 4.1

Another possible extensive frame is shown in Figure 4.27, where Bill moves first. □

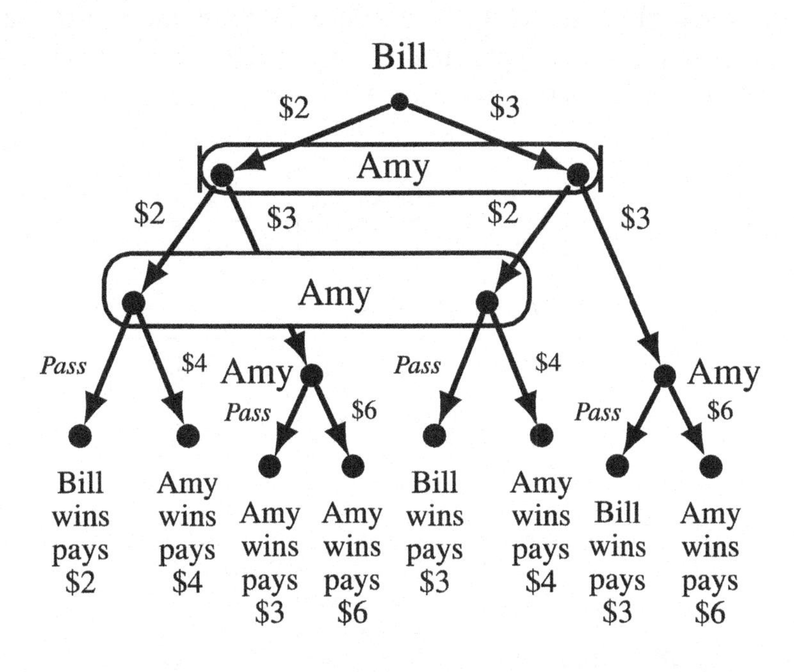

Figure 4.27: Another possible game-frame for Exercise 4.1

Solution to Exercise 4.2. The extensive form is shown in Figure 4.28 The extensive form is shown in Figure 4.28 (since the players are selfish and greedy we can take a player's utility of an outcome to be the profit of that player at that outcome). □

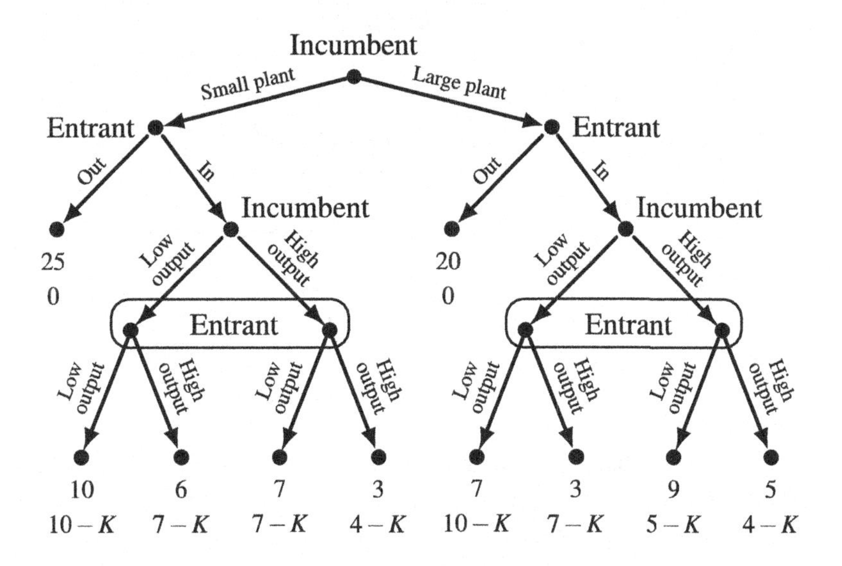

Figure 4.28: The extensive-form game for Exercise 4.2

Solution to Exercise 4.3. The strategic form is shown in Figure 4.29.
Amy's strategy (x, y, w, z) means: at the beginning I bid $\$x$, at the non-trivial information set on the left I choose y, at the singleton node in the middle I choose w and at the singleton node on the right I choose z. The numbers are bid amounts and P stands for "Pass". □

AMY \ BILL	Bid \$2	Bid \$3
$\$2, P, P, P$	Bill wins, pays \$2	Bill wins, pays \$3
$\$2, P, P, \$6$	Bill wins, pays \$2	Bill wins, pays \$3
$\$2, P, \$6, P$	Bill wins, pays \$2	Bill wins, pays \$3
$\$2, P, \$6, \$6$	Bill wins, pays \$2	Bill wins, pays \$3
$\$2, \$4, P, P$	Amy wins, pays \$4	Amy wins, pays \$4
$\$2, \$4, P, \$6$	Amy wins, pays \$4	Amy wins, pays \$4
$\$2, \$4, \$6, P$	Amy wins, pays \$4	Amy wins, pays \$4
$\$2, \$4, \$6, \$6$	Amy wins, pays \$4	Amy wins, pays \$4
$\$3, P, P, P$	Amy wins, pays \$3	Bill wins, pays \$3
$\$3, P, P, \$6$	Amy wins, pays \$3	Amy wins, pays \$6
$\$3, P, \$6, P$	Amy wins, pays \$6	Bill wins, pays \$3
$\$3, P, \$6, \$6$	Amy wins, pays \$6	Amy wins, pays \$6
$\$3, \$4, P, P$	Amy wins, pays \$3	Bill wins, pays \$3
$\$3, \$4, P, \$6$	Amy wins, pays \$3	Amy wins, pays \$6
$\$3, \$4, \$6, P$	Amy wins, pays \$6	Bill wins, pays \$3
$\$3, \$4, \$6, \$6$	Amy wins, pays \$6	Amy wins, pays \$6

Figure 4.29: The strategic form for Exercise 4.3

Solution to Exercise 4.4.

(a) The potential entrant has four information sets, hence a strategy has to specify what she would do in each of the four situations. A possible strategy is: "if the incumbent chooses a small plant I stay out, if the incumbent chooses a large plant I enter, if small plant and I entered then I choose low output, if large plant and I entered then I choose high output".

(b) The potential entrant has $2^4 = 16$ strategies.

(c) The strategic form is shown in Figure 4.30.

(d) For the case where $K = 2$ the Nash equilibria are highlighted in Figure 4.30. □

ENTRANT \ INCUMBENT	SLL	SLH	SHL	SHH	LLL	LLH	LHL	LHH
OOLL	0, 25	0, 25	0, 25	0, 25	0, 20	0, 20	0, 20	0, 20
OOLH	0, 25	0, 25	0, 25	0, 25	0, 20	0, 20	0, 20	0, 20
OOHL	0, 25	0, 25	0, 25	0, 25	0, 20	0, 20	0, 20	0, 20
OOHH	0, 25	0, 25	0, 25	0, 25	0, 20	0, 20	0, 20	0, 20
OILL	0, 25	0, 25	0, 25	0, 25	10-K, 7	5-K, 9	10-K, 7	5-K, 9
OILH	0, 25	0, 25	0, 25	0, 25	7-K, 3	4-K, 5	7-K, 3	4-K, 5
OIHL	0, 25	0, 25	0, 25	0, 25	10-K, 7	5-K, 9	10-K, 7	5-K, 9
OIHH	0, 25	0, 25	0, 25	0, 25	7-K, 3	4-K, 5	7-K, 3	4-K, 5
IOLL	10-K, 10	10-K, 10	7-K, 7	7-K, 7	0, 20	0, 20	0, 20	0, 20
IOLH	10-K, 10	10-K, 10	7-K, 7	7-K, 7	0, 20	0, 20	0, 20	0, 20
IOHL	7-K, 6	7-K, 6	4-K, 3	4-K, 3	0, 20	0, 20	0, 20	0, 20
IOHH	7-K, 6	7-K, 6	4-K, 3	4-K, 3	0, 20	0, 20	0, 20	0, 20
IILL	10-K, 10	10-K, 10	7-K, 7	7-K, 7	10-K, 7	5-K, 9	10-K, 7	5-K, 9
IILH	10-K, 10	10-K, 10	7-K, 7	7-K, 7	7-K, 3	4-K, 5	7-K, 3	4-K, 5
IIHL	7-K, 6	7-K, 6	4-K, 3	4-K, 3	10-K, 7	5-K, 9	10-K, 7	5-K, 9
IIHH	7-K, 6	7-K, 6	4-K, 3	4-K, 3	7-K, 3	4-K, 5	7-K, 3	4-K, 5

Figure 4.30: The strategic form for Exercise 4.4

Solution to Exercise 4.5. There are no proper subgames. □

Solution to Exercise 4.6. There are no proper subgames. □

Solution to Exercise 4.7.
(a) Only one proper subgame: it starts at Player 2's node.
(b) Since it is the only subgame, it is minimal. □

Solution to Exercise 4.8. The game under consideration is shown in Figure 4.31.

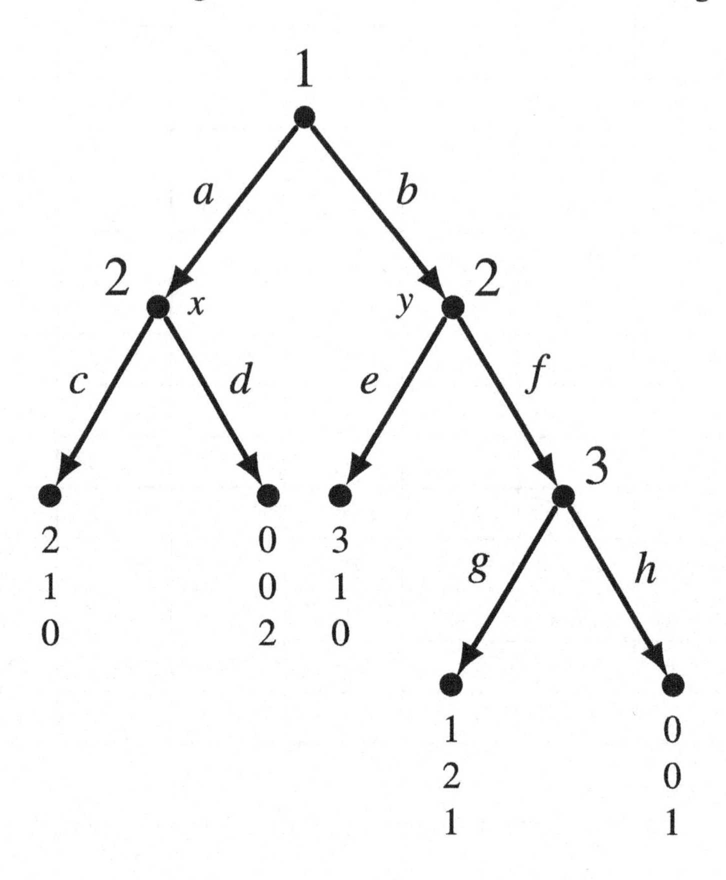

Figure 4.31: The game considered in Exercise 4.8

(a) There are three proper subgames: one starting at node x, one starting at node y and one starting at the node of Player 3.
(b) Two: the one starting at node x and the one starting at the decision node of Player 3. In a perfect-information game a minimal proper subgame is one that starts at a decision node followed only by terminal nodes. □

Solution to Exercise 4.9. The strategic form is shown in Figure 4.32.
The Nash equilibria are: $(A,(G,C),E)$ and $(B,(H,C),F)$.
The extensive-form game has two proper subgames. The one on the left has a unique Nash equilibrium, (G,E), and the one on the right has a unique Nash equilibrium, C.

Hence the game reduces to the game shown in Figure 4.33. In that game A is the unique optimal choice. Hence there is only one subgame-perfect equilibrium, namely $(A,(G,C),E)$. □

Player 3: E

Player 1 \ Player 2	GC	GD	HC	HD
A	2 2 4	2 2 4	2 0 1	2 0 1
B	1 4 3	4 3 2	1 4 3	4 3 2

Player 3: F

Player 1 \ Player 2	GC	GD	HC	HD
A	4 5 3	4 5 3	0 6 0	0 6 0
B	1 4 3	4 3 2	1 4 3	4 3 2

Figure 4.32: The strategic form for Exercise 4.9

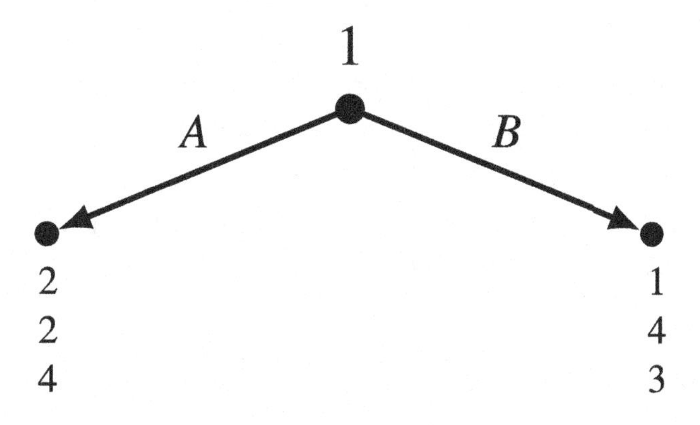

Figure 4.33: The reduced extensive-form game for Exercise 4.9

Solution to Exercise 4.10. Consider first the subgame that starts at Player 2's decision node following choice A of Player 1. The strategic-form of this game is shown in Figure 4.34 (where only the payoff of Players 2 and 3 are shown). The unique Nash equilibrium is (E,H).

Player 2 \ Player 3	F	G	H
C	0 0	2 0	1 1
D	1 0	1 2	2 1
E	0 1	0 1	3 2

Figure 4.34: The strategic-form of subgame that starts at Player 2's decision node following choice A of Player 1

Now consider the subgame that starts at Player 2's decision node following choice B of Player 1. The strategic-form of this game is shown in Figure 4.35. This game has two Nash equilibria: (L,P) and (M,R).

Player 2 \ Player 3	P	R
L	2 1	1 0
M	0 0	1 1

Figure 4.35: The strategic-form of subgame that starts at Player 2's decision node following choice B of Player 1

Thus there are two subgame-perfect equilibria of the entire game:

1. Player 1's strategy: A; Player 2's strategy: E if A and L if B; Player 3's strategy: H if A and P if B.
2. Player 1's strategy: B; Player 2's strategy: E if A and M if B; Player 3's strategy: H if A and R if B.

□

Solution to Exercise 4.11. Given the players' preferences, we can assign the following utilities to the outcomes:

outcome	Amy's utility	Bill's utility
Amy wins and pays \$3	2	0
Amy wins and pays \$4	1	0
Amy wins and pays \$6	-1	0
Bill wins and pays \$2	0	2
Bill wins and pays \$3	0	1

In the extensive form there are only two proper subgames: they start at Amy's singleton information sets (the two nodes on the right).

- In both subgames Amy will choose to pass (since she values the object at \$5 and is thus not willing to pay \$6).
- Replacing Amy's left singleton node with payoffs of 2 for Amy and 0 for Bill, and Amy's right singleton node with payoffs of 0 for Amy and 1 for Bill, we get a reduced game whose associated strategic form is as follows (in Amy's strategy the first component is the initial bet and the second component is her choice at her information set following Bill's choices after Amy's bet of \$2):

		Bill: \$2	Bill: \$3
Amy	\$2, pass	0 , 2	0 , 1
	\$2, \$4	1 , 0	1 , 0
	\$3, pass	2 , 0	0 , 1
	\$3, \$4	2 , 0	0 , 1

This game has one Nash equilibrium: ((\$2,\$4),\$3)). Thus the initial extensive-form game has one subgame-perfect equilibrium, which is as follows: Amy's strategy is (\$2, \$4, pass, pass), Bill's strategy is \$3. The corresponding outcome is: Amy wins the auction and pays \$4. □

Solution to Exercise 4.12.

(a) The extensive-form game is shown in Figure 4.36.

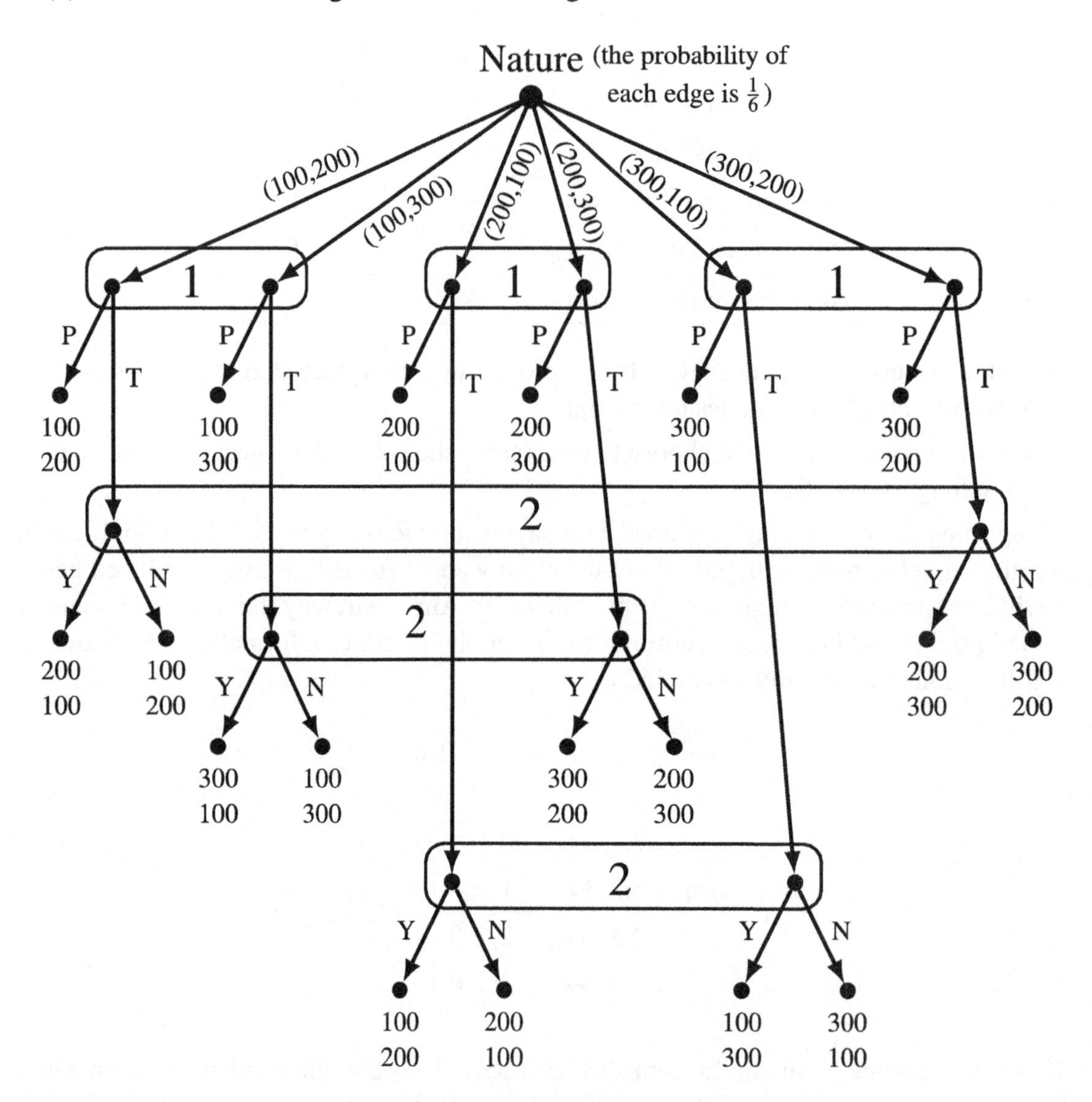

Figure 4.36: The extensive-form game for Exercise 4.12

(b) Player 1's strategies are the same as in Example 4.3. Player 2 now has 8 strategies. Each strategy has to specify how to reply to Player 1's proposal depending on the sum he (Player 2) has. Thus one possible strategy is: if I have \$100 I say No, if I have \$200 I say Yes and if I have \$300 I say No. □

Solution to Exercise 4.13.

(a) The extensive-form game is shown in Figure 4.37

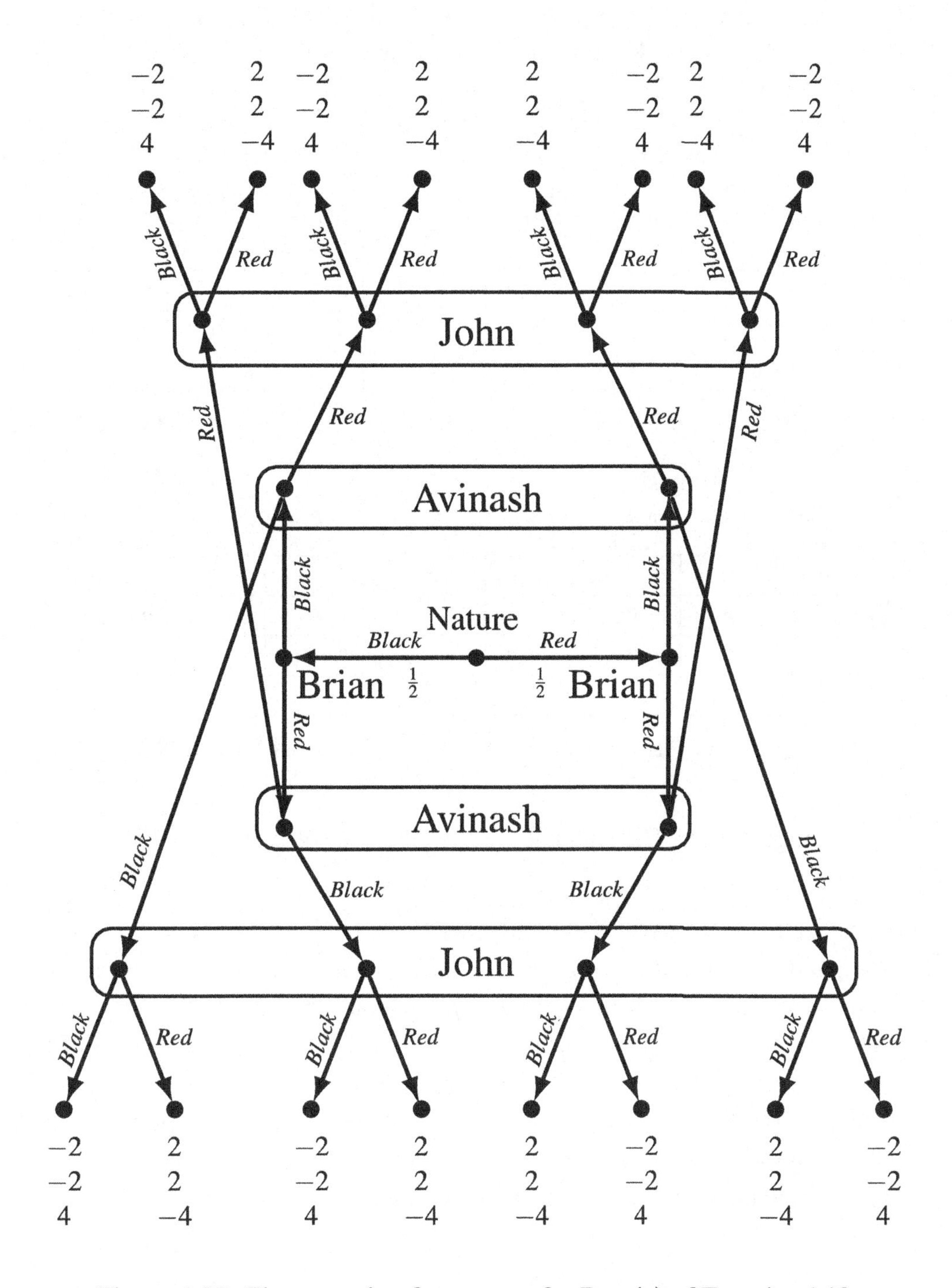

Figure 4.37: The extensive-form game for Part **(a)** of Exercise 4.13

(b) Each player has two information sets, two choices at each information set, hence four strategies.
The strategic form is shown in Figure 4.38
(interpretation: for Avinash "if B, R, if R, B" means "if Brian tells me B then I say R and if Brian tells me R then I say B"; similarly for the other strategies and for the other players).

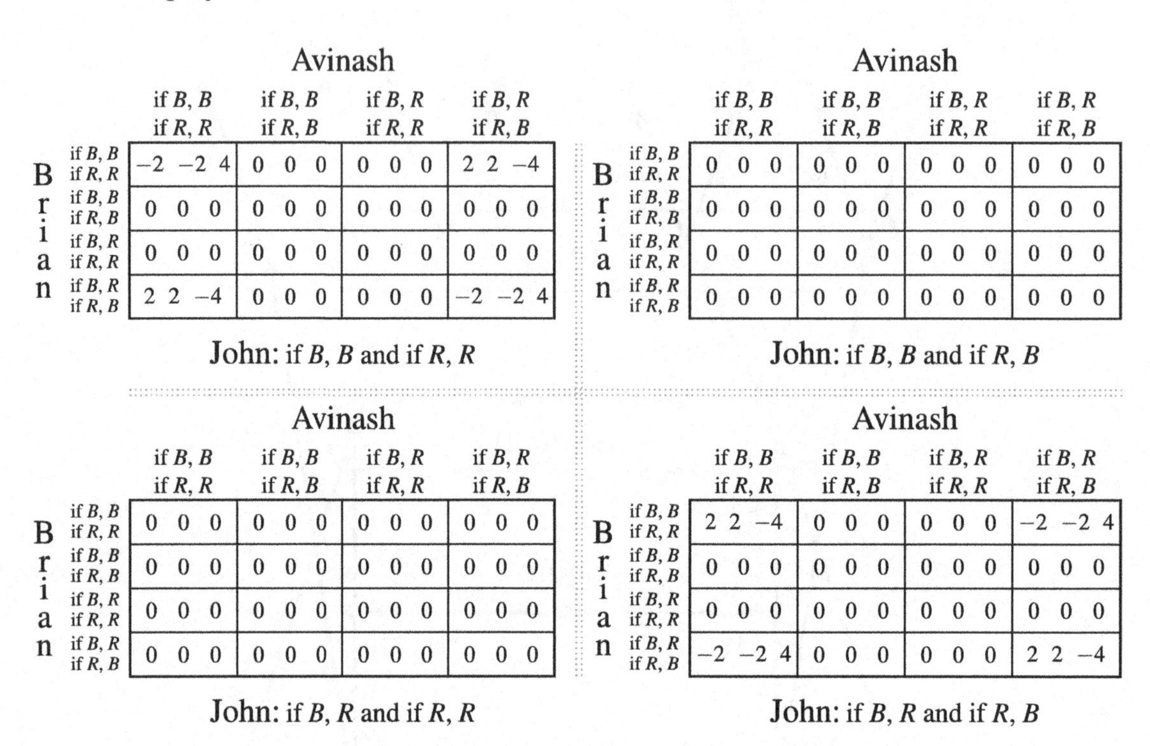

John: if B, B and if R, R

Brian \ Avinash	if B,B if R,R	if B,B if R,B	if B,R if R,R	if B,R if R,B
if B,B if R,R	−2 −2 4	0 0 0	0 0 0	2 2 −4
if B,B if R,B	0 0 0	0 0 0	0 0 0	0 0 0
if B,R if R,R	0 0 0	0 0 0	0 0 0	0 0 0
if B,R if R,B	2 2 −4	0 0 0	0 0 0	−2 −2 4

John: if B, B and if R, B

Brian \ Avinash	if B,B if R,R	if B,B if R,B	if B,R if R,R	if B,R if R,B
if B,B if R,R	0 0 0	0 0 0	0 0 0	0 0 0
if B,B if R,B	0 0 0	0 0 0	0 0 0	0 0 0
if B,R if R,R	0 0 0	0 0 0	0 0 0	0 0 0
if B,R if R,B	0 0 0	0 0 0	0 0 0	0 0 0

John: if B, R and if R, R

Brian \ Avinash	if B,B if R,R	if B,B if R,B	if B,R if R,R	if B,R if R,B
if B,B if R,R	0 0 0	0 0 0	0 0 0	0 0 0
if B,B if R,B	0 0 0	0 0 0	0 0 0	0 0 0
if B,R if R,R	0 0 0	0 0 0	0 0 0	0 0 0
if B,R if R,B	0 0 0	0 0 0	0 0 0	0 0 0

John: if B, R and if R, B

Brian \ Avinash	if B,B if R,R	if B,B if R,B	if B,R if R,R	if B,R if R,B
if B,B if R,R	2 2 −4	0 0 0	0 0 0	−2 −2 4
if B,B if R,B	0 0 0	0 0 0	0 0 0	0 0 0
if B,R if R,R	0 0 0	0 0 0	0 0 0	0 0 0
if B,R if R,B	−2 −2 4	0 0 0	0 0 0	2 2 −4

Figure 4.38: The strategic form for Part **(b)** of Exercise 4.13

How can we fill in the payoffs without spending more than 24 hours on this problem? There is a quick way of doing it. First of all, when John's strategy is to guess Black, no matter what Avinash says, he has a 50% chance of being right and a 50% chance of being wrong. Thus his expected payoff is $\frac{1}{2}(4)+\frac{1}{2}(-4)=0$ and the expected payoff of each of the other two players is $\frac{1}{2}(2)+\frac{1}{2}(-2)=0$. This explains why the second table is filled with the same payoff vector, namely $(0,0,0)$. The same reasoning applies to the case where when John's strategy is to guess Red, no matter what Avinash says (leading to the third table, filled with the same payoff vector $(0,0,0)$).

For the remaining strategies of John's, one can proceed as follows:
Start with the two colors, B and R. Under B write T (for true) if Brian's strategy says "if B then B" and write F (for false) if Brian's strategy says "if B then R"; similarly, under R write T (for true) if Brian's strategy says "if R then R" and write F (for false) if Brian's strategy says "if R then B".

In the next row, in the B column rewrite what is in the previous row if Avinash's strategy says "if B then B" and change a T into an F or an F into a T if Avinash's strategy says "if B then R". Similarly for the R column. Now repeat the same for John (in the B column a T remains a T and an F remains an F is John's strategy is "if B then B", while a T is changed into an F and an F is changed into a T if John's strategy is "if B then R").

Now in each column the payoffs are $(-2,-2,4)$ if the last row has a T and $(2,2,-4)$ if the last row has an F. The payoffs are then given by $\frac{1}{2}$ the payoff in the left column plus $\frac{1}{2}$ the payoff in the right column. For example, for the cell in the second row, third column of the third table we have the calculations shown in Figure 4.39. □

	B	R
Brian's strategy: if B, B and if R, B	T	F
Avinash's strategy: if B, R and if R, R	F	T
John's strategy: if B, R and if R, R	F	T
Payoffs:	$(2,2,-4)$	$(-2,-2,4)$

Expected payoffs: $\frac{1}{2}(2,2,-4)+\frac{1}{2}(-2,-2,4)=(0,0,0)$

Figure 4.39: The calculations for the expected payoffs

Solution to Exercise 4.14.

(a) The extensive-form representation of the simplified poker game is shown in Figure 4.40 (the top number is Yvonne's net take in dollars and the bottom number is Zoe's net take).

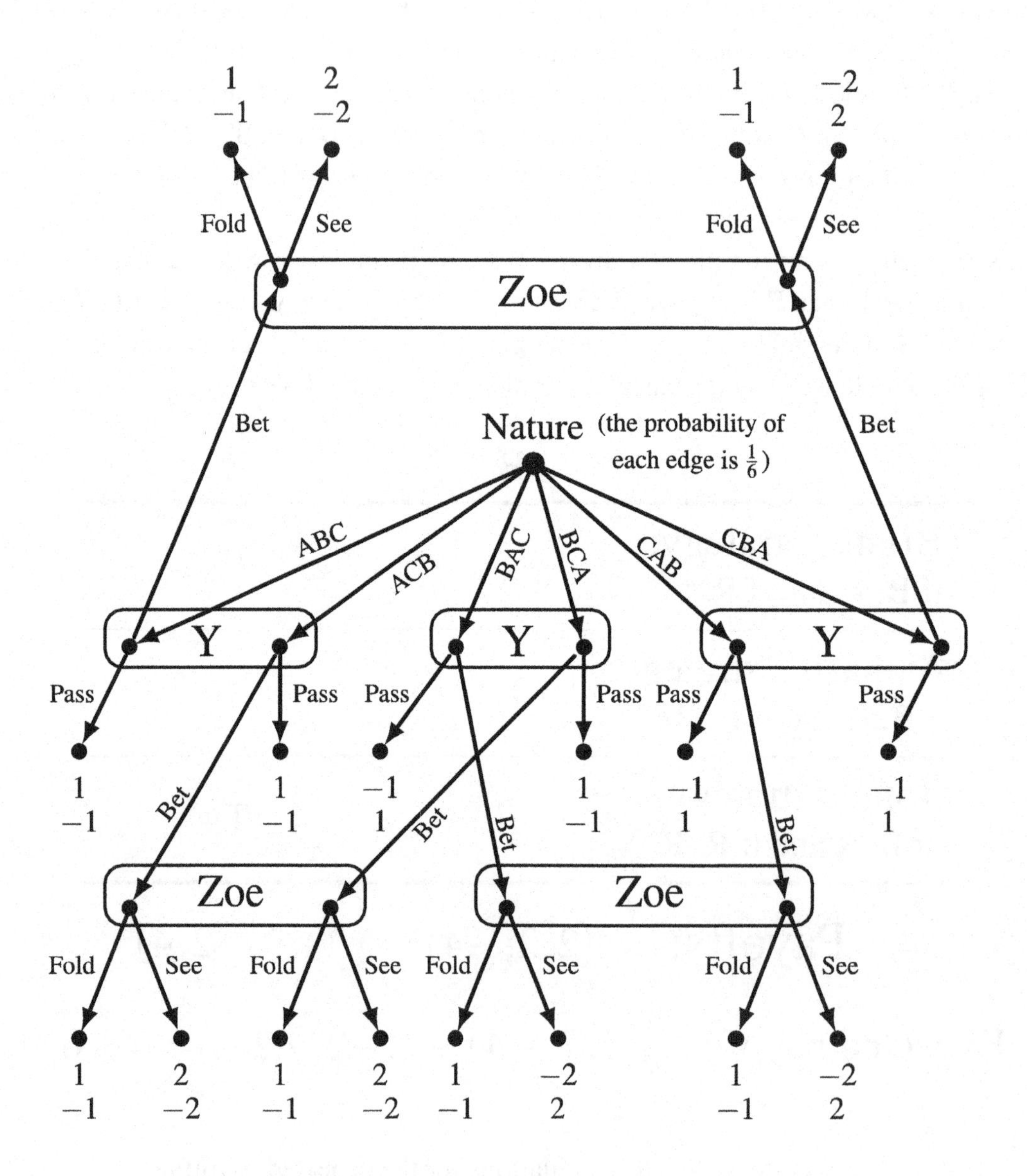

Figure 4.40: The extensive-form game for Exercise 4.14

(b) Yvonne has eight strategies (three information sets, two choices at each information set, thus $2 \times 2 \times 2 = 8$ possible strategies).

(c) Similarly, Zoe has eight strategies.

(d) Yvonne uses the strategy "If A pass, if B pass, if C bet" and Zoe uses the strategy "If A fold, if B fold, if C fold"). The table below shows how to compute the expected net payoff for Yvonne. Zoe's expected net payoff is the negative of that.

Top card is:	A	A	B	B	C	C
Second card is:	B	C	A	C	A	B
Probability:	$\frac{1}{6}$	$\frac{1}{6}$	$\frac{1}{6}$	$\frac{1}{6}$	$\frac{1}{6}$	$\frac{1}{6}$
Yvonne's action:	*pass*	*pass*	*pass*	*pass*	*bet*	*bet*
Zoe's action:	−−	−−	−−	−−	*fold*	*fold*
Yvonne's payoff:	1	1	−1	1	1	1

Yvonne's expected payoff: $\frac{1}{6}(1+1-1+1+1+1) = \frac{4}{6}$.

(e) Yvonne uses the strategy "If A pass, if B pass, if C bet" and Zoe uses the strategy "see with any card". The table below shows how to compute the expected net payoff for Yvonne. Zoe's expected net payoff is the negative of that.

Top card is:	A	A	B	B	C	C
Second card is:	B	C	A	C	A	B
Probability:	$\frac{1}{6}$	$\frac{1}{6}$	$\frac{1}{6}$	$\frac{1}{6}$	$\frac{1}{6}$	$\frac{1}{6}$
Yvonne's action:	*pass*	*pass*	*pass*	*pass*	*bet*	*bet*
Zoe's action:	−−	−−	−−	−−	*see*	*see*
Yvonne's payoff:	1	1	−1	1	−2	−2

Yvonne's expected payoff: $\frac{1}{6}(1+1-1+1-2-2) = -\frac{2}{6}$.

(f) The strategic form is shown in Figure 4.41.

(g) Let $\rhd$ denote weak dominance, that is, $a \rhd b$ means that a weakly dominates b.
FOR YVONNE (row player): 3rd row $\rhd 1^{st}$ row, $6^{th} \rhd 4^{th}$, $7^{th} \rhd 4^{th}$, $7^{th} \rhd 5^{th}$, $2^{nd} \rhd 8^{th}$.
FOR ZOE (column player): 1st col $\rhd$ 5th col, $3^{rd} \rhd 1^{st}$, $3^{rd} \rhd 4^{th}$, $3^{rd} \rhd 5^{th}$, $3^{rd} \rhd 7^{th}$, $3^{rd} \rhd 8^{th}$, $2^{nd} \rhd 8^{th}$, $4^{th} \rhd 5^{th}$, $4^{th} \rhd 8^{th}$, $6^{th} \rhd 2^{nd}$, $6^{th} \rhd 4^{th}$, $6^{th} \rhd 5^{th}$, $6^{th} \rhd 7^{th}$, $6^{th} \rhd 8^{th}$, $7^{th} \rhd 4^{th}$, $7^{th} \rhd 5^{th}$, $7^{th} \rhd 8^{th}$.

YVONNE \ ZOE	If A fold, If B fold, If C fold	If A see, If B see, If C see	If A see, If B fold, If C fold	If A fold, If B see, If C fold	If A fold, If B fold, If C see	If A see, If B see, If C fold	If A see, If B fold, If C see	If A fold, If B see, If C see
If A pass, if B pass if C pass	0, 0	0, 0	0, 0	0, 0	0, 0	0, 0	0, 0	0, 0
If A bet, if B bet if C bet	1, −1	0, 0	0, 0	4/6, −4/6	8/6, −8/6	−2/6, 2/6	2/6, −2/6	1, −1
If A bet, if B pass if C pass	0, 0	2/6, −2/6	0, 0	1/6, −1/6	1/6, −1/6	1/6, −1/6	1/6, −1/6	2/6, −2/6
If A pass, if B bet if C pass	2/6, −2/6	0, 0	−1/6, 1/6	2/6, −2/6	3/6, −3/6	−1/6, 1/6	0, 0	3/6, −3/6
If A pass, if B pass if C bet	4/6, −4/6	−2/6, 2/6	1/6, −1/6	1/6, −1/6	4/6, −4/6	−2/6, 2/6	1/6, −1/6	1/6, −1/6
If A bet, if B bet if C pass	2/6, −2/6	2/6, −2/6	−1/6, 1/6	3/6, −3/6	4/6, −4/6	0, 0	1/6, −1/6	5/6, −5/6
If A bet, if B pass if C bet	4/6, −4/6	0, 0	1/6, −1/6	2/6, −2/6	5/6, −5/6	−1/6, 1/6	2/6, −2/6	3/6, −3/6
if A Pass, If B or C, Bet,	1, −1	−2/6, 2/6	0, 0	3/6, −3/6	7/6, −7/6	−3/6, 3/6	1/6, −1/6	4/6, −4/6

Figure 4.41: The strategic form for Part **(f)** of Exercise 4.14

(h) Eliminating rows 1, 4, 5 and 8 and all columns except 3 and 6 we are left with the reduced game shown below:

		Zoe: See only if A	Zoe: See only with A or B
Yvonne	Bet always	0 , 0	$-\frac{2}{6}$, $\frac{2}{6}$
	Bet only if A	0 , 0	$\frac{1}{6}$, $-\frac{1}{6}$
	Bet only if A or B	$-\frac{1}{6}$, $\frac{1}{6}$	0 , 0
	Bet only if A or C	$\frac{1}{6}$, $-\frac{1}{6}$	$-\frac{1}{6}$, $\frac{1}{6}$

In this reduced game, the second row dominates the first and the third. Eliminating them we are led to the reduced game shown below, which is a remarkable simplification of the original strategic form:

		Zoe: See only if A	Zoe: See only with A or B
Yvonne	Bet only if A	0 , 0	$\frac{1}{6}$, $-\frac{1}{6}$
	Bet only if A or C	$\frac{1}{6}$, $-\frac{1}{6}$	$-\frac{1}{6}$, $\frac{1}{6}$

□

Solution to Exercise 4.15.

(a) The extensive-form is shown in Figure 4.42.

(b) In the auction subgame for every player it is a weakly dominant strategy to bid his own value. Thus a natural Nash equilibrium for this subgame is the dominant-strategy equilibrium (although there are other Nash equilibria and one could choose any one of the alternative Nash equilibria).
Let $p_j = \max\{p_1, \ldots, p_n\}$ be the highest value and $p_k = \max\{p_1, \ldots, p_n\} \setminus \{p_j\}$ be the second highest value. Then the auction, if it takes place, will be won by Player j and he will pay p_k. Hence there are three cases.

Case 1: $p > p_k$. In this case Player G will sell to the Chinese (and the strategy of Player i in the subgame is to bid p_i), G's payoff is p and the payoff of Player i is 0.

Case 2: $p < p_k$. In this case Player G announces the auction, the strategy of Player i in the subgame is to bid p_i, the winner is Player j and he pays p_k, so that the payoff of G is p_k, the payoff of player j is $p_j - p_k$ and the payoff of every other player is 0.

Case 3: $p = p_k$. In this case there are two subgame-perfect equilibria: one as in Case 1 and the other as in Case 2 and G is indifferent between the two.

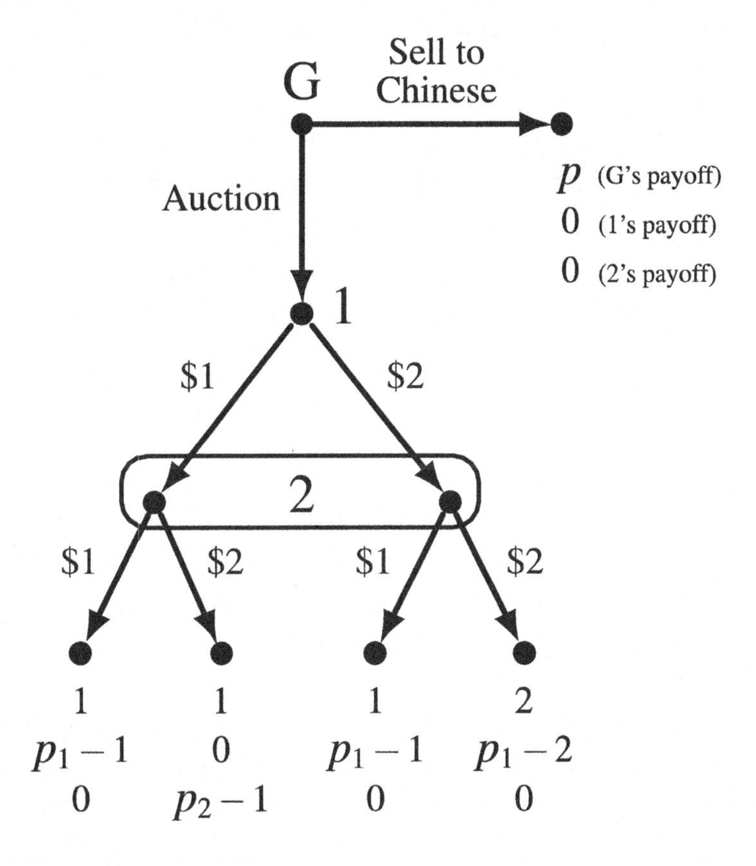

Figure 4.42: The extensive form for Part **(a)** of Exercise 4.15

(c) The extensive-form game is shown in Figure 4.43.

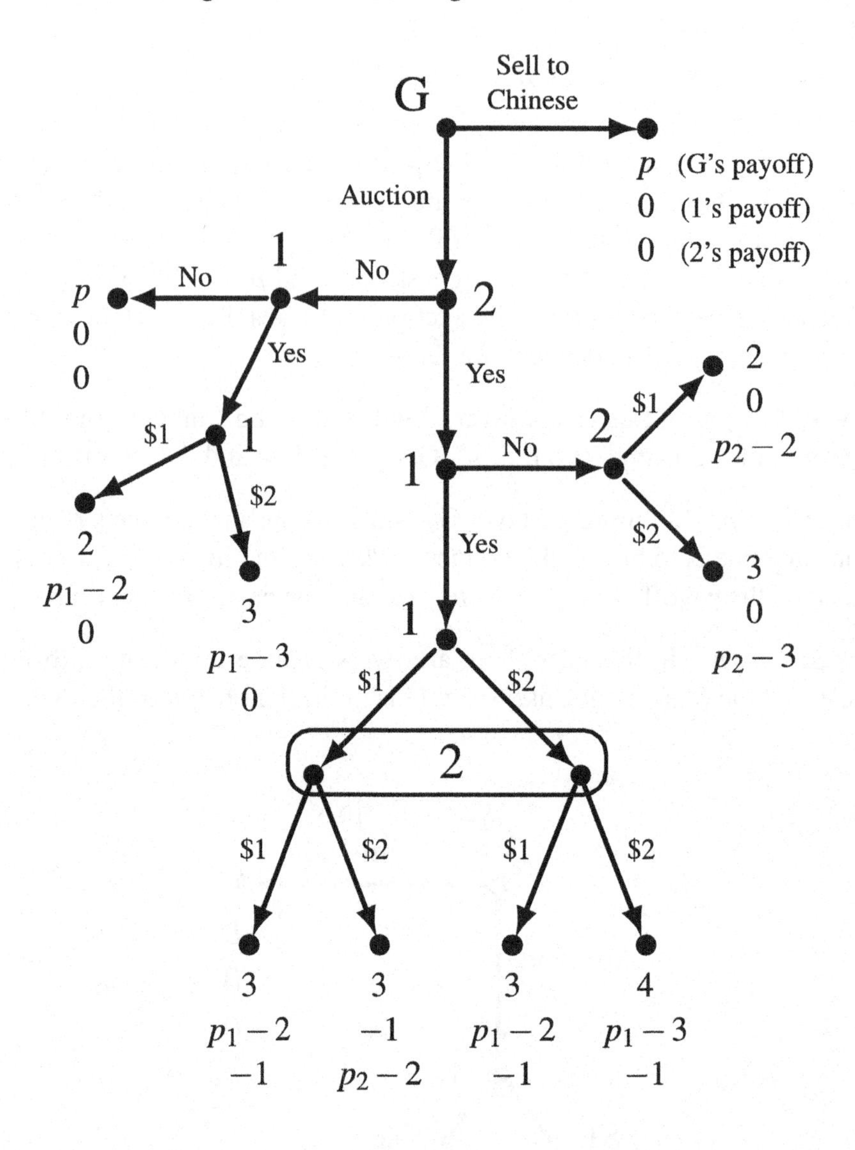

Figure 4.43: The extensive-form game for Part **(c)** of Exercise 4.15

(d) In the simultaneous subgame after both players have said Yes, the participation fee paid is a sunk cost and for every player bidding the true value is a weakly dominant strategy. Thus the outcome there is as follows: Player 1 bids p_1, gets the palace by paying p_2, G's payoff is (p_2+2), 1's payoff is (p_1-p_2-1) and Player 2's payoff is -1.

In the subgames where one player said No and the other said Yes the optimal choice is obviously $x=1$, with payoffs of 2 for Player G, 0 for the player who said No and p_i-2 for the player who said Yes.

Thus the game reduces to the one shown in Panel A of Figure 4.44.

By assumption, $p_1 > p_2+1 > 2$, so that $p_1-p_2-1 > 0$ and $p_1-2 > 0$. Thus at the bottom node and at the left node Player 1 prefers Yes to No.

Thus the game reduces to the one shown in Panel B of Figure 4.44.

Hence Player 2 will say No. The subgame-perfect equilibrium is as follows: (1) if $p > 2$ then player G will sell to the Chinese (and the choices off the equilibrium path are as explained above) and the payoffs are $(p,0,0)$; (2) if $p < 2$ then G chooses to auction, 2 says No, 1 says Yes and then offers \$1 and the payoffs are $(2, p_1 - 2, 0)$ (and the choices off the equilibrium path are as explained above); (3) if $p = 2$ then there are two equilibria: one as in (1) and the other as in (2).

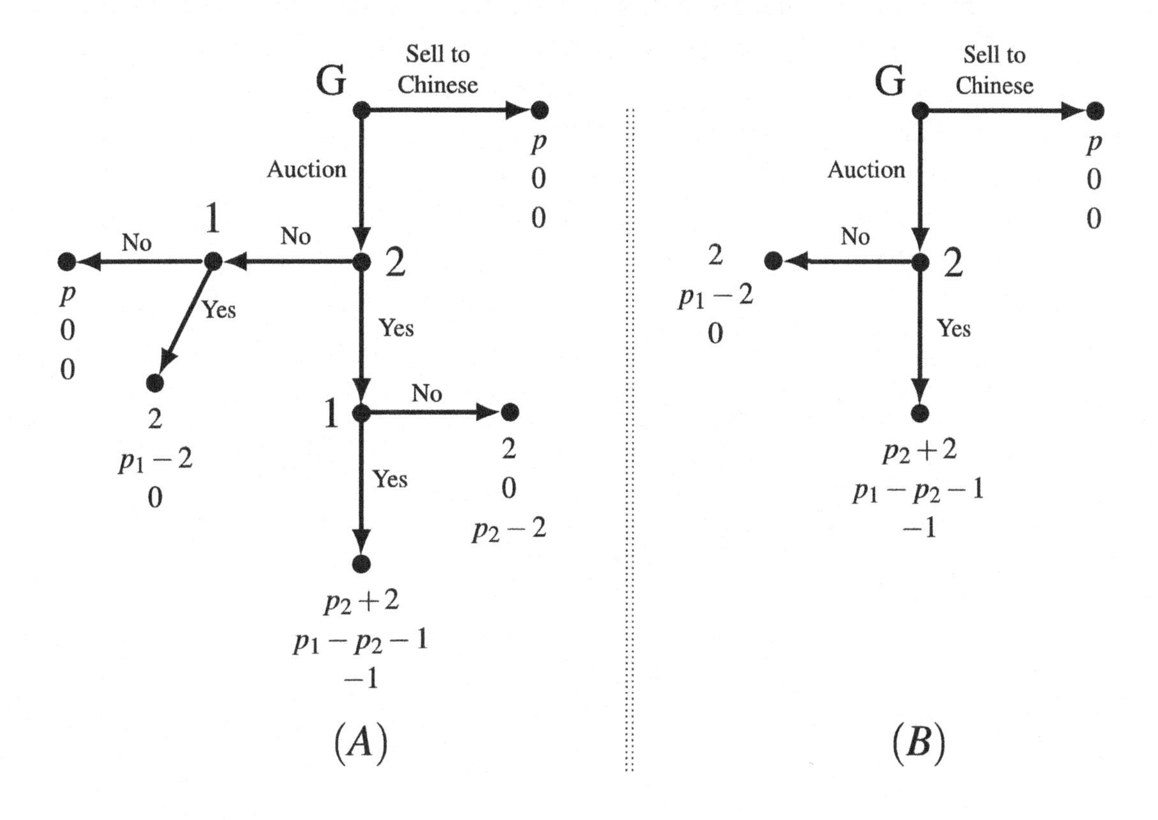

Figure 4.44: The extensive-form game for Part (d) of Exercise 4.15

(e) When the loser is given the fraction a of the amount paid by the winner (that is, the loser is given the fraction a of his own bid), **it is no longer true that bidding one's true value is a dominant strategy**. In fact, (p_1, p_2) is not even a Nash equilibrium any more. To see this, imagine that Player 1's true value is 10 and Player 2's true value is 6 and $a = 50\%$. Then if Player 1 bids 10 and 2 bids 6, Player 2 ends up losing the auction but being given \$3, while if he increased his bid to 8 then he would still lose the auction but receive \$4. This shows that there cannot be a Nash equilibrium where Player 2 bids less than Player 1. Now there are several Nash equilibria of the auction, for example, all pairs (b_1, b_2) with $b_1 = b_2 = b$ and $p_2 \leq b < p_1$ provided that $p_1 - b \geq a(b-1)$, that is, $b \leq \frac{p_1+a}{1+a}$ (but there are more: for example all pairs (b_1, b_2) with $b_1 = b_2 = b$ and $b < p_2$ provided that $p_1 - b \geq a(b-1)$ and $ab \geq p_2 - b$)). Thus to find a subgame-perfect equilibrium of the game one first has to select a Nash equilibrium of the auction game and then apply backward induction to see if the players would want to say Yes or No to the auction, etc.

(f) Let us start by considering the perfect-information game that is played if the Chinese says Yes. This is a game similar to the one discussed in Example 3.2 (Chapter 3, Section 3.5). We first determine the losing positions. Whoever has to move when the sum is 36 cannot win. Thus 36 is a losing position. Working backwards, the losing positions are 32, 28, 24, 20, 16, 12, 8, 4 and 0. Thus the first player (= player G) starts from a losing position: whatever his initial choice, he can be made to choose the second time when the sum is 4, and then 8, etc. Hence **the second player (= the Chinese) has a winning strategy**, which is as follows: if Player G just chose n, then choose $(4-n)$. If the Chinese says Yes and then follows this strategy he can guarantee that he will buy the palace for \$50. Thus the subgame-perfect equilibrium of this game is: the Chinese says Yes and uses the winning strategy in the ensuing game, while for Player G we can pick any arbitrary choices (so that, in fact, there are many subgame-perfect equilibria, but they share the same outcome). □

II

Games with Cardinal Payoffs

5. Expected Utility Theory

5.1 Money lotteries and attitudes to risk

The introduction of chance moves gives rise to probabilistic outcomes, which we called lotteries. In Chapter 4 we restricted attention to lotteries whose outcomes are sums of money (money lotteries) and to one possible way of ranking such lotteries, based on the notion of risk neutrality. In this section we will continue to focus on money lotteries and define other possible attitudes to risk.[1]

As before, we restrict attention to finite lotteries. Recall that a money lottery is a probability distribution of the form

$$\begin{pmatrix} \$x_1 & \$x_2 & \dots & \$x_n \\ p_1 & p_2 & \dots & p_n \end{pmatrix}$$

(with $0 \leq p_i \leq 1$, for all $i = 1, 2, ..., n$, and $p_1 + p_2 + \dots + p_n = 1$) and that (Definition 4.2.2, Chapter 4) its expected value is the number $(x_1 p_1 + x_2 p_2 + \dots + x_n p_n)$.
If L is a money lottery, we denote by $\mathbb{E}[L]$ the expected value of L. Thus, for example, if

$$L = \begin{pmatrix} \$30 & \$45 & \$90 \\ \frac{1}{3} & \frac{5}{9} & \frac{1}{9} \end{pmatrix} \quad \text{then} \quad \mathbb{E}[L] = \tfrac{1}{3}(30) + \tfrac{5}{9}(45) + \tfrac{1}{9}(90) = 45.$$

Recall also (Definition 4.2.3, Chapter 4) that a person is said to be *risk neutral* if she considers a money lottery to be just as good as its expected value for certain. For example, a risk-neutral person would consider getting \$45 with certainty to be just as good as playing lottery $L = \begin{pmatrix} \$30 & \$45 & \$90 \\ \frac{1}{3} & \frac{5}{9} & \frac{1}{9} \end{pmatrix}$.
We can now consider different attitudes to risk, besides risk neutrality.

[1] In the next section we will consider more general lotteries, where the outcomes need not be sums of money, and introduce the theory of expected utility.

Definition 5.1.1 Let L be a money lottery and consider the choice between L and getting $\$\mathbb{E}[L]$ (the expected value of L) for certain . Then

- An individual who prefers $\$\mathbb{E}[L]$ for certain to L is said to be *risk averse.*
- An individual who is indifferent between $\$\mathbb{E}[L]$ for certain and L is said to be *risk neutral.*
- An individual who prefers L to $\$\mathbb{E}[L]$ for certain is said to be *risk loving*.

Note that if an individual is risk **neutral**, has transitive preferences over money lotteries and prefers more money to less, then we can tell how that individual ranks any two money lotteries. For example, how would a risk neutral individual rank the two lotteries $L_1 = \begin{pmatrix} \$30 & \$45 & \$90 \\ \frac{1}{3} & \frac{5}{9} & \frac{1}{9} \end{pmatrix}$ and $L_2 = \begin{pmatrix} \$5 & \$100 \\ \frac{3}{5} & \frac{2}{5} \end{pmatrix}$?
Since $\mathbb{E}[L_1] = 45$ and the individual is risk neutral, $L_1 \sim \$45$; since $\mathbb{E}[L_2] = 43$ and the individual is risk neutral, $\$43 \sim L_2$; since the individual prefers more money to less, $\$45 \succ \$43$; thus, by transitivity, $L_1 \succ L_2$.

On the other hand, knowing that an individual is risk **averse**, has transitive preferences over money lotteries and prefers more money to less is not sufficient to predict how she will choose between two arbitrary money lotteries.
For example, as we will see later (see Exercise 5.11), it is possible that one risk-averse individual will prefer $L_3 = \begin{pmatrix} \$28 \\ 1 \end{pmatrix}$ (whose expected value is 28) to $L_4 = \begin{pmatrix} \$10 & \$50 \\ \frac{1}{2} & \frac{1}{2} \end{pmatrix}$ (whose expected value is 30), while another risk-averse individual will prefer L_4 to L_3.

Similarly, knowing that an individual is risk **loving**, has transitive preferences over money lotteries and prefers more money to less is not sufficient to predict how she will choose between two arbitrary money lotteries.

R Note that "rationality" does not, and should not, dictate whether an individual should be risk neutral, risk averse or risk loving: an individual's attitude to risk is merely a reflection of that individual's preferences. It is a generally accepted principle that *de gustibus non est disputandum* (in matters of taste, there can be no disputes). According to this principle, there is no such thing as an irrational preference and thus there is no such thing as an irrational attitude to risk. From an empirical point of view, however, most people reveal through their choices (e.g. the decision to buy insurance) that they are risk averse, at least when the stakes are high.

As noted above, with the exception of risk-neutral individuals, even if we restrict attention to money lotteries we are not able to say much – in general – about how an individual would choose among lotteries. What we need is a theory of "rational" preferences over lotteries that (1) is general enough to cover lotteries whose outcomes are not necessarily sums of money and (2) is capable of accounting for different attitudes to risk in the case of money lotteries. One such theory is the theory of expected utility, to which we now turn.

Test your understanding of the concepts introduced in this section, by going through the exercises in Section 5.4.1 at the end of this chapter.

5.2 Expected utility: theorems

The theory of expected utility was developed by the founders of game theory, namely John von Neumann and Oskar Morgenstern, in their 1944 book *Theory of Games and Economic Behavior*. In a rather unconventional way, we shall first (in this section) state the main result of the theory (which we split into two theorems) and then (in the following section) explain the assumptions (or axioms) behind that result. The reader who is not interested in understanding the conceptual foundations of expected utility theory, but wants to understand what the theory says and how it can be used, can study this section and skip the next.

Let O be a set of *basic outcomes*. Note that a basic outcome need not be a sum of money: it could be the state of an individual's health, or whether the individual under consideration receives an award, or whether it will rain on the day of her planned outdoor party, etc.

Let $\mathscr{L}(O)$ be the set of *simple lotteries* (or probability distributions) over O. We will assume throughout that O is a finite set: $O = \{o_1, o_2, ..., o_m\}$ $(m \geq 1)$.

Thus, an element of $\mathscr{L}(O)$ is of the form $\begin{pmatrix} o_1 & o_2 & \cdots & o_m \\ p_1 & p_2 & \cdots & p_m \end{pmatrix}$ with $0 \leq p_i \leq 1$, for all $i = 1, 2, ..., m$, and $p_1 + p_2 + ... + p_m = 1$.

We will use the symbol L (with or without subscript) to denote an element of $\mathscr{L}(O)$, that is, a simple lottery. Lotteries are used to represent situations of uncertainty. For example, if $m = 4$ and the individual faces the lottery $L = \begin{pmatrix} o_1 & o_2 & o_3 & o_4 \\ \frac{2}{5} & 0 & \frac{1}{5} & \frac{2}{5} \end{pmatrix}$ then she knows that, eventually, the outcome will be one and only one of o_1, o_2, o_3, o_4, but does not know which one; furthermore, she is able to quantify her uncertainty by assigning probabilities to these outcomes.

We interpret these probabilities either as objectively obtained from relevant (past) data or as subjective estimates by the individual. For example, an individual who is considering whether or not to insure her bicycle against theft for the following 12 months knows that there are two relevant basic outcomes: either the bicycle will be stolen or it will not be stolen. Furthermore, she can look up data on past bicycle thefts in her area and use the proportion of bicycles that were stolen as an "objective" estimate of the probability that her bicycle will be stolen. Alternatively, she can use a more subjective estimate: for example she might use a lower probability of theft than suggested by the data because she knows herself to be very conscientious and – unlike other people – to always lock her bicycle when left unattended.

The assignment of *zero probability* to a particular basic outcome is taken to be an expression of *belief, not impossibility*: the individual is confident that the outcome will not arise, but she cannot rule out that outcome on logical grounds or by appealing to the laws of nature.

Among the elements of $\mathscr{L}(O)$ there are the degenerate lotteries that assign probability 1 to one basic outcome: for example, if $m = 4$ one degenerate lottery is $\begin{pmatrix} o_1 & o_2 & o_3 & o_4 \\ 0 & 0 & 1 & 0 \end{pmatrix}$.

To simplify the notation we will often denote degenerate lotteries as basic outcomes, that is, instead of writing $\begin{pmatrix} o_1 & o_2 & o_3 & o_4 \\ 0 & 0 & 1 & 0 \end{pmatrix}$ we will simply write o_3.

Thus, in general, the degenerate lottery $\begin{pmatrix} o_1 & \cdots & o_{i-1} & o_i & o_{i+1} & \cdots & o_m \\ 0 & 0 & 0 & 1 & 0 & 0 & 0 \end{pmatrix}$ will be denoted by o_i. As another simplification, we will often omit those outcomes that are assigned zero probability. For example, if $m = 4$, the lottery $\begin{pmatrix} o_1 & o_2 & o_3 & o_4 \\ \frac{1}{3} & 0 & \frac{2}{3} & 0 \end{pmatrix}$ will be written more simply as $\begin{pmatrix} o_1 & o_3 \\ \frac{1}{3} & \frac{2}{3} \end{pmatrix}$.

In this chapter we shall call the individual under consideration the Decision-Maker, or *DM* for short. The theory of expected utility assumes that the *DM* has a complete and transitive ranking $\succsim$ of the elements of $\mathscr{L}(O)$ (indeed, this is one of the axioms listed in the next section). As in Chapter 2, the interpretation of $L \succsim L'$ is that the *DM* considers L to be at least as good as L'. By completeness, given any two lotteries L and L', either $L \succ L'$ (the *DM* prefers L to L') or $L' \succ L$ (the *DM* prefers L' to L) or $L \sim L'$ (the *DM* is indifferent between L and L'). Furthermore, by transitivity, for any three lotteries L_1, L_2 and L_3, if $L_1 \succsim L_2$ and $L_2 \succsim L_3$, then $L_1 \succsim L_3$. Besides completeness and transitivity, a number of other "rationality" constraints are postulated on the ranking $\succsim$ of the elements of $\mathscr{L}(O)$; these constraints are the so-called Expected Utility Axioms and are discussed in the next section.

Definition 5.2.1 A ranking $\succsim$ of the elements of $\mathscr{L}(O)$ that satisfies the Expected Utility Axioms (listed in the next section) is called a *von Neumann-Morgenstern ranking*.

The following two theorems are the key results in the theory of expected utility.

Theorem 5.2.1 [von Neumann-Morgenstern, 1944]. Let $O = \{o_1, o_2, ..., o_m\}$ be a set of basic outcomes and let $\mathscr{L}(O)$ be the set of simple lotteries over O. If $\succsim$ is a von Neumann-Morgenstern ranking of the elements of $\mathscr{L}(O)$ then there exists a function $U : O \to \mathbb{R}$, called a *von Neumann-Morgenstern utility function*,indexvon Neumann-Morgenstern!utility function that assigns a number (called *utility*) to every basic outcome and is such that, for any two lotteries $L = \begin{pmatrix} o_1 & o_2 & \cdots & o_m \\ p_1 & p_2 & \cdots & p_m \end{pmatrix}$ and $L' = \begin{pmatrix} o_1 & o_2 & \cdots & o_m \\ q_1 & q_2 & \cdots & q_m \end{pmatrix}$,

$$L \succ L' \text{ if and only if } \mathbb{E}[U(L)] > \mathbb{E}[U(L')], \text{ and}$$
$$L \sim L' \text{ if and only if } \mathbb{E}[U(L)] = \mathbb{E}[U(L')], \text{ where}$$
$$U(L) = \begin{pmatrix} U(o_1) & U(o_2) & \cdots & U(o_m) \\ p_1 & p_2 & \cdots & p_m \end{pmatrix}, U(L') = \begin{pmatrix} U(o_1) & U(o_2) & \cdots & U(o_m) \\ q_1 & q_2 & \cdots & q_m \end{pmatrix},$$

$\mathbb{E}[U(L)]$ is the expected value of the lottery $U(L)$ and $\mathbb{E}[U(L')]$ is the expected value of the lottery $U(L')$, that is,

$$\mathbb{E}[U(L)] = p_1 U(o_1) + p_2 U(o_2) + ... + p_m U(o_m), \text{ and}$$
$$\mathbb{E}[U(L')] = q_1 U(o_1) + q_2 U(o_2) + ... + q_m U(o_m).$$

$\mathbb{E}[U(L)]$ is called the *expected utility* of lottery L (and $\mathbb{E}[U(L')]$ the expected utility of lottery L'). We say that any function $U : O \to \mathbb{R}$ that satisfies the property that, for any two lotteries L and L', $L \succsim L'$ if and only if $\mathbb{E}[U(L)] \geq \mathbb{E}[U(L')]$ *represents the preferences* (or ranking) $\succsim$.

Before we comment on Theorem 5.2.1 we give an example of how one can use it. Theorem 5.2.1 sometimes allows us to predict an individual's choice between two lotteries C and D if we know how that individual ranks two different lotteries A and B. For example, suppose that we observe that Susan is faced with the choice between lotteries A and B below and she says that she prefers A to B:

$$A = \begin{pmatrix} o_1 & o_2 & o_3 \\ 0 & 0.25 & 0.75 \end{pmatrix} \qquad B = \begin{pmatrix} o_1 & o_2 & o_3 \\ 0.2 & 0 & 0.8 \end{pmatrix}$$

With this information we can predict which of the following two lotteries C and D she will choose, if she has von Neumann-Morgenstern preferences:

$$C = \begin{pmatrix} o_1 & o_2 & o_3 \\ 0.8 & 0 & 0.2 \end{pmatrix} \qquad D = \begin{pmatrix} o_1 & o_2 & o_3 \\ 0 & 1 & 0 \end{pmatrix} = o_2.$$

Let U be a von Neumann-Morgenstern utility function whose existence is guaranteed by Theorem 5.2.1. Let $U(o_1) = a$, $U(o_2) = b$ and $U(o_3) = c$ (where a, b and c are numbers). Then, since Susan prefers A to B, the expected utility of A must be greater than the expected utility of B: $0.25b + 0.75c > 0.2a + 0.8c$. This inequality is equivalent to $0.25b > 0.2a + 0.05c$ or, dividing both sides by 0.25, $b > 0.8a + 0.2c$. It follows from this and Theorem 5.2.1 that Susan prefers D to C, because the expected utility of D is b and the expected utility of C is $0.8a + 0.2c$. Note that, in this example, we merely used the fact that a von Neumann-Morgenstern utility function *exists*, even though we do not know what the values of this function are.

Theorem 5.2.1 is an example of a "representation theorem" and is a generalization of a similar result for the case of the ranking of a finite set of basic outcomes O. It is not difficult to prove that if $\succsim$ is a complete and transitive ranking of O then there exists a function $U : O \to \mathbb{R}$, called a utility function (see Chapter 2), such that, for any two basic outcomes $o, o' \in O$, $U(o) \geq U(o')$ if and only if $o \succsim o'$. Now, it is quite possible that an individual has a complete and transitive ranking of O, is fully aware of her ranking and yet she is not able to answer the question "what is your utility function?", perhaps because she has never heard about utility functions. A utility function is a *tool* that we can use to represent her ranking, nothing more than that. The same applies to von Neumann-Morgenstern rankings: Theorem 5.2.1 tells us that if an individual has a von Neumann-Morgenstern ranking of the set of lotteries $\mathscr{L}(O)$ then there exists a von Neumann-Morgenstern utility function that we can use to represent her preferences, but it would not make sense for us to ask the individual "what is your von Neumann-Morgenstern utility function?" (indeed this was a question that could not even be conceived before von Neumann and Morgenstern stated and proved Theorem 5.2.1 in 1944!).

Theorem 5.2.1 tells us that a von Neumann-Morgenstern utility function exists; the next theorem can be used to actually construct such a function, by asking the individual to answer a few questions, formulated in a way that is fully comprehensible to her (without using the word 'utility'). The theorem says that, although there are many utility functions that represent a given von Neumann-Morgenstern ranking, once you know one function you "know them all", in the sense that there is a simple operation that transforms one function into the other.

Theorem 5.2.2 [von Neumann-Morgenstern, 1944].
Let $\succsim$ be a von Neumann-Morgenstern ranking of the set of basic lotteries $\mathscr{L}(O)$, where $O = \{o_1, o_2, ..., o_m\}$. Then the following are true.

(A) If $U : O \to \mathbb{R}$ is a von Neumann-Morgenstern utility function that represents $\succsim$, then, for any two real numbers a and b, with $a > 0$, the function $V : O \to \mathbb{R}$ defined by $V(o_i) = aU(o_i) + b$ (for every $i = 1, \dots, m$) is also a von Neumann-Morgenstern utility function that represents $\succsim$.

(B) If $U : O \to \mathbb{R}$ and $V : O \to \mathbb{R}$ are two von Neumann-Morgenstern utility functions that represent $\succsim$, then there exist two real numbers a and b, with $a > 0$, such that $V(o_i) = aU(o_i) + b$ (for every $i = 1, \dots, m$).

Proof. The proof of Part A of Theorem 5.2.2 is very simple. Let a and b be two numbers, with $a > 0$. The hypothesis is that $U : O \to \mathbb{R}$ is a von Neumann-Morgenstern utility function that represents $\succsim$, that is, that, for any two lotteries

$$L = \begin{pmatrix} o_1 & \dots & o_m \\ p_1 & \dots & p_m \end{pmatrix} \quad \text{and} \quad L' = \begin{pmatrix} o_1 & \dots & o_m \\ q_1 & \dots & q_m \end{pmatrix},$$

$$L \succsim L' \text{ if and only if } p_1U(o_1) + ... + p_mU(o_m) \geq q_1U(o_1) + ... + q_mU(o_m) \quad (5.1)$$

Multiplying both sides of the inequality in (5.1) by $a > 0$ and adding $(p_1 + \cdots + p_m)b$ to the left-hand side and $(q_1 + \cdots + q_m)b$ to the right-hand side we obtain

$$p_1 [aU(o_1) + b] + ... + p_m [aU(o_m) + b] \geq q_1 [aU(o_1) + b] + ... + q_m [aU(o_m) + b] \quad (5.2)$$

Defining $V(o_i) = aU(o_i) + b$, it follows from (5.1) and (5.2) that

$$L \succsim L' \text{ if and only if } p_1V(o_1) + ... + p_mV(o_m) \geq q_1V(o_1) + ... + q_mV(o_m),$$

that is, the function V is a von Neumann-Morgenstern utility function that represents the ranking $\succsim$. The proof of Part B will be given later, after introducing more notation and some observations. ■

Suppose that the *DM* has a von Neumann-Morgenstern ranking of the set of lotteries $\mathscr{L}(O)$. Since among the lotteries there are the degenerate ones that assign probability 1 to a single basic outcome, it follows that the *DM* has a complete and transitive ranking of the basic outcomes. We shall write o_{best} for a best basic outcome, that is, a basic outcome which is at least as good as any other basic outcome ($o_{best} \succsim o$, for every $o \in O$) and o_{worst} for a worst basic outcome, that is, a basic outcome such that every other outcome is at least as good as it ($o \succsim o_{worst}$, for every $o \in O$). Note that there may be several best outcomes (then the *DM* would be indifferent among them) and several worst outcomes; then o_{best} will denote an arbitrary best outcome and o_{worst} an arbitrary worst outcome. We shall assume throughout that the *DM* is not indifferent among all the outcomes, that is, we shall assume that $o_{best} \succ o_{worst}$.

We now show that, in virtue of Theorem 5.2.2, among the von Neumann-Morgenstern utility functions that represent a given von Neumann-Morgenstern ranking $\succsim$ of $\mathscr{L}(O)$, there is one that assigns the value 1 to the best basic outcome(s) and the value 0 to the worst basic outcome(s). To see this, consider an arbitrary von Neumann-Morgenstern

utility function $F: O \to \mathbb{R}$ that represents $\succsim$ and define $G: O \to \mathbb{R}$ as follows: for every $o \in O$, $G(o) = F(o) - F(o_{worst})$.
Then, by Theorem 5.2.2 (with $a = 1$ and $b = -F(o_{worst})$), G is also a utility function that represents $\succsim$ and, by construction, $G(o_{worst}) = F(o_{worst}) - F(o_{worst}) = 0$; note also that, since $o_{best} \succ o_{worst}$, it follows that $G(o_{best}) > 0$.
Finally, define $U: O \to \mathbb{R}$ as follows: for every $o \in O$, $U(o) = \frac{G(o)}{G(o_{best})}$.
Then, by Theorem 5.2.2 (with $a = \frac{1}{G(o_{best})}$ and $b = 0$), U is a utility function that represents $\succsim$ and, by construction, $U(o_{worst}) = 0$ and $U(o_{best}) = 1$.
For example, if there are six basic outcomes and the ranking of the basic outcomes is $o_3 \sim o_6 \succ o_1 \succ o_4 \succ o_2 \sim o_5$, then one can take as o_{best} either o_3 or o_6 and as o_{worst} either o_2 or o_5; furthermore, if F is given by $\begin{array}{cccccc} o_1 & o_2 & o_3 & o_4 & o_5 & o_6 \\ 2 & -2 & 8 & 0 & -2 & 8 \end{array}$ then G is the function $\begin{array}{cccccc} o_1 & o_2 & o_3 & o_4 & o_5 & o_6 \\ 4 & 0 & 10 & 2 & 0 & 10 \end{array}$ and U is the function $\begin{array}{cccccc} o_1 & o_2 & o_3 & o_4 & o_5 & o_6 \\ 0.4 & 0 & 1 & 0.2 & 0 & 1 \end{array}$.

Definition 5.2.2 Let $U: O \to \mathbb{R}$ be a utility function that represents a given von Neumann-Morgenstern ranking $\succsim$ of the set of lotteries $\mathscr{L}(O)$. We say that U is *normalized* if $U(o_{worst}) = 0$ and $U(o_{best}) = 1$.

The transformations described above show how to normalize any given utility function. Armed with the notion of a normalized utility function we can now complete the proof of Theorem 5.2.2.

Proof of Part B of Theorem 5.2.2. Let $F: O \to \mathbb{R}$ and $G: O \to \mathbb{R}$ be two von Neumann-Morgenstern utility functions that represent a given von Neumann-Morgenstern ranking of $\mathscr{L}(O)$.
Let $U: O \to \mathbb{R}$ be the normalization of F and $V: O \to \mathbb{R}$ be the normalization of G. First we show that it must be that $U = V$, that is, $U(o) = V(o)$ for every $o \in O$.
Suppose, by contradiction, that there is an $\hat{o} \in O$ such that $U(\hat{o}) \neq V(\hat{o})$. Without loss of generality we can assume that $U(\hat{o}) > V(\hat{o})$.
Construct the following lottery: $L = \begin{pmatrix} o_{best} & o_{worst} \\ \hat{p} & 1-\hat{p} \end{pmatrix}$ with $\hat{p} = U(\hat{o})$ (recall that U is normalized and thus takes on values in the interval from 0 to 1).
Then $\mathbb{E}[U(L)] = \mathbb{E}[V(L)] = U(\hat{o})$. Hence, according to U it must be that $\hat{o} \sim L$ (this follows from Theorem 5.2.1), while according to V it must be (again, by Theorem 5.2.1) that $L \succ \hat{o}$ (since $\mathbb{E}[V(L)] = U(\hat{o}) > V(\hat{o})$). Then U and V cannot be two representations of the same ranking. Now let $a_1 = \frac{1}{F(o_{best}) - F(o_{worst})}$ and $b_1 = -\frac{F(o_{worst})}{F(o_{best}) - F(o_{worst})}$.
Note that $a_1 > 0$. Then it is easy to verify that, for every $o \in O$, $U(o) = a_1 F(o) + b_1$.
Similarly let $a_2 = \frac{1}{G(o_{best}) - G(o_{worst})}$ and $b_2 = -\frac{G(o_{worst})}{G(o_{best}) - G(o_{worst})}$; again, $a_2 > 0$ and, for every $o \in O$, $V(o) = a_2 G(o) + b_2$. We can invert the latter transformation and obtain that, for every $o \in O$, $G(o) = \frac{V(o)}{a_2} - \frac{b_2}{a_2}$.
Thus, we can transform F into U, which – as proved above – is the same as V, and then transform V into G thus obtaining the following transformation of F into G: $G(o) = aF(o) + b$ where $a = \frac{a_1}{a_2} > 0$ and $b = \frac{b_1 - b_2}{a_2}$. ■

R Theorem 5.2.2 is often stated as follows: a utility function that represents a von Neumann-Morgenstern ranking $\succsim$ of $\mathscr{L}(O)$ is *unique up to a positive affine transformation*.[2] Because of Theorem 5.2.2, a von Neumann-Morgenstern utility function is usually referred to as a *cardinal* utility function.

Theorem 5.2.1 guarantees the existence of a utility function that represents a given von Neumann-Morgenstern ranking $\succsim$ of $\mathscr{L}(O)$ and Theorem 5.2.2 characterizes the set of such functions. Can one actually construct a utility function that represents a given ranking? The answer is affirmative: if there are m basic outcomes one can construct an individual's von Neumann-Morgenstern utility function by asking her at most $(m-1)$ questions. The first question is "what is your ranking of the basic outcomes?". Then we can construct the normalized utility function by first assigning the value 1 to the best outcome(s) and the value 0 to the worst outcome(s). This leaves us with at most $(m-2)$ values to determine. For this we appeal to one of the axioms discussed in the next section, namely the Continuity Axiom, which says that, for every basic outcome o_i there is a probability $p_i \in [0,1]$ such that the *DM* is indifferent between o_i for certain and the lottery that gives a best outcome with probability p_i and a worst outcome with probability $(1-p_i)$: $o_i \sim \begin{pmatrix} o_{best} & o_{worst} \\ p_i & 1-p_i \end{pmatrix}$. Thus, for each basic outcome o_i for which a utility has not been determined yet, we should ask the individual to tell us the value of p_i such that $o_i \sim \begin{pmatrix} o_{best} & o_{worst} \\ p_i & 1-p_i \end{pmatrix}$; then we can set $U_i(o_i) = p_i$, because the expected utility of the lottery $\begin{pmatrix} o_{best} & o_{worst} \\ p_i & 1-p_i \end{pmatrix}$ is $p_i U_i(o_{best}) + (1-p_i)\,U_i(o_{worst}) = p_i(1) + (1-p_i)0 = p_i$.

Example 5.1 Suppose that there are five basic outcomes, that is, $O = \{o_1, o_2, o_3, o_4, o_5\}$ and the *DM*, who has von Neumann-Morgenstern preferences, tells us that her ranking of the basic outcomes is as follows: $o_2 \succ o_1 \sim o_5 \succ o_3 \sim o_4$.

- Then we can begin by assigning utility 1 to the best outcome o_2 and utility 0 to the worst outcomes o_3 and o_4: $\begin{pmatrix} \text{outcome:} & o_1 & o_2 & o_3 & o_4 & o_5 \\ \text{utility:} & ? & 1 & 0 & 0 & ? \end{pmatrix}$.
- There is only one value left to be determined, namely the utility of o_1 (which is also the utility of o_5, since $o_1 \sim o_5$).
- To find this value, we ask the *DM* to tell us what value of p makes her indifferent between the lottery $L = \begin{pmatrix} o_2 & o_3 \\ p & 1-p \end{pmatrix}$ and outcome o_1 with certainty.
- Suppose that her answer is: 0.4. Then her normalized von Neumann-Morgenstern utility function is $\begin{pmatrix} \text{outcome:} & o_1 & o_2 & o_3 & o_4 & o_5 \\ \text{utility:} & 0.4 & 1 & 0 & 0 & 0.4 \end{pmatrix}$. Knowing this, we can predict her choice among any set of lotteries over these five basic outcomes.

■

Test your understanding of the concepts introduced in this section, by going through the exercises in Section 5.4.2 at the end of this chapter.

[2] An affine transformation is a function $f : \mathbb{R} \to \mathbb{R}$ of the form $f(x) = ax + b$ with $a, b \in \mathbb{R}$. The affine transformation is positive if $a > 0$.

5.3 Expected utility: the axioms

We can now turn to the list of rationality axioms proposed by von Neumann and Morgenstern. This section makes heavy use of mathematical notation and, as mentioned in the previous section, if the reader is not interested in understanding in what sense the theory of expected utility captures the notion of rationality, he/she can skip it without affecting his/her ability to understand the rest of this book.

Let $O = \{o_1, o_2, ..., o_m\}$ be the set of basic outcomes and $\mathscr{L}(O)$ the set of simple lotteries, that is, the set of probability distributions over O. Let $\succsim$ be a binary relation on $\mathscr{L}(O)$. We say that $\succsim$ is a von Neumann-Morgenstern ranking of $\mathscr{L}(O)$ if it satisfies the following four axioms or properties.

Axiom 1 [Completeness and transitivity]. $\succsim$ is complete (for every two lotteries L and L' either $L \succsim L'$ or $L' \succsim L$ or both) and transitive (for any three lotteries L_1, L_2 and L_3, if $L_1 \succsim L_2$ and $L_2 \succsim L_3$ then $L_1 \succsim L_3$).

As noted in the previous section, Axiom 1 implies that there is a complete and transitive ranking of the basic outcomes. Recall that o_{best} denotes a best basic outcome and o_{worst} denotes a worst basic outcome and that we are assuming that $o_{best} \succ o_{worst}$, that is, that the *DM* is not indifferent among all the basic outcomes.

Axiom 2 [Monotonicity]. $\begin{pmatrix} o_{best} & o_{worst} \\ p & 1-p \end{pmatrix} \succsim \begin{pmatrix} o_{best} & o_{worst} \\ q & 1-q \end{pmatrix}$ if and only if $p \geq q$.

Axiom 3 [Continuity]. For every basic outcome o_i there is a $p_i \in [0,1]$ such that $o_i \sim \begin{pmatrix} o_{best} & o_{worst} \\ p_i & 1-p_i \end{pmatrix}$.

Before we introduce the last axiom we need to define a compound lottery.

Definition 5.3.1 A *compound lottery* is a lottery of the form $\begin{pmatrix} x_1 & x_2 & \cdots & x_r \\ p_1 & p_2 & \cdots & p_r \end{pmatrix}$ where each x_i is either an element of O or an element of $\mathscr{L}(O)$.

For example, let $m = 4$.

Then $L = \begin{pmatrix} o_1 & o_2 & o_3 & o_4 \\ \frac{2}{5} & 0 & \frac{1}{5} & \frac{2}{5} \end{pmatrix}$ is a simple lottery (an element of $\mathscr{L}(O)$),

while $C = \begin{pmatrix} \begin{pmatrix} o_1 & o_2 & o_3 & o_4 \\ \frac{1}{3} & \frac{1}{6} & \frac{1}{3} & \frac{1}{6} \end{pmatrix} & o_1 & \begin{pmatrix} o_1 & o_2 & o_3 & o_4 \\ \frac{1}{5} & 0 & \frac{1}{5} & \frac{3}{5} \end{pmatrix} \\ \frac{1}{2} & \frac{1}{4} & \frac{1}{4} \end{pmatrix}$ is a compound lottery.[3]

[3] With $r = 3$, $x_1 = \begin{pmatrix} o_1 & o_2 & o_3 & o_4 \\ \frac{1}{3} & \frac{1}{6} & \frac{1}{3} & \frac{1}{6} \end{pmatrix}$, $x_2 = o_1$, $x_3 = \begin{pmatrix} o_1 & o_2 & o_3 & o_4 \\ \frac{1}{5} & 0 & \frac{1}{5} & \frac{3}{5} \end{pmatrix}$, $p_1 = \frac{1}{2}, p_2 = \frac{1}{4}$ and $p_3 = \frac{1}{4}$.

The compound lottery $C = \begin{pmatrix} \begin{pmatrix} o_1 & o_2 & o_3 & o_4 \\ \frac{1}{3} & \frac{1}{6} & \frac{1}{3} & \frac{1}{6} \end{pmatrix} & o_1 & \begin{pmatrix} o_1 & o_2 & o_3 & o_4 \\ \frac{1}{5} & 0 & \frac{1}{5} & \frac{3}{5} \end{pmatrix} \\ \\ \frac{1}{2} & \frac{1}{4} & \frac{1}{4} \end{pmatrix}$

can be viewed graphically as a tree, as shown in Figure 5.1.

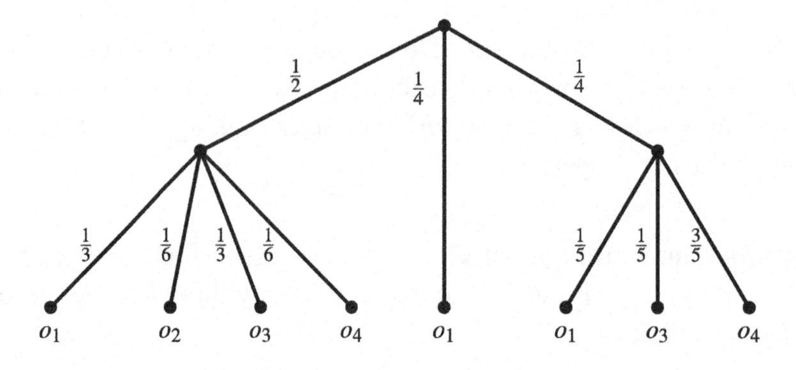

Figure 5.1: A compound lottery

Definition 5.3.2 Given a compound lottery $C = \begin{pmatrix} x_1 & x_2 & \dots & x_r \\ p_1 & p_2 & \dots & p_r \end{pmatrix}$ the *corresponding simple lottery* $L(C) = \begin{pmatrix} o_1 & o_2 & \dots & o_m \\ q_1 & q_2 & \dots & q_m \end{pmatrix}$ is defined as follows. First of all, for $i = 1, \dots, m$ and $j = 1, \dots, r$, define

$$o_i(x_j) = \begin{cases} 1 & \text{if } x_j = o_i \\ 0 & \text{if } x_j = o_k \text{ with } k \neq i \\ s_i & \text{if } x_j = \begin{pmatrix} o_1 & \dots & o_{i-1} & o_i & o_{i+1} & \dots & o_m \\ s_1 & \dots & s_{i-1} & s_i & s_{i+1} & \dots & s_m \end{pmatrix} \end{cases}$$

Then $q_i = \sum_{j=1}^{r} p_j o_i(x_j)$.

Continuing the above example where

$$C = \begin{pmatrix} \begin{pmatrix} o_1 & o_2 & o_3 & o_4 \\ \frac{1}{3} & \frac{1}{6} & \frac{1}{3} & \frac{1}{6} \end{pmatrix} & o_1 & \begin{pmatrix} o_1 & o_2 & o_3 & o_4 \\ \frac{1}{5} & 0 & \frac{1}{5} & \frac{3}{5} \end{pmatrix} \\ \\ \frac{1}{2} & \frac{1}{4} & \frac{1}{4} \end{pmatrix}$$

(see Figure 5.1) we have that $r = 3$, $x_1 = \begin{pmatrix} o_1 & o_2 & o_3 & o_4 \\ \frac{1}{3} & \frac{1}{6} & \frac{1}{3} & \frac{1}{6} \end{pmatrix}$, $x_2 = o_1$ and $x_3 = \begin{pmatrix} o_1 & o_2 & o_3 & o_4 \\ \frac{1}{5} & 0 & \frac{1}{5} & \frac{3}{5} \end{pmatrix}$, so that

$$o_1(x_1) = \tfrac{1}{3}, \quad o_1(x_2) = 1, \quad \text{and} \quad o_1(x_3) = \tfrac{1}{5}$$

and thus $q_1 = \frac{1}{2}\left(\frac{1}{3}\right) + \frac{1}{4}(1) + \frac{1}{4}\left(\frac{1}{5}\right) = \frac{28}{60}$. Similarly, $q_2 = \frac{1}{2}\left(\frac{1}{6}\right) + \frac{1}{4}(0) + \frac{1}{4}(0) = \frac{1}{12} = \frac{5}{60}$,

$q_3 = \frac{1}{2}\left(\frac{1}{3}\right) + \frac{1}{4}(0) + \frac{1}{4}\left(\frac{1}{5}\right) = \frac{13}{60}$ and $q_4 = \frac{1}{2}\left(\frac{1}{6}\right) + \frac{1}{4}(0) + \frac{1}{4}\left(\frac{3}{5}\right) = \frac{14}{60}$. These numbers correspond to multiplying the probabilities along the edges of the tree of Figure 5.1 leading to an outcome, as shown in Figure 5.2 and then adding up the probabilities of each outcome, as shown in Figure 5.3. Thus, the simple lottery $L(C)$ that corresponds to C is $L(C) = \begin{pmatrix} o_1 & o_2 & o_3 & o_4 \\ \frac{28}{60} & \frac{5}{60} & \frac{13}{60} & \frac{14}{60} \end{pmatrix}$, namely the lottery shown in Figure 5.3.

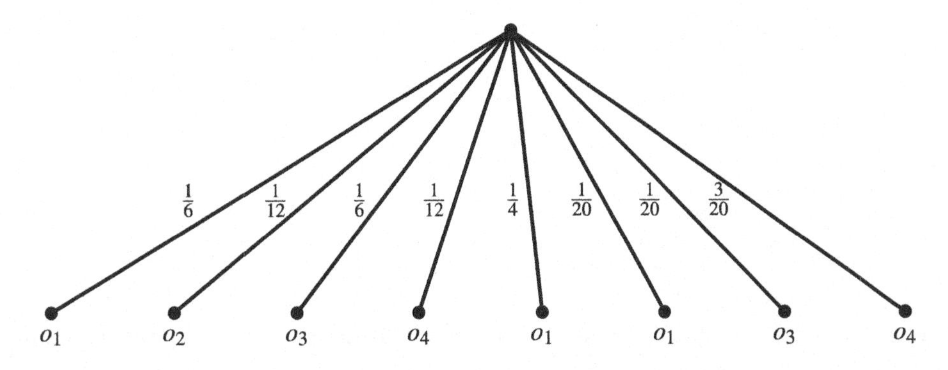

Figure 5.2: Simplification of Figure 5.1 obtained by merging paths into simple edges and associating with the simple edges the products of the probabilities along the path.

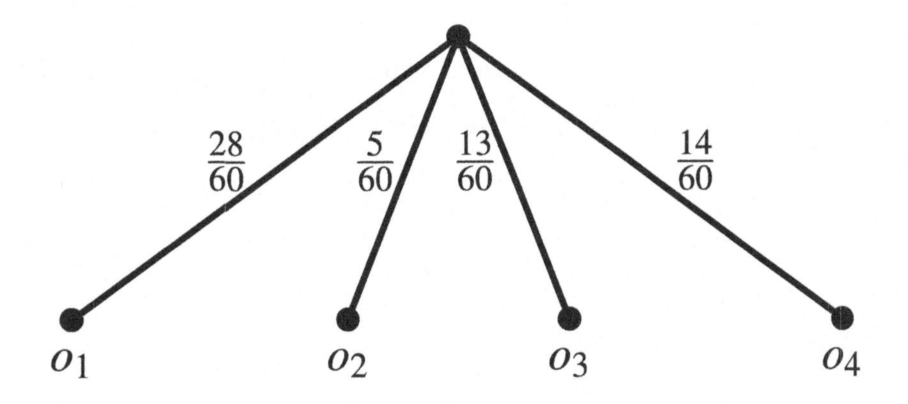

Figure 5.3: Simplification of Figure 5.2 obtained by adding, for each outcome, the probabilities of that outcome.

Axiom 4 [Independence or substitutability]. Consider an arbitrary basic outcome o_i and an arbitrary simple lottery $L = \begin{pmatrix} o_1 & \cdots & o_{i-1} & o_i & o_{i+1} & \cdots & o_m \\ p_1 & \cdots & p_{i-1} & p_i & p_{i+1} & \cdots & p_m \end{pmatrix}$. If $\hat{L}$ is a simple lottery such that $o_i \sim \hat{L}$, then $L \sim M$ where M is the simple lottery corresponding to the compound lottery $C = \begin{pmatrix} o_1 & \cdots & o_{i-1} & \hat{L} & o_{i+1} & \cdots & o_m \\ p_1 & \cdots & p_{i-1} & p_i & p_{i+1} & \cdots & p_m \end{pmatrix}$ obtained by replacing o_i with $\hat{L}$ in L.

We can now prove the first theorem of the previous section.

Proof of Theorem 5.2.1. To simplify the notation, throughout this proof we will assume that we have renumbered the basic outcomes in such a way that $o_{best} = o_1$ and $o_{worst} = o_m$. First of all, for every basic outcome o_i, let $u_i \in [0,1]$ be such that $o_i \sim \begin{pmatrix} o_1 & o_m \\ u_i & 1-u_i \end{pmatrix}$. The existence of such a value u_i is guaranteed by the Continuity Axiom (Axiom 3); clearly $u_1 = 1$ and $u_m = 0$. Now consider an arbitrary lottery

$$L_1 = \begin{pmatrix} o_1 & \dots & o_m \\ p_1 & \dots & p_m \end{pmatrix}.$$

First we show that

$$L_1 \sim \begin{pmatrix} o_1 & o_m \\ \sum_{i=1}^{m} p_i u_i & 1 - \sum_{i=1}^{m} p_i u_i \end{pmatrix} \tag{5.3}$$

This is done through a repeated application of the Independence Axiom (Axiom 4), as follows. Consider the compound lottery

$$\mathscr{C}_2 = \begin{pmatrix} o_1 & \begin{pmatrix} o_1 & o_m \\ u_2 & 1-u_2 \end{pmatrix} & o_3 & \dots & o_m \\ p_1 & p_2 & p_3 & \dots & p_m \end{pmatrix}$$

obtained by replacing o_2 in lottery L_1 with the lottery $\begin{pmatrix} o_1 & o_m \\ u_2 & 1-u_2 \end{pmatrix}$ that the *DM* considers to be just as good as o_2. The simple lottery corresponding to $\mathscr{C}_2$ is

$$L_2 = \begin{pmatrix} o_1 & o_3 & \dots & o_{m-1} & o_m \\ p_1 + p_2 u_2 & p_3 & \dots & p_{m-1} & p_m + p_2(1-u_2) \end{pmatrix}.$$

Note that o_2 is assigned probability 0 in L_2 and thus we have omitted it. By Axiom 4, $L_1 \sim L_2$. Now apply the same argument to L_2: let

$$\mathscr{C}_3 = \begin{pmatrix} o_1 & \begin{pmatrix} o_1 & o_m \\ u_3 & 1-u_3 \end{pmatrix} & \dots & o_m \\ p_1 + p_2 u_2 & p_3 & \dots & p_m + p_2(1-u_2) \end{pmatrix}$$

whose corresponding simple lottery is

$$L_3 = \begin{pmatrix} o_1 & \dots & o_m \\ p_1 + p_2 u_2 + p_3 u_3 & \dots & p_m + p_2(1-u_2) + p_3(1-u_3) \end{pmatrix}.$$

Note, again, that o_3 is assigned probability zero in L_3. By Axiom 4, $L_2 \sim L_3$; thus, by transitivity (since $L_1 \sim L_2$ and $L_2 \sim L_3$) we have that $L_1 \sim L_3$. Repeating this argument we get that $L_1 \sim L_{m-1}$, where

$$L_{m-1} = \begin{pmatrix} o_1 & o_m \\ p_1 + p_2 u_2 + \dots + p_{m-1} u_{m-1} & p_m + p_2(1-u_2) + \dots + p_{m-1}(1-u_{m-1}) \end{pmatrix}.$$

Since $u_1 = 1$ (so that $p_1u_1 = p_1$) and $u_m = 0$ (so that $p_mu_m = 0$),

$$p_1 + p_2u_2 + \ldots + p_{m-1}u_{m-1} = \sum_{i=1}^{m} p_iu_i$$

and

$$p_2(1-u_2) + \ldots + p_{m-1}(1-u_{m-1}) + p_m = \sum_{i=2}^{m} p_i - \sum_{i=2}^{m-1} p_iu_i = p_1 + \sum_{i=2}^{m} p_i - \sum_{i=2}^{m-1} p_iu_i - p_1$$

$$= _{(\text{since } u_1=1 \text{ and } u_m=0)} \sum_{i=1}^{m} p_i - \sum_{i=2}^{m-1} p_iu_i - p_1u_1 - p_mu_m = _{\left(\text{since } \sum_{i=1}^{m} p_i=1\right)} 1 - \sum_{i=1}^{m} p_iu_i$$

Thus, $L_{m-1} = \begin{pmatrix} o_1 & o_m \\ \sum_{i=1}^{m} p_iu_i & 1-\sum_{i=1}^{m} p_iu_i \end{pmatrix}$, which proves (5.3). Now define the following utility function $U : \{o_1, \ldots, o_m\} \to [0,1]$: $U(o_i) = u_i$, where, as before, for every basic outcome o_i, $u_i \in [0,1]$ is such that $o_i \sim \begin{pmatrix} o_1 & o_m \\ u_i & 1-u_i \end{pmatrix}$. Consider two arbitrary lotteries $L = \begin{pmatrix} o_1 & \cdots & o_m \\ p_1 & \cdots & p_m \end{pmatrix}$ and $L' = \begin{pmatrix} o_1 & \cdots & o_m \\ q_1 & \cdots & q_m \end{pmatrix}$. We want to show that $L \succsim L'$ if and only if $\mathbb{E}[U(L)] \geq \mathbb{E}[U(L')]$, that is, if and only if $\sum_{i=1}^{m} p_iu_i \geq \sum_{i=1}^{m} q_iu_i$. By (5.3), $L \sim M$, where $M = \begin{pmatrix} o_1 & o_m \\ \sum_{i=1}^{m} p_iu_i & 1-\sum_{i=1}^{m} p_iu_i \end{pmatrix}$ and also $L' \sim M'$, where $M' = \begin{pmatrix} o_1 & o_m \\ \sum_{i=1}^{m} q_iu_i & 1-\sum_{i=1}^{m} q_iu_i \end{pmatrix}$.
Thus, by transitivity of $\succsim$, $L \succsim L'$ if and only if $M \succsim M'$; by the Monotonicity Axiom (Axiom 2), $M \succsim M'$ if and only if $\sum_{i=1}^{m} p_iu_i \geq \sum_{i=1}^{m} q_iu_i$. ■

The following example, known as the *Allais paradox*, suggests that one should view expected utility theory as a "prescriptive" or "normative" theory (that is, as a theory about how rational people should choose) rather than as a descriptive theory (that is, as a theory about the actual behavior of individuals). In 1953 the French economist Maurice Allais published a paper regarding a survey he had conducted in 1952 concerning a hypothetical decision problem. Subjects "with good training in and knowledge of the theory of probability, so that they could be considered to behave rationally" were asked to rank the following pairs of lotteries:

$$A = \begin{pmatrix} \$5 \text{ Million} & \$0 \\ \frac{89}{100} & \frac{11}{100} \end{pmatrix} \quad \textit{versus} \quad B = \begin{pmatrix} \$1 \text{ Million} & \$0 \\ \frac{90}{100} & \frac{10}{100} \end{pmatrix}$$

and

$$C = \begin{pmatrix} \$5 \text{ Million} & \$1 \text{ Million} & \$0 \\ \frac{89}{100} & \frac{10}{100} & \frac{1}{100} \end{pmatrix} \quad \textit{versus} \quad D = \begin{pmatrix} \$1 \text{ Million} \\ 1 \end{pmatrix}.$$

Most subjects reported the following ranking: $A \succ B$ and $D \succ C$. Such ranking violates the axioms of expected utility. To see this, let $O = \{o_1, o_2, o_3\}$ with $o_1 = \$5$ Million, $o_2 = \$1$ Million and $o_3 = \$0$. Let us assume that the individual in question prefers more money to less, so that $o_1 \succ o_2 \succ o_3$ and has a von Neumann-Morgenstern ranking of the lotteries over $\mathscr{L}(O)$. Let $u_2 \in (0,1)$ be such that $D \sim \begin{pmatrix} \$5 \text{ Million} & \$0 \\ u_2 & 1-u_2 \end{pmatrix}$ (the existence of such u_2 is guaranteed by the Continuity Axiom). Then, since $D \succ C$, by transitivity

$$\begin{pmatrix} \$5 \text{ Million} & \$0 \\ u_2 & 1-u_2 \end{pmatrix} \succ C. \tag{5.4}$$

Let C' be the simple lottery corresponding to the compound lottery

$$\begin{pmatrix} \$5 \text{ Million} & \begin{pmatrix} \$5 \text{ Million} & \$0 \\ u_2 & 1-u_2 \end{pmatrix} & \$0 \\ \frac{89}{100} & \frac{10}{100} & \frac{1}{100} \end{pmatrix}.$$

Then $C' = \begin{pmatrix} \$5 \text{ Million} & \$0 \\ \frac{89}{100} + \frac{10}{100}u_2 & 1 - \left(\frac{89}{100} + \frac{10}{100}u_2\right) \end{pmatrix}.$

By the Independence Axiom, $C \sim C'$ and thus, by (5.4) and transitivity,

$$\begin{pmatrix} \$5 \text{ Million} & \$0 \\ u_2 & 1-u_2 \end{pmatrix} \succ \begin{pmatrix} \$5 \text{ Million} & \$0 \\ \frac{89}{100} + \frac{10}{100}u_2 & 1 - \left(\frac{89}{100} + \frac{10}{100}u_2\right) \end{pmatrix}.$$

Hence, by the Monotonicity Axiom, $u_2 > \frac{89}{100} + \frac{10}{100}u_2$, that is,

$$u_2 > \frac{89}{90}. \tag{5.5}$$

Let B' be the simple lottery corresponding to the following compound lottery, constructed from B by replacing the basic outcome '\$1 Million' with $\begin{pmatrix} \$5 \text{ Million} & \$0 \\ u_2 & 1-u_2 \end{pmatrix}$:

$$\begin{pmatrix} \begin{pmatrix} \$5 \; Million & \$0 \\ u_2 & 1-u_2 \end{pmatrix} & \$0 \\ \frac{90}{100} & \frac{10}{100} \end{pmatrix}.$$

Then

$$B' = \begin{pmatrix} \$5 \text{ Million} & \$0 \\ \frac{90}{100}u_2 & 1 - \frac{90}{100}u_2 \end{pmatrix}.$$

By the Independence Axiom, $B \sim B'$; thus, since $A \succ B$, by transitivity, $A \succ B'$ and therefore, by the Monotonicity Axiom, $\frac{89}{100} > \frac{90}{100}u_2$, that is, $u_2 < \frac{89}{90}$, contradicting (5.5).

Thus, if one finds the expected utility axioms compelling as axioms of rationality, then one cannot consistently express a preference for A over B and also a preference for D over C.

Another well-known example is the *Ellsberg paradox*. Suppose that you are told that an urn contains 30 red balls and 60 more balls that are either blue or yellow. You don't know how many blue or how many yellow balls there are, but the number of blue balls plus the number of yellow ball equals 60 (they could be all blue or all yellow or any combination of the two). The balls are well mixed so that each individual ball is as likely to be drawn as any other. You are given a choice between the bets A and B, where

A = you get \$100 if you pick a red ball and nothing otherwise,
B = you get \$100 if you pick a blue ball and nothing otherwise.

Many subjects in experiments state a strict preference for A over B: $A \succ B$. Consider now the following bets:

C = you get \$100 if you pick a red or yellow ball and nothing otherwise,
D = you get \$100 if you pick a blue or yellow ball and nothing otherwise.

Do the axioms of expected utility constrain your ranking of C and D? Many subjects in experiments state the following ranking: $A \succ B$ and $D \succsim C$. All such people violate the axioms of expected utility. The fraction of red balls in the urn is $\frac{30}{90} = \frac{1}{3}$. Let p_2 be the fraction of blue balls and p_3 the fraction of yellow balls (either of these can be zero: all we know is that $p_2 + p_3 = \frac{60}{90} = \frac{2}{3}$). Then A, B, C and D can be viewed as the following lotteries:

$$A = \begin{pmatrix} \$100 & \$0 \\ \frac{1}{3} & p_2 + p_3 \end{pmatrix}, \quad B = \begin{pmatrix} \$100 & \$0 \\ p_2 & \frac{1}{3} + p_3 \end{pmatrix}$$

$$C = \begin{pmatrix} \$100 & \$0 \\ \frac{1}{3} + p_3 & p_2 \end{pmatrix}, \quad D = \begin{pmatrix} \$100 & \$0 \\ p_2 + p_3 = \frac{2}{3} & \frac{1}{3} \end{pmatrix}$$

Let U be the normalized von Neumann-Morgenstern utility function that represents the individual's ranking; then $U(\$100) = 1$ and $U(0) = 0$. Thus,

$$\mathbb{E}[U(A)] = \frac{1}{3}, \quad \mathbb{E}[U(B)] = p_2, \quad \mathbb{E}[U(C)] = \frac{1}{3} + p_3, \quad \text{and} \quad \mathbb{E}[U(D)] = p_2 + p_3 = \frac{2}{3}.$$

Hence, $A \succ B$ if and only if $\frac{1}{3} > p_2$, which implies that $p_3 > \frac{1}{3}$, so that $\mathbb{E}[U(C)] = \frac{1}{3} + p_3 > \mathbb{E}[U(D)] = \frac{2}{3}$ and thus $C \succ D$ (similarly, $B \succ A$ if and only if $\frac{1}{3} < p_2$, which implies that $\mathbb{E}[U(C)] < \mathbb{E}[U(D)]$ and thus $D \succ C$).

> Test your understanding of the concepts introduced in this section, by going through the exercises in Section 5.4.3 at the end of this chapter.

5.4 Exercises

The solutions to the following exercises are given in Section 5.5 at the end of this chapter.

5.4.1 Exercises for Section 5.1: Money lotteries and attitudes to risk

Exercise 5.1 What is the expected value of the following lottery?

$$\begin{pmatrix} 24 & 12 & 48 & 6 \\ \frac{1}{6} & \frac{2}{6} & \frac{1}{6} & \frac{2}{6} \end{pmatrix}$$

■

Exercise 5.2 Consider the following lottery:

$$\begin{pmatrix} o_1 & o_2 & o_3 \\ \frac{1}{4} & \frac{1}{2} & \frac{1}{4} \end{pmatrix}$$

where

- o_1 = you get an invitation to have dinner at the White House,
- o_2 = you get (for free) a puppy of your choice
- o_3 = you get \$600.

What is the expected value of this lottery? ■

Exercise 5.3 Consider the following money lottery

$$L = \begin{pmatrix} \$10 & \$15 & \$18 & \$20 & \$25 & \$30 & \$36 \\ \frac{3}{12} & \frac{1}{12} & 0 & \frac{3}{12} & \frac{2}{12} & 0 & \frac{3}{12} \end{pmatrix}$$

(a) What is the expected value of the lottery?

(b) Ann prefers more money to less and has transitive preferences. She says that, between getting \$20 for certain and playing the above lottery, she would prefer \$20 for certain. What is her attitude to risk?

(c) Bob prefers more money to less and has transitive preferences. He says that, given the same choice as Ann, he would prefer playing the lottery. What is his attitude to risk?

■

Exercise 5.4 Sam has a debilitating illness and has been offered two mutually exclusive courses of action: (1) take some well-known drugs which have been tested for a long time and (2) take a new experimental drug. If he chooses (1) then for certain his pain will be reduced to a bearable level. If he chooses (2) then he has a 50% chance of being completely cured and a 50% chance of no benefits from the drug and possibly some harmful side effects. He chose (1). What is his attitude to risk? ■

5.4.2 Exercises for Section 5.2: Expected utility theory

Exercise 5.5 Ben is offered a choice between the following two money lotteries: $A = \begin{pmatrix} \$4,000 & \$0 \\ 0.8 & 0.2 \end{pmatrix}$ and $B = \begin{pmatrix} \$3,000 \\ 1 \end{pmatrix}$. He says he strictly prefers B to A. Which of the following two lotteries, C and D, will Ben choose if he satisfies the axioms of expected utility and prefers more money to less?
$C = \begin{pmatrix} \$4,000 & \$0 \\ 0.2 & 0.8 \end{pmatrix}$, $D = \begin{pmatrix} \$3,000 & \$0 \\ 0.25 & 0.75 \end{pmatrix}$. ■

Exercise 5.6 There are three basic outcomes, o_1, o_2 and o_3. Ann satisfies the axioms of expected utility and her preferences over lotteries involving these three outcomes can be represented by the following von Neumann-Morgenstern utility function:
$V(o_2) = a > V(o_1) = b > V(o_3) = c$. Normalize the utility function. ■

Exercise 5.7 Consider the following lotteries:

$$L_1 = \begin{pmatrix} \$3000 & \$500 \\ \frac{5}{6} & \frac{1}{6} \end{pmatrix}, \quad L_2 = \begin{pmatrix} \$3000 & \$500 \\ \frac{2}{3} & \frac{1}{3} \end{pmatrix},$$

$$L_3 = \begin{pmatrix} \$3000 & \$2000 & \$1000 & \$500 \\ \frac{1}{4} & \frac{1}{4} & \frac{1}{4} & \frac{1}{4} \end{pmatrix}, \quad L_4 = \begin{pmatrix} \$2000 & \$1000 \\ \frac{1}{2} & \frac{1}{2} \end{pmatrix}.$$

Jennifer says that she is indifferent between lottery L_1 and getting \$2,000 for certain. She is also indifferent between lottery L_2 and getting \$1,000 for certain. Finally, she says that between L_3 and L_4 she would chose L_3.
Is she rational according to the theory of expected utility? [Assume that she prefers more money to less.] ■

Exercise 5.8 Consider the following basic outcomes:

- o_1 = a Summer internship at the White House,
- o_2 = a free one-week vacation in Europe,
- o_3 = \$800,
- o_4 = a free ticket to a concert.

Rachel says that her ranking of these outcomes is $o_1 \succ o_2 \succ o_3 \succ o_4$. She also says that **(1)** she is indifferent between $\begin{pmatrix} o_2 \\ 1 \end{pmatrix}$ and $\begin{pmatrix} o_1 & o_4 \\ \frac{4}{5} & \frac{1}{5} \end{pmatrix}$ and **(2)** she is indifferent between $\begin{pmatrix} o_3 \\ 1 \end{pmatrix}$ and $\begin{pmatrix} o_1 & o_4 \\ \frac{1}{2} & \frac{1}{2} \end{pmatrix}$. If she satisfies the axioms of expected utility theory, which of the two lotteries $L_1 = \begin{pmatrix} o_1 & o_2 & o_3 & o_4 \\ \frac{1}{8} & \frac{2}{8} & \frac{3}{8} & \frac{2}{8} \end{pmatrix}$ and $L_2 = \begin{pmatrix} o_1 & o_2 & o_3 \\ \frac{1}{5} & \frac{3}{5} & \frac{1}{5} \end{pmatrix}$ will she choose? ■

Exercise 5.9 Consider the following lotteries: $L_1 = \begin{pmatrix} \$30 & \$28 & \$24 & \$18 & \$8 \\ \frac{2}{10} & \frac{1}{10} & \frac{1}{10} & \frac{2}{10} & \frac{4}{10} \end{pmatrix}$ and $L_2 = \begin{pmatrix} \$30 & \$28 & \$8 \\ \frac{1}{10} & \frac{4}{10} & \frac{5}{10} \end{pmatrix}$.

(a) Which lottery would a risk neutral person choose?

(b) Paul's von Neumann-Morgenstern utility-of-money function is $U(m) = ln(m)$, where ln denotes the natural logarithm. Which lottery would Paul choose?

■

Exercise 5.10 There are five basic outcomes. Jane has a von Neumann-Morgenstern ranking of the set of lotteries over the set of basic outcomes that can be represented by either of the following utility functions U and V: $\begin{pmatrix} & o_1 & o_2 & o_3 & o_4 & o_5 \\ U: & 44 & 170 & -10 & 26 & 98 \\ V: & 32 & 95 & 5 & 23 & 59 \end{pmatrix}$.

(a) Show how to normalize each of U and V and verify that you get the same normalized utility function.

(b) Show how to transform U into V with a positive affine transformation of the form $x \mapsto ax + b$ with $a, b \in \mathbb{R}$ and $a > 0$.

■

Exercise 5.11 Consider the following lotteries: $L_3 = \begin{pmatrix} \$28 \\ 1 \end{pmatrix}$, $L_4 = \begin{pmatrix} \$10 & \$50 \\ \frac{1}{2} & \frac{1}{2} \end{pmatrix}$.

(a) Ann has the following von Neumann-Morgenstern utility function: $U_{Ann}(\$m) = \sqrt{m}$. How does she rank the two lotteries?

(b) Bob has the following von Neumann-Morgenstern utility function: $U_{Bob}(\$m) = 2m - \frac{m^4}{100^3}$. How does he rank the two lotteries?

(c) Verify that both Ann and Bob are risk averse, by determining what they would choose between lottery L_4 and its expected value for certain.

■

5.4.3 Exercises for Section 5.3: Expected utility axioms

Exercise 5.12 Let $O = \{o_1, o_2, o_3, o_4\}$. Find the simple lottery corresponding to the following compound lottery

$$\begin{pmatrix} \begin{pmatrix} o_1 & o_2 & o_3 & o_4 \\ \frac{2}{5} & \frac{1}{10} & \frac{3}{10} & \frac{1}{5} \end{pmatrix} & o_2 & \begin{pmatrix} o_1 & o_3 & o_4 \\ \frac{1}{5} & \frac{1}{5} & \frac{3}{5} \end{pmatrix} & \begin{pmatrix} o_2 & o_3 \\ \frac{1}{3} & \frac{2}{3} \end{pmatrix} \\ \frac{1}{8} & \frac{1}{4} & \frac{1}{8} & \frac{1}{2} \end{pmatrix}$$

■

Exercise 5.13 Let $O = \{o_1, o_2, o_3, o_4\}$. Suppose that the *DM* has a von Neumann-Morgenstern ranking of $\mathscr{L}(O)$ and states the following indifference:

$o_1 \sim \begin{pmatrix} o_2 & o_4 \\ \frac{1}{4} & \frac{3}{4} \end{pmatrix}$ and $o_2 \sim \begin{pmatrix} o_3 & o_4 \\ \frac{3}{5} & \frac{2}{5} \end{pmatrix}$.

Find a lottery that the *DM* considers just as good as $L = \begin{pmatrix} o_1 & o_2 & o_3 & o_4 \\ \frac{1}{3} & \frac{2}{9} & \frac{1}{9} & \frac{1}{3} \end{pmatrix}$.

Do not add any information to what is given above (in particular, do not make any assumptions about which outcome is best and which is worst). ■

Exercise 5.14 — ⋆⋆⋆ **Challenging Question** ⋆⋆⋆.

Would you be willing to pay more in order to reduce the probability of dying within the next hour from one sixth to zero or from four sixths to three sixths? Unfortunately, this is not a hypothetical question: you accidentally entered the office of a mad scientist and have been overpowered and tied to a chair. The mad scientist has put six glasses in front of you, numbered 1 to 6, and tells you that one of them contains a deadly poison and the other five contain a harmless liquid. He says that he is going to roll a die and make you drink from the glass whose number matches the number that shows from the rolling of the die. You beg to be exempted and he asks you "what is the largest amount of money that you would be willing to pay to replace the glass containing the poison with one containing a harmless liquid?". Interpret this question as "what sum of money x makes you indifferent between (1) leaving the poison in whichever glass contains it and rolling the die, and (2) reducing your wealth by $\$x$ and rolling the die after the poison has been replaced by a harmless liquid". Your answer is: $\$X$. Then he asks you "suppose that instead of one glass with poison there had been four glasses with poison (and two with a harmless liquid); what is the largest amount of money that you would be willing to pay to replace one glass with poison with a glass containing a harmless liquid (and thus roll the die with 3 glasses with poison and 3 with a harmless liquid)?". Your answer is: $\$Y$. Show that if $X > Y$ then you do not satisfy the axioms of Expected Utility Theory. [Hint: think about what the basic outcomes are; assume that you do not care about how much money is left in your estate if you die and that, when alive, you prefer more money to less.] ■

5.5 Solutions to Exercises

Solution to Exercise 5.1 The expected value of the lottery $\begin{pmatrix} 24 & 12 & 48 & 6 \\ \frac{1}{6} & \frac{2}{6} & \frac{1}{6} & \frac{2}{6} \end{pmatrix}$ is $\frac{1}{6}(24) + \frac{2}{6}(12) + \frac{1}{6}(48) + \frac{2}{6}(6) = 18$. □

Solution to Exercise 5.2 This was a trick question! There is no expected value because the basic outcomes are not numbers. □

Solution to Exercise 5.3

(a) The expected value of the lottery

$$L = \begin{pmatrix} \$10 & \$15 & \$18 & \$20 & \$25 & \$30 & \$36 \\ \frac{3}{12} & \frac{1}{12} & 0 & \frac{3}{12} & \frac{2}{12} & 0 & \frac{3}{12} \end{pmatrix}$$

is $\mathbb{E}[L] = \frac{3}{12}(10) + \frac{1}{12}(15) + (0)(18) + \frac{3}{12}(20) + \frac{2}{12}(25) + (0)(30) + \frac{3}{12}(36) = \frac{263}{12} =$ \$21.92

(b) Since Ann prefers more money to less, she prefers \$21.92 to \$20 (\$21.92 $\succ$ \$20). She said that she prefers \$20 to lottery L (\$20 $\succ L$). Thus, since her preferences are transitive, she prefers \$21.92 to lottery L (\$21.92 $\succ L$). Hence, she is risk averse.

(c) The answer is: we cannot tell. First of all, since Bob prefers more money to less, he prefers \$21.92 to \$20 (\$21.92 $\succ$ \$20). Bob could be risk neutral, because a risk neutral person would be indifferent between L and \$21.92 ($L \sim$ \$21.92); since Bob prefers \$21.92 to \$20 and has transitive preferences, if risk neutral he would prefer L to \$20.

However, Bob could also be risk loving: a risk-loving person prefers L to \$21.92 ($L \succ$ \$21.92) and we know that he prefers \$21.92 to \$20; thus, by transitivity, if risk loving, he would prefer L to \$20.

But Bob could also be risk averse: he could consistently prefer \$21.92 to L and L to \$20 (for example, he could consider L to be just as good as \$20.50). □

Solution to Exercise 5.4 Just like Exercise 5.2, this was a trick question! Here the basic outcomes are not sums of money but states of health. Since the described choice is not one between money lotteries, the definitions of risk aversion/neutrality/love are not applicable. □

Solution to Exercise 5.5 Since Ben prefers B to A, he must prefer D to C.

Proof. Let U be a von Neumann-Morgenstern utility function that represents Ben's preferences.

- Let $U(\$4{,}000) = a, U(\$3{,}000) = b$ and $U(\$0) = c$.
- Since Ben prefers more money to less, $a > b > c$.
- Then $\mathbb{E}[U(A)] = 0.8U(\$4{,}000) + 0.2U(\$0) = 0.8a + 0.2c$ and $\mathbb{E}[U(B)] = U(\$3{,}000) = b$.
- Since Ben prefers B to A, it must be that $b > 0.8a + 0.2c$.

Let us now compare C and D: $\mathbb{E}[U(C)] = 0.2a + 0.8c$ and $\mathbb{E}[U(D)] = 0.25b + 0.75c$.

- Since $b > 0.8a + 0.2c$, $0.25b > 0.25(0.8a + 0.2c) = 0.2a + 0.05c$ and thus, adding $0.75c$ to both sides, we get that $0.25b + 0.75c > 0.2a + 0.8c$, that is, $\mathbb{E}[U(D)] > \mathbb{E}[U(C)]$, so that $D \succ C$.

Note that the proof would have been somewhat easier if we had taken the normalized utility function, so that $a = 1$ and $c = 0$. □

Solution to Exercise 5.6 Define the function U as follows:
$U(x) = \frac{1}{a-c}V(x) - \frac{c}{a-c} = \frac{V(x)-c}{a-c}$ (note that, by hypothesis, $a > c$ and thus $\frac{1}{a-c} > 0$).
Then U represents the same preferences as V.
Then $U(o_2) = \frac{V(o_2)-c}{a-c} = \frac{a-c}{a-c} = 1$, $U(o_1) = \frac{V(o_1)-c}{a-c} = \frac{b-c}{a-c}$, and $U(o_3) = \frac{V(o_3)-c}{a-c} = \frac{c-c}{a-c} = 0$.
Note that, since $a > b > c$, $0 < \frac{b-c}{a-c} < 1$. □

Solution to Exercise 5.7 We can take the set of basic outcomes to be $\{\$3000, \$2000, \$1000, \$500\}$. Suppose that there is a von Neumann-Morgenstern utility function U that represents Jennifer's preferences. We can normalize it so that $U(\$3000) = 1$ and $U(\$500) = 0$.
- Since Jennifer is indifferent between L_1 and \$2000, $U(\$2000) = \frac{5}{6}$ (since the expected utility of L_1 is $\frac{5}{6}(1) + \frac{1}{6}(0) = \frac{5}{6}$).
- Since she is indifferent between L_2 and \$1000, $U(\$1000) = \frac{2}{3}$ (since the expected utility of L_2 is $\frac{2}{3}(1) + \frac{1}{3}(0) = \frac{2}{3}$).
Thus, $\mathbb{E}[U(L_3)] = \frac{1}{4}(1) + \frac{1}{4}\left(\frac{5}{6}\right) + \frac{1}{4}\left(\frac{2}{3}\right) + \frac{1}{4}(0) = \frac{5}{8}$ and $\mathbb{E}[U(L_4)] = \frac{1}{2}\left(\frac{5}{6}\right) + \frac{1}{2}\left(\frac{2}{3}\right) = \frac{3}{4}$.
Since $\frac{3}{4} > \frac{5}{8}$, Jennifer should prefer L_4 to L_3. Hence, she is not rational according to the theory of expected utility. □

Solution to Exercise 5.8 Normalize her utility function so that $U(o_1) = 1$ and $U(o_4) = 0$.
Since Rachel is indifferent between $\begin{pmatrix} o_2 \\ 1 \end{pmatrix}$ and $\begin{pmatrix} o_1 & o_4 \\ \frac{4}{5} & \frac{1}{5} \end{pmatrix}$, we have that $U(o_2) = \frac{4}{5}$.
Similarly, since she is indifferent between $\begin{pmatrix} o_3 \\ 1 \end{pmatrix}$ and $\begin{pmatrix} o_1 & o_4 \\ \frac{1}{2} & \frac{1}{2} \end{pmatrix}$, $U(o_3) = \frac{1}{2}$.
Then the expected utility of $L_1 = \begin{pmatrix} o_1 & o_2 & o_3 & o_4 \\ \frac{1}{8} & \frac{2}{8} & \frac{3}{8} & \frac{2}{8} \end{pmatrix}$ is $\frac{1}{8}(1) + \frac{2}{8}(\frac{4}{5}) + \frac{3}{8}(\frac{1}{2}) + \frac{2}{8}(0) = \frac{41}{80} = 0.5125$,
while the expected utility of $L_2 = \begin{pmatrix} o_1 & o_2 & o_3 \\ \frac{1}{5} & \frac{3}{5} & \frac{1}{5} \end{pmatrix}$ is $\frac{1}{5}(1) + \frac{3}{5}(\frac{4}{5}) + \frac{1}{5}(\frac{1}{2}) =. \frac{39}{50} = 0.78$.
Hence, she prefers L_2 to L_1. □

Solution to Exercise 5.9

(a) The expected value of L_1 is $\frac{2}{10}(30) + \frac{1}{10}(28) + \frac{1}{10}(24) + \frac{2}{10}(18) + \frac{4}{10}(8) = 18$ and the expected value of L_2 is $\frac{1}{10}(30) + \frac{4}{10}(28) + \frac{5}{10}8 = 18.2$.

Hence, a risk-neutral person would prefer L_2 to L_1.

(b) The expected utility of L_1 is $\frac{1}{5}\ln(30) + \frac{1}{10}\ln(28) + \frac{1}{10}\ln(24) + \frac{1}{5}\ln(18) + \frac{2}{5}\ln(8) = 2.741$ while the expected utility of L_2 is $\frac{1}{10}\ln(30) + \frac{2}{5}\ln(28) + \frac{1}{2}\ln(8) = 2.713$.
Thus, Paul would choose L_1 (since he prefers L_1 to L_2). □

Solution to Exercise 5.10

(a) To normalize U first add 10 to each value and then divide by 180. Denote the normalization of U by $\overline{U}$.
Then

$$\overline{U}: \begin{array}{ccccc} o_1 & o_2 & o_3 & o_4 & o_5 \\ \frac{54}{180}=0.3 & \frac{180}{180}=1 & \frac{0}{180}=0 & \frac{36}{180}=0.2 & \frac{108}{180}=0.6 \end{array}$$

To normalize V first subtract 5 from each value and then divide by 90. Denote the normalization of V by $\overline{V}$.
Then

$$\overline{V}: \begin{array}{ccccc} o_1 & o_2 & o_3 & o_4 & o_5 \\ \frac{27}{90}=0.3 & \frac{90}{90}=1 & \frac{0}{90}=0 & \frac{18}{90}=0.2 & \frac{54}{90}=0.6 \end{array}$$

(b) The transformation is of the form $V(o) = aU(o) + b$. To find the values of a and b plug in two sets of values and solve the system of equations $\begin{cases} 44a + b = 32 \\ 170a + b = 95 \end{cases}$.
The solution is $a = \frac{1}{2}$, $b = 10$. Thus, $V(o) = \frac{1}{2}U(o) + 10$. □

Solution to Exercise 5.11

(a) Ann prefers L_3 to L_4 ($L_3 \succ_{Ann} L_4$). In fact, $\mathbb{E}[U_{Ann}(L_3)] = \sqrt{28} = 5.2915$ while $\mathbb{E}[U_{Ann}(L_4)] = \frac{1}{2}\sqrt{10} + \frac{1}{2}\sqrt{50} = 5.1167$.

(b) Bob prefers L_4 to L_3 ($L_4 \succ_{Bob} L_3$). In fact, $\mathbb{E}[U_{Bob}(L_3)] = 2(28) - \frac{28^4}{100^3} = 55.3853$ while $\mathbb{E}[U_{Bob}(L_4)] = \frac{1}{2}\left[2(10) - \frac{10^4}{100^3}\right] + \frac{1}{2}\left[2(50) - \frac{50^4}{100^3}\right] = 56.87$.

(c) The expected value of lottery L_4 is $\frac{1}{2}10 + \frac{1}{2}50 = 30$; thus, a risk-averse person would strictly prefer \$30 with certainty to the lottery L_4. We saw in part **(a)** that for Ann the expected utility of lottery L_4 is 5.1167; the utility of \$30 is $\sqrt{30} = 5.4772$. Thus, Ann would indeed choose \$30 for certain over the lottery L_4. We saw in part **(b)** that for Bob the expected utility of lottery L_4 is 56.87; the utility of \$30 is $2(30) - \frac{30^4}{100^3} = 59.19$. Thus, Bob would indeed choose \$30 for certain over the lottery L_4. □

Solution to Exercise 5.12 The simple lottery is $\begin{pmatrix} o_1 & o_2 & o_3 & o_4 \\ \frac{18}{240} & \frac{103}{240} & \frac{95}{240} & \frac{24}{240} \end{pmatrix}$. For example, the probability of o_2 is computed as follows: $\frac{1}{8}\left(\frac{1}{10}\right) + \frac{1}{4}(1) + \frac{1}{8}(0) + \frac{1}{2}\left(\frac{1}{3}\right) = \frac{103}{240}$. □

Solution to Exercise 5.13 Using the stated indifference, use lottery L to construct the compound lottery $\begin{pmatrix} \begin{pmatrix} o_2 & o_4 \\ \frac{1}{4} & \frac{3}{4} \end{pmatrix} & \begin{pmatrix} o_3 & o_4 \\ \frac{3}{5} & \frac{2}{5} \end{pmatrix} & o_3 & o_4 \\ \frac{1}{3} & \frac{2}{9} & \frac{1}{9} & \frac{1}{3} \end{pmatrix}$, whose corresponding simple lottery is $L' = \begin{pmatrix} o_1 & o_2 & o_3 & o_4 \\ 0 & \frac{1}{12} & \frac{11}{45} & \frac{121}{180} \end{pmatrix}$. Then, by the Independence Axiom, $L \sim L'$. □

Solution to Exercise 5.14 Let W be your initial wealth. The basic outcomes are:

1. you do not pay any money, do not die and live to enjoy your wealth W (denote this outcome by A_0),
2. you pay \$$Y$, do not die and live to enjoy your remaining wealth $W - Y$ (call this outcome A_Y),
3. you pay \$$X$, do not die and live to enjoy your remaining wealth $W - X$ (call this outcome A_X),
4. you die (call this outcome D); this could happen because **(a)** you do not pay any money, roll the die and drink the poison or **(b)** you pay \$$Y$, roll the die and drink the poison; we assume that you are indifferent between these two outcomes.

Since, by hypothesis, $X > Y$, your ranking of these outcomes must be $A_0 \succ A_Y \succ A_X \succ D$. If you satisfy the von Neumann-Morgenstern axioms, then your preferences can be represented by a von Neumann-Morgenstern utility function U defined on the set of basic outcomes. We can normalize your utility function by setting $U(A_0) = 1$ and $U(D) = 0$. Furthermore, it must be that

$$U(A_Y) > U(A_X). \tag{5.6}$$

The maximum amount \$$P$ that you are willing to pay is that amount that makes you indifferent between **(1)** rolling the die with the initial number of poisoned glasses and **(2)** giving up \$$P$ and rolling the die with one less poisoned glass.

Thus – based on your answers – you are indifferent between the two lotteries

$$\begin{pmatrix} D & A_0 \\ \frac{1}{6} & \frac{5}{6} \end{pmatrix} \text{ and } \begin{pmatrix} A_X \\ 1 \end{pmatrix}$$

and you are indifferent between the two lotteries:

$$\begin{pmatrix} D & A_0 \\ \frac{4}{6} & \frac{2}{6} \end{pmatrix} \text{ and } \begin{pmatrix} D & A_Y \\ \frac{3}{6} & \frac{3}{6} \end{pmatrix}.$$

Thus,

$$\underbrace{\tfrac{1}{6}U(D) + \tfrac{5}{6}U(A_0)}_{=\frac{1}{6}0+\frac{5}{6}1=\frac{5}{6}} = U(A_X) \text{ and } \underbrace{\tfrac{4}{6}U(D) + \tfrac{2}{6}U(A_0)}_{=\frac{4}{6}0+\frac{2}{6}1=\frac{2}{6}} = \underbrace{\tfrac{3}{6}U(D) + \tfrac{3}{6}U(A_Y)}_{=\frac{3}{6}0+\frac{3}{6}U(A_Y)}.$$

Hence, $U(A_X) = \frac{5}{6}$ and $U(A_Y) = \frac{2}{3} = \frac{4}{6}$, so that $U(A_X) > U(A_Y)$, contradicting (5.6). □

6. Strategic-form Games

6.1 Strategic-form games with cardinal payoffs

At the end of Chapter 4 we discussed the possibility of incorporating random events in extensive-form games by means of chance moves. The introduction of chance moves gives rise to probabilistic outcomes and thus to the issue of how a player might rank such outcomes. Random events can also occur in strategic-form games, as shown in Figure 6.1, which represents the simple first-price auction of Example 6.1 below.

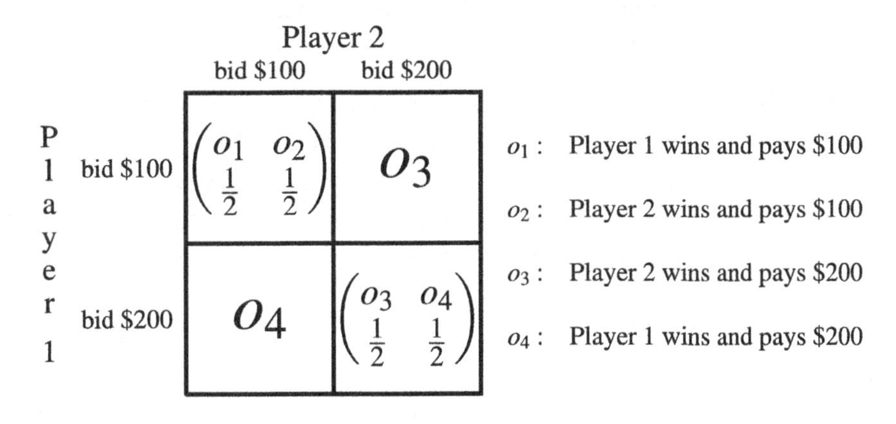

Figure 6.1: A game-frame in strategic form representing Example 6.1

Example 6.1 Two players simultaneously submit a bid for a painting. Only two bids are possible: \$100 and \$200. If one player bids \$200 and the other \$100 then the high bidder wins the painting and has to pay her own bid. If the two players bid the same amount then a fair coin is tossed and if the outcome is Heads the winner is Player 1 (who then has to pay her own bid) while if the outcome is Tails the winner is Player 2 (who then has to pay his own bid). ■

Suppose that Player 1 ranks the basic outcomes as follows: $o_1 \succ_1 o_4 \succ_1 o_2 \sim_1 o_3$, that is, she prefers winning to not winning; conditional on winning, she prefers to pay less and, conditional on not winning, she is indifferent as to how much Player 2 pays. Suppose also that Player 1 believes that Player 2 is going to submit a bid of \$100 (perhaps she has been informed of this by somebody spying on Player 2). What should we expect Player 1 to do? Knowing her ranking of the basic outcomes is of no help, because we need to know how she ranks the probabilistic outcome $\begin{pmatrix} o_1 & o_2 \\ \frac{1}{2} & \frac{1}{2} \end{pmatrix}$ relative to the basic outcome o_4.

The theory of expected utility introduced in Chapter 5 provides one possible answer to the question of how players rank probabilistic outcomes. With the aid of expected utility theory we can now generalize the definition of strategic-form game. First we generalize the notion of game-frame in strategic form (Definition 2.1.1, Chapter 2) by allowing probabilistic outcomes, or lotteries, to be associated with strategy profiles. In the following definition, the bulleted items coincide with the first three items of Definition 2.1.1 (Chapter 2); the modified item is the last one, preceded by the symbol ★.

Definition 6.1.1 A *game-frame in strategic form* is a quadruple $\langle I, (S_i)_{i\in I}, O, f\rangle$ where:

- $I = \{1,\ldots,n\}$ is a set of *players* ($n \geq 2$).
- For every Player $i \in I$, S_i is the set of *strategies* (or choices) of Player i. As before, we denote by $S = S_1 \times \cdots \times S_n$ the set of *strategy profiles*.
- O is a set of *basic outcomes*.

★ $f : S \to \mathscr{L}(O)$ is a function that associates with every strategy profile s a lottery over the set of basic outcomes O (as in Chapter 5, we denote by $\mathscr{L}(O)$ the set of lotteries, or probability distributions, over O).

If, for every $s \in S$, $f(s)$ is a degenerate lottery (that is, a basic outcome) then we are back to Definition 2.1.1 (Chapter 2).

From a game-frame one obtains a game by adding, for every player $i \in I$, a von Neumann-Morgenstern ranking $\succsim_i$ of the elements of $\mathscr{L}(O)$. It is more convenient to represent such a ranking by means of a von Neumann-Morgenstern utility function $U_i : O \to \mathbb{R}$. We denote by $\mathbb{E}[U_i(f(s))]$ the expected utility of lottery $f(s) \in \mathscr{L}(O)$ for Player i. The following definition mirrors Definition 2.1.2 of Chapter 2.

Definition 6.1.2 A *game in strategic form with cardinal payoffs* is a quintuple $\langle I, (S_i)_{i\in I}, O, f, (\succsim_i)_{i\in I}\rangle$ where:

- $\langle I, (S_i)_{i\in I}, O, f\rangle$ is a game-frame in strategic form (Definition 6.1.1) and
- for every Player $i \in I$, $\succsim_i$ is a von Neumann-Morgenstern ranking of the set of lotteries $\mathscr{L}(O)$.

If we represent each ranking $\succsim_i$ by means of a von Neumann-Morgenstern utility function U_i and define $\pi_i : S \to \mathbb{R}$ by $\pi_i(s) = \mathbb{E}[U_i(f(s))]$, then $\langle I, (S_1,...,S_n), (\pi_1,...,\pi_n)\rangle$ is called a *reduced-form game in strategic form with cardinal payoffs* ('reduced-form' because some information is lost, namely the specification of the possible outcomes). The function $\pi_i : S \to \mathbb{R}$ is called the *von Neumann-Morgenstern payoff function of Player i.*

For example, consider the first-price auction of Example 6.1 whose game-frame in strategic form was shown in Figure 6.1. Let $O = \{o_1, o_2, o_3, o_4\}$ and suppose that Player 1 has a von Neumann-Morgenstern ranking of $\mathscr{L}(O)$ that is represented by the following von Neumann-Morgenstern utility function U_1 (note that the implied ordinal ranking of the basic outcomes is indeed $o_1 \succ_1 o_4 \succ_1 o_2 \sim_1 o_3$):

$$\begin{array}{rcccc} \text{outcome:} & o_1 & o_2 & o_3 & o_4 \\ U_1: & 4 & 1 & 1 & 2 \end{array}$$

Then, for Player 1, the expected utility of lottery $\begin{pmatrix} o_1 & o_2 \\ \frac{1}{2} & \frac{1}{2} \end{pmatrix}$ is 2.5 and the expected utility of lottery $\begin{pmatrix} o_3 & o_4 \\ \frac{1}{2} & \frac{1}{2} \end{pmatrix}$ is 1.5.

Suppose also that Player 2 has (somewhat spiteful) preferences represented by the following von Neumann-Morgenstern utility function U_2:

$$\begin{array}{rcccc} \text{outcome:} & o_1 & o_2 & o_3 & o_4 \\ U_2: & 1 & 6 & 4 & 5 \end{array}$$

Thus, for Player 2, the expected utility of lottery $\begin{pmatrix} o_1 & o_2 \\ \frac{1}{2} & \frac{1}{2} \end{pmatrix}$ is 3.5 and the expected utility of lottery $\begin{pmatrix} o_3 & o_4 \\ \frac{1}{2} & \frac{1}{2} \end{pmatrix}$ is 4.5. Then we can represent the game in reduced form as shown in Figure 6.2.

		Player 2	
		\$100	\$200
Player 1	\$100	2.5 3.5	1 4
	\$200	2 5	1.5 4.5

Figure 6.2: A cardinal game in reduced form based on the game-frame of 6.1

The game of Figure 6.2 does not have any Nash equilibria. However, we will show in the next section that if we extend the notion of strategy, by allowing players to choose randomly, then the game of Figure 6.2 does have a Nash equilibrium.

Test your understanding of the concepts introduced in this section, by going through the exercises in Section 6.5.1 at the end of this chapter.

6.2 Mixed strategies

Definition 6.2.1 Consider a game in strategic form with cardinal payoffs and recall that S_i denotes the set of strategies of Player i. From now on, we shall call S_i the set of *pure strategies* of Player i. We assume that S_i is a finite set (for every $i \in I$). A *mixed strategy* of Player i is a probability distribution over the set of pure strategies S_i. The set of mixed strategies of Player i is denoted by Σ_i.

R Since among the mixed strategies of Player i there are the degenerate strategies that assign probability 1 to a pure strategy, the set of mixed strategies includes the set of pure strategies (viewed as degenerate probability distributions).

For example, one possible mixed strategy for Player 1 in the game of Figure 6.2 is $\begin{pmatrix} \$100 & \$200 \\ \frac{1}{3} & \frac{2}{3} \end{pmatrix}$. The traditional interpretation of a mixed strategy is in terms of objective randomization: the player, instead of choosing a pure strategy herself, delegates the choice to a random device.[1] For example, Player 1 choosing the mixed strategy $\begin{pmatrix} \$100 & \$200 \\ \frac{1}{3} & \frac{2}{3} \end{pmatrix}$ is interpreted as a decision to let, say, a die determine whether she will bid \$100 or \$200: Player 1 will roll a die and if the outcome is 1 or 2 then she will bid \$100, while if the outcome is 3, 4, 5 or 6 then she will bid \$200. Suppose that Player 1 chooses this mixed strategy and Player 2 chooses the mixed strategy $\begin{pmatrix} \$100 & \$200 \\ \frac{3}{5} & \frac{2}{5} \end{pmatrix}$. Since the players rely on independent random devices, this pair of mixed strategies gives rise to the following probabilistic outcome:

$$\begin{pmatrix} \text{strategy profile} & (\$100,\$100) & (\$100,\$200) & (\$200,\$100) & (\$200,\$200) \\ \text{outcome} & \begin{pmatrix} o_1 & o_2 \\ \frac{1}{2} & \frac{1}{2} \end{pmatrix} & o_3 & o_4 & \begin{pmatrix} o_3 & o_4 \\ \frac{1}{2} & \frac{1}{2} \end{pmatrix} \\ \text{probability} & \frac{1}{3}\left(\frac{3}{5}\right) = \frac{3}{15} & \frac{1}{3}\left(\frac{2}{5}\right) = \frac{2}{15} & \frac{2}{3}\left(\frac{3}{5}\right) = \frac{6}{15} & \frac{2}{3}\left(\frac{2}{5}\right) = \frac{4}{15} \end{pmatrix}$$

If the two players have von Neumann-Morgenstern preferences, then – by the Compound Lottery Axiom (Chapter 5) – they will view the above as the following lottery:

$$\begin{pmatrix} \text{outcome} & o_1 & o_2 & o_3 & o_4 \\ \text{probability} & \frac{3}{30} & \frac{3}{30} & \frac{8}{30} & \frac{16}{30} \end{pmatrix}.$$

[1]An alternative interpretation of mixed strategies in terms of beliefs will be discussed in Chapter 10 and in Part V (Chapters 14-16).

Using the von Neumann-Morgenstern utility functions postulated in the previous section, namely

$$\begin{array}{lcccc} \textit{outcome}: & o_1 & o_2 & o_3 & o_4 \\ U_1: & 4 & 1 & 1 & 2 \end{array} \quad \text{and} \quad \begin{array}{lcccc} \textit{outcome}: & o_1 & o_2 & o_3 & o_4 \\ U_2: & 1 & 6 & 4 & 5 \end{array}$$

the lottery $\begin{pmatrix} o_1 & o_2 & o_3 & o_4 \\ \frac{3}{30} & \frac{3}{30} & \frac{8}{30} & \frac{16}{30} \end{pmatrix}$ has an expected utility of

For Player 1: $\frac{3}{30}(4)+\frac{3}{30}(1)+\frac{8}{30}(1)+\frac{16}{30}(2)=\frac{55}{30}$

For Player 2: $\frac{3}{30}(1)+\frac{3}{30}(6)+\frac{8}{30}(4)+\frac{16}{30}(5)=\frac{133}{30}$.

Thus we can define the payoffs of the two players from this mixed strategy profile by

$$\Pi_1\left[\begin{pmatrix} \$100 & \$200 \\ \frac{1}{3} & \frac{2}{3} \end{pmatrix}, \begin{pmatrix} \$100 & \$200 \\ \frac{3}{5} & \frac{2}{5} \end{pmatrix}\right] = \frac{55}{30}$$

$$\Pi_2\left[\begin{pmatrix} \$100 & \$200 \\ \frac{1}{3} & \frac{2}{3} \end{pmatrix}, \begin{pmatrix} \$100 & \$200 \\ \frac{3}{5} & \frac{2}{5} \end{pmatrix}\right] = \frac{133}{30}$$

Note that we can calculate these payoffs in a different – but equivalent – way by using the reduced-form game of Figure 6.2, as follows:

$$\begin{pmatrix} \text{strategy profile} & (\$100,\$100) & (\$100,\$200) & (\$200,\$100) & (\$200,\$200) \\ \text{expected utilities} & (2.5,3.5) & (1,4) & (2,5) & (1.5,4.5) \\ \text{probability} & \frac{1}{3}\left(\frac{3}{5}\right)=\frac{3}{15} & \frac{1}{3}\left(\frac{2}{5}\right)=\frac{2}{15} & \frac{2}{3}\left(\frac{3}{5}\right)=\frac{6}{15} & \frac{2}{3}\left(\frac{2}{5}\right)=\frac{4}{15} \end{pmatrix}$$

so that the expected payoff of Player 1 is

$$\tfrac{3}{15}(2.5)+\tfrac{2}{15}(1)+\tfrac{6}{15}(2)+\tfrac{4}{15}(1.5)=\tfrac{55}{30}$$

and the expected payoff of Player 2 is

$$\tfrac{3}{15}(3.5)+\tfrac{2}{15}(4)+\tfrac{6}{15}(5)+\tfrac{4}{15}(4.5)=\tfrac{133}{30}.$$

The previous example provides the rationale for the following definition. First some notation.

- Let $\sigma_i \in \Sigma_i$ be a mixed strategy of Player i; then, for every pure strategy $s_i \in S_i$ of Player i, we denote by $\sigma_i(s_i)$ the probability that σ_i assigns to s_i.[2]
- Let Σ be the set of mixed-strategy profiles, that is, $\Sigma = \Sigma_1 \times \cdots \times \Sigma_n$.
- Consider a mixed-strategy profile $\sigma = (\sigma_1, ..., \sigma_n) \in \Sigma$ and a pure-strategy profile $s = (s_1, \ldots, s_n) \in S$; then we denote by $\sigma(s)$ the product of the probabilities $\sigma_i(s_i)$, that is, $\sigma(s) = \prod_{i=1}^{n} \sigma_i(s_i) = \sigma_1(s_1) \times ... \times \sigma_n(s_n)$.[3]

Definition 6.2.2 Consider a reduced-form game in strategic form with cardinal payoffs $G = \langle I, (S_1, ..., S_n), (\pi_1, ..., \pi_n) \rangle$ (Definition 6.1.2), where, for every Player $i \in I$, the set of pure strategies S_i is finite. Then the *mixed-strategy extension of* G is the reduced-form game in strategic form $\langle I, (\Sigma_1, ..., \Sigma_n), (\Pi_1, ..., \Pi_n) \rangle$ where, for every Player $i \in I$,

- Σ_i is the set of mixed strategies of Player i in G (that is, Σ_i is the set of probability distributions over S_i).
- The payoff function $\Pi_i : \Sigma \to \mathbb{R}$ is defined by $\Pi_i(\sigma) = \sum_{s \in S} \sigma(s)\, \pi_i(s)$.[a]

[a] In the above example, if $\sigma_1 = \begin{pmatrix} \$100 & \$200 \\ \frac{1}{3} & \frac{2}{3} \end{pmatrix}$ and $\sigma_2 = \begin{pmatrix} \$100 & \$200 \\ \frac{3}{5} & \frac{2}{5} \end{pmatrix}$

then $\Pi_1(\sigma_1, \sigma_2) = \frac{3}{15}(2.5) + \frac{2}{15}(1) + \frac{6}{15}(2) + \frac{4}{15}(1.5) = \frac{55}{30}$.

Definition 6.2.3 Fix a reduced-form game in strategic form with cardinal payoffs $G = \langle I, (S_1, \ldots, S_n), (\pi_1, \ldots, \pi_n) \rangle$ (Definition 6.1.2), where, for every player $i \in I$, the set of pure strategies S_i is finite. A *Nash equilibrium in mixed-strategies of* G is a Nash equilibrium of the mixed-strategy extension of G.

[2] In the above example, if $\sigma_1 = \begin{pmatrix} \$100 & \$200 \\ \frac{1}{3} & \frac{2}{3} \end{pmatrix}$ then $\sigma_1(\$200) = \frac{2}{3}$.

[3] In the above example, if $\sigma = (\sigma_1, \sigma_2)$ with $\sigma_1 = \begin{pmatrix} \$100 & \$200 \\ \frac{1}{3} & \frac{2}{3} \end{pmatrix}$ and $\sigma_2 = \begin{pmatrix} \$100 & \$200 \\ \frac{3}{5} & \frac{2}{5} \end{pmatrix}$ then $\sigma_1(\$200) = \frac{2}{3}$, $\sigma_2(\$100) = \frac{3}{5}$ and thus $\sigma((\$200, \$100)) = \frac{2}{3}\left(\frac{3}{5}\right) = \frac{6}{15}$.

For example, consider the reduced-form game of Figure 6.3 (which reproduces Figure 6.2: with all the payoffs multiplied by 10; this corresponds to representing the preferences of the players with different utility functions that are a obtained from the ones used above by multiplying them by 10).

Is $\sigma = (\sigma_1, \sigma_2)$ with $\sigma_1 = \begin{pmatrix} \$100 & \$200 \\ \frac{1}{3} & \frac{2}{3} \end{pmatrix}$ and $\sigma_2 = \begin{pmatrix} \$100 & \$200 \\ \frac{3}{5} & \frac{2}{5} \end{pmatrix}$ a mixed-strategy Nash equilibrium of this game?

		Player 2	
		\$100	\$200
Player 1	\$100	25 35	10 40
	\$200	20 50	15 45

Figure 6.3: The game of Figure 6.2 with the payoffs multiplied by 10

The payoff of Player 1 is

$$\Pi_1(\sigma_1, \sigma_2) = \tfrac{3}{15}(25) + \tfrac{2}{15}(10) + \tfrac{6}{15}(20) + \tfrac{4}{15}(15) = \tfrac{55}{3}.$$

If Player 1 switched from $\sigma_1 = \begin{pmatrix} \$100 & \$200 \\ \frac{1}{3} & \frac{2}{3} \end{pmatrix}$ to $\hat{\sigma}_1 = \begin{pmatrix} \$100 & \$200 \\ 1 & 0 \end{pmatrix}$, that is, to the pure strategy \$100, then Player 1's payoff would be larger:

$$\Pi_1(\hat{\sigma}_1, \sigma_2) = \tfrac{3}{5}(25) + \tfrac{2}{5}(10) = 19.$$

Thus, since $19 > \frac{55}{3}$, it is not a Nash equilibrium.

John Nash (who shared the 1994 Nobel Memorial prize in economics with John Harsanyi and Reinhard Selten), proved the following theorem.

Theorem 6.2.1 — **Nash, 1951.** Every reduced-form game in strategic form with cardinal payoffs $\langle I, (S_1, \ldots, S_n), (\pi_1, \ldots, \pi_n) \rangle$ (Definition 6.1.2), where, for every Player $i \in I$, the set of pure strategies S_i is finite, has at least one Nash equilibrium in mixed-strategies.

We will not give the proof of this theorem, since it is rather complex (it requires the use of fixed-point theorems).

		Player 2	
		\$100	\$200
Player 1	\$100	25 35	10 40
	\$200	20 50	15 45

Going back to the game of Figure 6.3 reproduced above, let us verify that, on the other hand, $\sigma^* = (\sigma_1^*, \sigma_2^*)$ with $\sigma_1^* = \sigma_2^* = \begin{pmatrix} \$100 & \$200 \\ \frac{1}{2} & \frac{1}{2} \end{pmatrix}$ is a Nash equilibrium in mixed strategies. The payoff of Player 1 is

$$\Pi_1(\sigma_1^*, \sigma_2^*) = \tfrac{1}{4}(25) + \tfrac{1}{4}(10) + \tfrac{1}{4}(20) + \tfrac{1}{4}(15) = \tfrac{70}{4} = 17.5.$$

Could Player 1 obtain a larger payoff with some other mixed strategy $\sigma_1 = \begin{pmatrix} \$100 & \$200 \\ p & 1-p \end{pmatrix}$ for some $p \neq \frac{1}{2}$?

Fix an arbitrary $p \in [0,1]$ and let us compute Player 1's payoff if she uses the strategy $\sigma_1 = \begin{pmatrix} \$100 & \$200 \\ p & 1-p \end{pmatrix}$ against Player 2's mixed strategy $\sigma_2^* = \begin{pmatrix} \$100 & \$200 \\ \frac{1}{2} & \frac{1}{2} \end{pmatrix}$:

$$\Pi_1 \left[\begin{pmatrix} \$100 & \$200 \\ p & 1-p \end{pmatrix}, \begin{pmatrix} \$100 & \$200 \\ \frac{1}{2} & \frac{1}{2} \end{pmatrix} \right] = \tfrac{1}{2}p25 + \tfrac{1}{2}p10 + \tfrac{1}{2}(1-p)20 + \tfrac{1}{2}(1-p)15$$

$$= p\left(\tfrac{1}{2}25 + \tfrac{1}{2}10\right) + (1-p)\left(\tfrac{1}{2}20 + \tfrac{1}{2}15\right) = \tfrac{35}{2} = 17.5.$$

Thus if Player 2 uses the mixed strategy $\sigma_2^* = \begin{pmatrix} \$100 & \$200 \\ \frac{1}{2} & \frac{1}{2} \end{pmatrix}$, then *Player 1 gets the same payoff no matter what mixed strategy she employs.*

It follows that any mixed strategy of Player 1 is a best reply to $\sigma_2^* = \begin{pmatrix} \$100 & \$200 \\ \frac{1}{2} & \frac{1}{2} \end{pmatrix}$; in particular, $\sigma_1^* = \begin{pmatrix} \$100 & \$200 \\ \frac{1}{2} & \frac{1}{2} \end{pmatrix}$ is a best reply to $\sigma_2^* = \begin{pmatrix} \$100 & \$200 \\ \frac{1}{2} & \frac{1}{2} \end{pmatrix}$.

It is easy to verify that the same applies to Player 2: any mixed strategy of Player 2 is a best reply to Player 1's mixed strategy $\sigma_1^* = \begin{pmatrix} \$100 & \$200 \\ \frac{1}{2} & \frac{1}{2} \end{pmatrix}$. Hence $\sigma^* = (\sigma_1^*, \sigma_2^*)$ is indeed a Nash equilibrium in mixed strategies.

We will see in the next section that this "indifference" phenomenon is true in general.

R Since, among the mixed strategies of Player i there are the degenerate strategies that assign probability 1 to a pure strategy, every Nash equilibrium in pure strategies is also a Nash equilibrium in mixed strategies. That is, the set of mixed-strategy Nash equilibria includes the set of pure-strategy Nash equilibria.

Test your understanding of the concepts introduced in this section, by going through the exercises in Section 6.5.2 at the end of this chapter.

6.3 Computing the mixed-strategy Nash equilibria

How can we find the mixed-strategy equilibria of a given game? The first important observation is that if a pure strategy is strictly dominated by another pure strategy then it cannot be played with positive probability at a Nash equilibrium. Thus, for the purpose of finding Nash equilibria, one can delete all the strictly dominated strategies and focus on the resulting game. But then the same reasoning applies to the resulting game and one can delete all the strictly dominated strategies in that game, and so on. Thus we have the following observation.

R In order to find the mixed-strategy Nash equilibria of a game one can first apply the iterated deletion of strictly dominated strategies (IDSDS: Chapter 2) and then find the Nash equilibria of the resulting game (which can then be viewed as Nash equilibria of the original game where all the pure strategies that were deleted are assigned zero probability). Note, however, that – as we will see in Section 6.4 – one can perform more deletions than allowed by the IDSDS procedure.

For example, consider the game of Figure 6.4.

	Player 2					
Player 1	*E*		*F*		*G*	
A	2	4	3	3	6	0
B	4	0	2	4	4	2
C	3	3	4	2	3	1
D	3	6	1	1	2	6

Figure 6.4: A reduced-form game with cardinal payoffs

In this game there are no pure-strategy Nash equilibria; however, by Nash's theorem there will be at least one mixed-strategy equilibrium. To find it we can first note that, for Player 1, D is strictly dominated by B; deleting D we get a smaller game where, for Player 2, G is strictly dominated by F. Deleting G we are left with a smaller game where A is strictly dominated by C. Deleting A we are left with the game shown in Figure 6.5.

		Player 2	
		E	F
Player 1	B	4 0	2 4
	C	3 3	4 2

Figure 6.5: The result of applying the IDSDS procedure to the game of Figure 6.4

We will see that the game of Figure 6.5 has a unique Nash equilibrium in mixed strategies given by $\left[\begin{pmatrix} B & C \\ \frac{1}{5} & \frac{4}{5} \end{pmatrix}, \begin{pmatrix} E & F \\ \frac{2}{3} & \frac{1}{3} \end{pmatrix}\right]$.

Thus the game of Figure 6.4 has a unique Nash equilibrium in mixed strategies given by

$$\left[\begin{pmatrix} A & B & C & D \\ 0 & \frac{1}{5} & \frac{4}{5} & 0 \end{pmatrix}, \begin{pmatrix} E & F & G \\ \frac{2}{3} & \frac{1}{3} & 0 \end{pmatrix}\right].$$

Once we have simplified the game by applying the IDSDS procedure, in order to find the mixed-strategy Nash equilibria we can use the following result.
First we recall some notation that was introduced in Chapter 2. Given a mixed-strategy profile $\sigma = (\sigma_1, \ldots, \sigma_n)$ and a Player i, we denote by σ_{-i} the profile of strategies of the players other than i and use (σ_i, σ_{-i}) as an alternative notation for σ; furthermore, (τ_i, σ_{-i}) denotes the result of replacing σ_i with τ_i in σ, that is, $(\tau_i, \sigma_{-i}) = (\sigma_1, ..., \sigma_{i-1}, \tau_i, \sigma_{i+1}, \ldots, \sigma_n)$.

Theorem 6.3.1 Consider a reduced-form game in strategic form with cardinal payoffs.
- Suppose that $\sigma^* = (\sigma_1^*, ..., \sigma_n^*)$ is a Nash equilibrium in mixed strategies.
- Consider an arbitrary Player i.
- Let $\pi_i^* = \Pi_i(\sigma^*)$ be the payoff of Player i at this Nash equilibrium and let $s_{ij}, s_{ik} \in S_i$ be two pure strategies of Player i such that $\sigma_i^*(s_{ij}) > 0$ and $\sigma_i^*(s_{ik}) > 0$, that is, s_{ij} and s_{ik} are two pure strategies to which the mixed strategy σ_i^* of Player i assigns positive probability.
- Then $\Pi_i\left(s_{ij}, \sigma_{-i}^*\right) = \Pi_i\left(s_{ik}, \sigma_{-i}^*\right) = \pi_i^*$.
In other words, when the other players use the mixed-strategy profile σ_{-i}^*, Player i gets the same payoff no matter whether she plays the mixed strategy σ_i^* or the pure strategy s_{ij} or the pure strategy s_{ik}.

The details of the proof of Theorem 6.3.1 will be omitted, but the idea is simple: if s_{ij} and s_{ik} are two pure strategies to which the mixed strategy σ_i^* of Player i assigns positive probability and $\Pi_i\left((s_{ij}, \sigma_{-i}^*)\right) > \Pi_i\left((s_{ik}, \sigma_{-i}^*)\right)$, then Player i can increase her payoff from $\pi_i^* = \Pi_i(\sigma^*)$ to a larger number by reducing the probability of s_{ik} to zero and adding that probability to $\sigma_i^*(s_{ij})$, that is, by switching from σ_i^* to the mixed strategy $\hat{\sigma}_i$ obtained as follows: $\hat{\sigma}_i(s_{ik}) = 0$, $\hat{\sigma}_i(s_{ij}) = \sigma_i^*(s_{ij}) + \sigma_i^*(s_{ik})$ and, for every other $s_i \in S_i$, $\hat{\sigma}_i(s_i) = \sigma_i^*(s_i)$. But this would contradict the hypothesis that $\sigma^* = (\sigma_1^*, ..., \sigma_n^*)$ is a Nash equilibrium.

Let us now go back to the game of Figure 6.5, which is reproduced in Figure 6.5, and see how we can use Theorem 6.3.1 to find the Nash equilibrium in mixed strategies.

		Player 2	
		E	F
Player 1	B	4 0	2 4
	C	3 3	4 2

Figure 6.6: Copy of Figure 6.5

We want to find values of p and q, strictly between 0 and 1, such that

$$\left[\begin{pmatrix} B & C \\ p & 1-p \end{pmatrix}, \begin{pmatrix} E & F \\ q & 1-q \end{pmatrix}\right]$$

is a Nash equilibrium.

By Theorem 6.3.1, if Player 1 played the pure strategy B against $\begin{pmatrix} E & F \\ q & 1-q \end{pmatrix}$ she should get the same payoff as if she were to play the pure strategy C.

The former would give her a payoff of $4q+2(1-q)$ and the latter a payoff of $3q+4(1-q)$.

Thus we need q to be such that $4q+2(1-q) = 3q+4(1-q)$, that is, $q = \frac{2}{3}$.

When $q = \frac{2}{3}$, both B and C give Player 1 a payoff of $\frac{10}{3}$ and thus any mixture of B and C would also give the same payoff of $\frac{10}{3}$.

In other words, Player 1 is indifferent among all her mixed strategies and thus any mixed strategy is a best response to $\begin{pmatrix} E & F \\ \frac{2}{3} & \frac{1}{3} \end{pmatrix}$.

Similar reasoning for Player 2 reveals that, by Theorem 6.3.1, we need p to be such that $0p+3(1-p) = 4p+2(1-p)$, that is, $p = \frac{1}{5}$.

Against $\begin{pmatrix} B & C \\ \frac{1}{5} & \frac{4}{5} \end{pmatrix}$ any mixed strategy of Player 2 gives him the same payoff of $\frac{12}{5}$; thus any mixed strategy of Player 2 is a best reply to $\begin{pmatrix} B & C \\ \frac{1}{5} & \frac{4}{5} \end{pmatrix}$.

It follows that $\left[\begin{pmatrix} B & C \\ \frac{1}{5} & \frac{4}{5} \end{pmatrix}, \begin{pmatrix} E & F \\ \frac{2}{3} & \frac{1}{3} \end{pmatrix}\right]$ is a Nash equilibrium.

R It follows from Theorem 6.3.1, and was illustrated in the above example, that at a mixed strategy Nash equilibrium where Player i plays two or more pure strategies with positive probability, Player i does not have an incentive to use that mixed strategy: she would get the same payoff if, instead of randomizing, she played one of the pure strategies in the support of her mixed strategy (that is, if she increased the probability of any pure strategy from a positive number to 1).[4] *The only purpose of randomizing is to make the other player indifferent among two or more of his own pure strategies.*

		Player 2			
		D		*E*	
Player 1	*A*	3	0	0	2
	B	0	2	3	0
	C	2	0	2	1

Figure 6.7: A reduced-form game with cardinal payoffs

The "indifference" condition explained above provides a necessary, *but not sufficient*, condition for a mixed-strategy profile to be a Nash equilibrium. To see that the condition is not sufficient, consider the game of Figure 6.7 and the mixed-strategy profile $\left[\begin{pmatrix} A & B & C \\ \frac{1}{2} & \frac{1}{2} & 0 \end{pmatrix}, \begin{pmatrix} D & E \\ \frac{1}{2} & \frac{1}{2} \end{pmatrix}\right]$. Given that Player 2 plays the mixed strategy $\begin{pmatrix} D & E \\ \frac{1}{2} & \frac{1}{2} \end{pmatrix}$, Player 1 is indifferent between the two pure strategies that are in the support of her own mixed strategy, namely A and B: the payoff from playing A is 1.5 and so is the payoff from playing B (and 1.5 is also the payoff associated with the mixed strategy under consideration). However, the profile $\left[\begin{pmatrix} A & B & C \\ \frac{1}{2} & \frac{1}{2} & 0 \end{pmatrix}, \begin{pmatrix} D & E \\ \frac{1}{2} & \frac{1}{2} \end{pmatrix}\right]$ is not a Nash equilibrium, because Player 1 could get a payoff of 2 by switching to the pure strategy C.

We know from Theorem 6.2.1 that this game does have a mixed-strategy Nash equilibrium. How can we find it? Let us calculate the best response of Player 1 to every possible mixed strategy $\begin{pmatrix} D & E \\ q & 1-q \end{pmatrix}$ of Player 2 (with $q \in [0,1]$).

For Player 1 the payoff from playing A against $\begin{pmatrix} D & E \\ q & 1-q \end{pmatrix}$ is $3q$, the payoff from playing B is $3-3q$ and the payoff from playing C is constant and equal to 2. These functions are shown in Figure 6.8.

[4]The support of a mixed strategy is the set of pure strategies that are assigned positive probability by that mixed strategy.

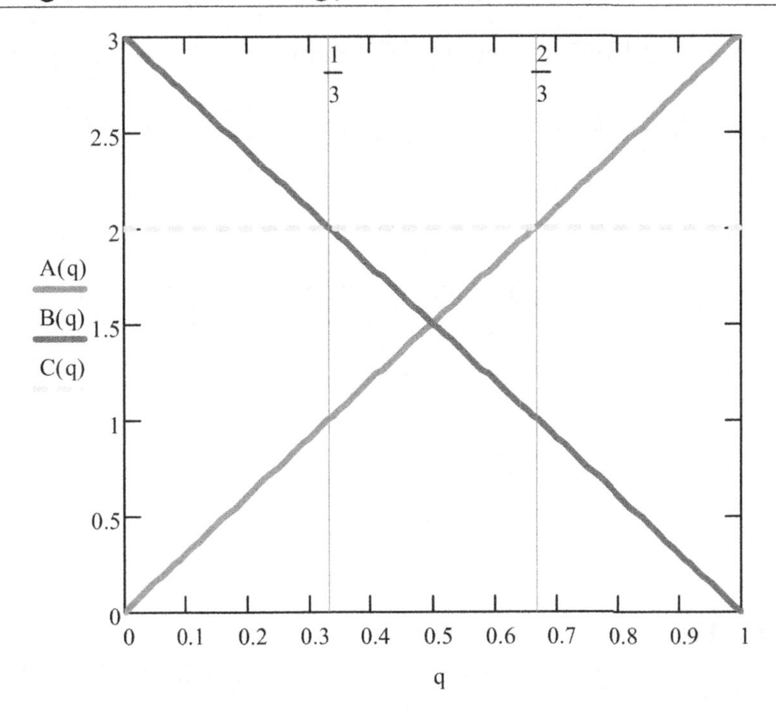

Figure 6.8: Player 1's payoff from each pure strategy against an arbitrary mixed strategy of Player 2

The upward-sloping line plots the function $A(q) = 3q$, the downward-sloping line plots the function $B(q) = 3 - 3q$ and the horizontal dashed line the function $C(q) = 2$.

The downward-sloping and horizontal lines intersect when $q = \frac{1}{3}$ and the upward-sloping and horizontal lines intersect when $q = \frac{2}{3}$.

The maximum payoff is given by the downward-sloping line up to $q = \frac{1}{3}$, then by the horizontal line up to $q = \frac{2}{3}$ and then by the upward-sloping line.

Thus the best reply function of Player 1 is as follows:

$$\text{Player1's best reply} = \begin{cases} B & \text{if } 0 \leq q < \frac{1}{3} \\ \begin{pmatrix} B & C \\ p & 1-p \end{pmatrix} \text{ for any } p \in [0,1] & \text{if } q = \frac{1}{3} \\ C & \text{if } \frac{1}{3} < q < \frac{2}{3} \\ \begin{pmatrix} A & C \\ p & 1-p \end{pmatrix} \text{ for any } p \in [0,1] & \text{if } q = \frac{2}{3} \\ A & \text{if } \frac{2}{3} < q \leq 1 \end{cases}$$

Hence if there is a mixed-strategy Nash equilibrium it is either of the form

$$\left[\begin{pmatrix} A & B & C \\ 0 & p & 1-p \end{pmatrix}, \begin{pmatrix} D & E \\ \frac{1}{3} & \frac{2}{3} \end{pmatrix}\right] \quad \text{or of the form} \quad \left[\begin{pmatrix} A & B & C \\ p & 0 & 1-p \end{pmatrix}, \begin{pmatrix} D & E \\ \frac{2}{3} & \frac{1}{3} \end{pmatrix}\right].$$

The latter cannot be a Nash equilibrium for any p, because when Player 1 plays B with probability 0, E strictly dominates D for Player 2 and thus Player 2's mixed strategy is not a best reply (E is the unique best reply). Thus the only candidate for a Nash equilibrium is of the form

$$\left[\begin{pmatrix} A & B & C \\ 0 & p & 1-p \end{pmatrix}, \begin{pmatrix} D & E \\ \frac{1}{3} & \frac{2}{3} \end{pmatrix}\right].$$

In this case, by Theorem 6.3.1, we need p to be such that Player 2 is indifferent between D and E: we need $2p = 1 - p$, that is, $p = \frac{1}{3}$. Hence the Nash equilibrium is

$$\left[\begin{pmatrix} A & B & C \\ 0 & \frac{1}{3} & \frac{2}{3} \end{pmatrix}, \begin{pmatrix} D & E \\ \frac{1}{3} & \frac{2}{3} \end{pmatrix}\right].$$

In games where the number of strategies or the number of players are larger than in the examples we have considered so far, finding the Nash equilibria involves lengthier calculations. However, computer programs have been developed that can be used to compute all the Nash equilibria of a finite game in a very short time.

Test your understanding of the concepts introduced in this section, by going through the exercises in Section 6.5.3 at the end of this chapter.

6.4 Strict dominance and rationalizability

We remarked in the previous section that a pure strategy that is strictly dominated by another pure strategy cannot be played with positive probability at a Nash equilibrium. Thus, when looking for a Nash equilibrium, one can first simplify the game by applying the IDSDS procedure (Chapter 2). When payoffs are cardinal (von Neumann-Morgenstern payoffs) it turns out that, *in a two-person game*, a pure strategy cannot be a best response to any mixed-strategy of the opponent not only when it is strictly dominated by another pure strategy but also when it is strictly dominated *by a mixed strategy*. To see this, consider the game of Figure 6.9.

		Player 2	
		D	E
Player 1	A	0 1	4 0
	B	1 2	1 4
	C	2 0	0 1

Figure 6.9: A strategic-form game with cardinal payoffs

The pure strategy B of Player 1 is not strictly dominated by another pure strategy and yet it cannot be a best reply to any mixed strategy of Player 2.

To see this, consider an arbitrary mixed strategy $\begin{pmatrix} D & E \\ q & 1-q \end{pmatrix}$ of Player 2 with $q \in [0,1]$. If Player 1 plays B against it, she gets a payoff of 1; if, instead, she plays the mixed strategy $\begin{pmatrix} A & B & C \\ \frac{1}{3} & 0 & \frac{2}{3} \end{pmatrix}$ then her payoff is $\frac{1}{3}4(1-q)+\frac{2}{3}2q = \frac{4}{3} > 1$.

Theorem 6.4.1 — Pearce, 1984. Consider a two-player reduced-form game in strategic form with cardinal payoffs, an arbitrary Player i and a pure strategy s_i of Player i. Then there is no mixed-strategy of the opponent to which s_i is a best response, if and only if s_i is strictly dominated by a mixed strategy σ_i of Player i (that is, there is a $\sigma_i \in \Sigma_i$ such that $\Pi_i(\sigma_i, \sigma_j) > \Pi_i(s_i, \sigma_j)$, for every $\sigma_j \in \Sigma_j$).

Note that, since the set of mixed strategies includes the set of pure strategies, strict dominance by a mixed strategy includes as a sub-case strict dominance by a pure strategy.

When the number of players is 3 or more, the generalization of Theorem 6.4.1 raises some subtle issues: see Exercise 6.14. However, we can appeal to the intuition behind Theorem 6.4.1 (see the remark below) to refine the IDSDS procedure for general n-player games with cardinal payoffs as follows.

Definition 6.4.1 — Cardinal IDSDS. The *Cardinal Iterated Deletion of Strictly Dominated Strategies* is the following algorithm. Given a finite n-player ($n \geq 2$) strategic-form game with cardinal payoffs G, let G^1 be the game obtained by removing from G, for every Player i, those pure strategies of Player i (if any) that are strictly dominated in G by some *mixed* strategy of Player i; let G^2 be the game obtained by removing from G^1, for every Player i, those pure strategies of Player i (if any) that are strictly dominated in G^1 by some mixed strategy of Player i, and so on. Let G^∞ be the output of this procedure. Since the initial game G is finite, G^∞ will be obtained in a finite number of steps. For every Player i, the pure strategies of Player i in G^∞ are called her *rationalizable strategies*.

Figure 6.10 illustrates this procedure as applied to the game in Panel (i).

Player 2

Player 1	D		E		F	
A	3	4	2	1	1	2
B	0	0	1	3	4	1
C	1	4	1	4	2	6

(i) The game $G^0 = G$

Player 2

Player 1	D		E		F	
A	3	4	2	1	1	2
B	0	0	1	3	4	1

(ii) The game G^1 after Step 1

Player 2

Player 1	D		E	
A	3	4	2	1
B	0	0	1	3

(iii) The game G^2 after Step 2

Player 2

Player 1	D		E	
A	3	4	2	1

(iv)a The game G^3 after Step 3

Player 2

Player 1	D	
A	3	4

(iv)b The game $G^4 = G^\infty$

Figure 6.10: Application of the cardinal IDSDS procedure

In the first step, the pure strategy C of Player 1 is deleted, because it is strictly dominated by the mixed strategy $\begin{pmatrix} A & B \\ \frac{1}{2} & \frac{1}{2} \end{pmatrix}$ thus yielding game G^1 shown in Panel (ii).

In the second step, the pure strategy F of Player 2 is deleted, because it is strictly dominated by the mixed strategy $\begin{pmatrix} D & E \\ \frac{1}{2} & \frac{1}{2} \end{pmatrix}$ thus yielding game G^2 shown in Panel (iii).

In the third step, B is deleted because it is strictly dominated by A thus yielding game G^3 shown in the top part of Panel (iv).

In the final step, E is deleted because it is strictly dominated by D so that the final output is the strategy profile (A,D).

Hence the only rationalizable strategies are A for Player 1 and D for Player 2.

Note that, in the game of Figure 6.10 Panel (i), since the only rationalizable strategy profile is (A,D), it follows that (A,D) is also the unique Nash equilibrium.

As noted in Chapter 2 the significance of the output of the IDSDS procedure is as follows. Consider game G in Panel (*i*) of Figure 5.9. Since, for Player 1, C is strictly dominated, if Player 1 is rational she will not play C. Thus, if Player 2 believes that Player 1 is rational then he believes that Player 1 will not play C, that is, he restricts attention to game G^1; since, in G^1, F is strictly dominated for Player 2, if Player 2 is rational he will not play F. It follows that if Player 1 believes that Player 2 is rational and that Player 2 believes that Player 1 is rational, then Player 1 restricts attention to game G^2 where rationality requires that Player 1 not play B, etc.

R Define a player to be rational if her chosen pure strategy is a best reply to her belief about what the opponent will do. In a two-player game a belief of Player 1 about what Player 2 will do can be expressed as a probability distribution over the set of pure strategies of Player 2; but this is the same object as a mixed strategy of Player 2. Thus, by Theorem 6.4.1, a rational Player 1 cannot choose a pure strategy that is strictly dominated by one of her own mixed strategies. The iterated reasoning outlined above can be captured by means of the notion of common knowledge of rationality. Indeed, it will be shown in Chapter 10 that if there is common knowledge of rationality then only rationalizable strategy profiles can be played. In a game with more than two players a belief of Player i about her opponents is no longer the same object as a mixed-strategy profile of the opponents, because a belief can allow for correlation in the behavior of the opponents, while the notion of mixed-strategy profile rules out such correlation (see Exercise 6.14).

R The iterated reasoning outlined above *requires that the von Neumann-Morgenstern preferences of both players be common knowledge between them.* For example, if Player 2 believes that Player 1 is rational but only knows her ordinal ranking of the outcomes, then Player 2 will not be able to deduce that it is irrational for Player 1 to play C and thus it cannot be irrational for him to play F. Expecting a player to know the von Neumann-Morgenstern preferences of another player is often (almost always?) very unrealistic! Thus one should be aware of the implicit assumptions that one makes (and one should question the assumptions made by others in their analyses).

Test your understanding of the concepts introduced in this section, by going through the exercises in Section 6.5.4 at the end of this chapter.

6.5 Exercises

6.5.1 Exercises for Section 6.1: Strategic-form games with cardinal payoffs

The answers to the following exercises are in Section 6.6 at the end of this chapter.

Exercise 6.1 Consider the following game-frame in strategic form, where o_1, o_2, o_3 and o_4 are basic outcomes:

		Player 2	
		c	d
Player	a	o_1	o_2
1	b	o_3	o_4

Both players satisfy the axioms of expected utility.
- The best outcome for Player 1 is o_3; she is indifferent between outcomes o_1 and o_4 and ranks them both as worst; she considers o_2 to be worse than o_3 and better than o_4; she is indifferent between o_2 with certainty and the lottery $\begin{pmatrix} o_3 & o_1 \\ 0.25 & 0.75 \end{pmatrix}$.
- The best outcome for Player 2 is o_4, which he considers to be just as good as o_1; he considers o_2 to be worse than o_1 and better than o_3; he is indifferent between o_2 with certainty and the lottery $\begin{pmatrix} o_1 & o_3 \\ 0.4 & 0.6 \end{pmatrix}$.

Find the normalized von Neumann-Morgenstern utility functions for the two players and write the corresponding reduced-form game. ■

Exercise 6.2 Consider the game-frame shown in Figure 6.11, where $o_1, \ldots, o_4$ are basic outcomes. Both players have von Neumann-Morgenstern rankings of the basic outcomes. The ranking of Player 1 can be represented by the following von Neumann-Morgenstern utility function:

outcome:	o_1	o_2	o_3	o_4
U_1 :	12	10	6	16

and the ranking of Player 2 can be represented by the following von Neumann-Morgenstern utility function:

outcome:	o_1	o_2	o_3	o_4
U_2 :	6	14	8	10

Write the corresponding reduced-form game. ■

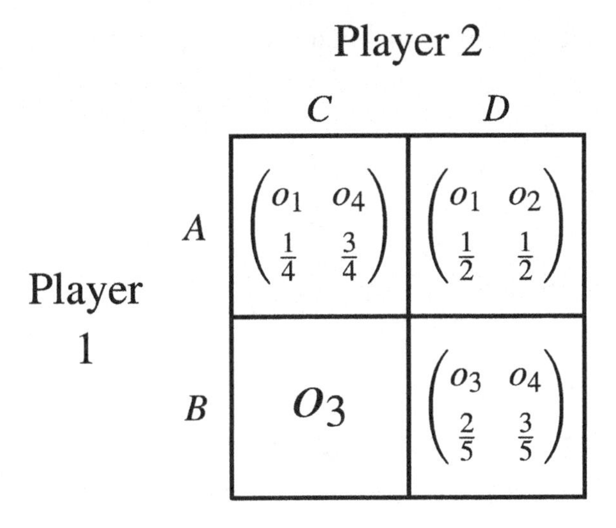

Figure 6.11: A game-frame in strategic form

6.5.2 Exercises for Section 6.2: Mixed strategies

The answers to the following exercises are in Section 6.6 at the end of this chapter.

		Player 2: D		Player 2: E	
Player 1	A	0	1	6	3
	B	4	4	2	0
	C	3	0	4	2

Figure 6.12: A strategic-form game with cardinal payoffs

Exercise 6.3 Consider the reduced-form game with cardinal payoffs shown in Figure 6.12.

(a) Calculate the players' payoffs from the mixed strategy profile

$$\left[\begin{pmatrix} A & B & C \\ \frac{1}{4} & \frac{3}{4} & 0 \end{pmatrix} \begin{pmatrix} D & E \\ \frac{1}{2} & \frac{1}{2} \end{pmatrix}\right].$$

(b) Is $\left[\begin{pmatrix} A & B & C \\ \frac{1}{4} & \frac{3}{4} & 0 \end{pmatrix} \begin{pmatrix} D & E \\ \frac{1}{2} & \frac{1}{2} \end{pmatrix}\right]$ a Nash equilibrium? ■

Exercise 6.4 Consider the following reduced-form game with cardinal payoffs:

		Player 2	
		D	E
Player 1	A	2 , 3	8 , 5
	B	6 , 6	4 , 2

Prove that $\left[\begin{pmatrix} A & B \\ \frac{2}{3} & \frac{1}{3} \end{pmatrix}\begin{pmatrix} D & E \\ \frac{1}{2} & \frac{1}{2} \end{pmatrix}\right]$ is a Nash equilibrium. ■

6.5.3 Exercises for Section 6.3: Computing the mixed-strategy Nash equilibria

The answers to the following exercises are in Section 6.6 at the end of this chapter.

Exercise 6.5 Consider again the game of Exercise 6.1.

(a) Find the mixed-strategy Nash equilibrium.

(b) Calculate the payoffs of both players at the Nash equilibrium.

■

Exercise 6.6 Find the Nash equilibrium of the game of Exercise 6.2. ■

Exercise 6.7 Find all the mixed-strategy Nash equilibria of the game of Exercise 6.4 and calculate the payoffs of both players at every Nash equilibrium. ■

Exercise 6.8 Find the mixed-strategy Nash equilibria of the following game:

		Player 2	
		L	R
	T	1 , 4	4 , 3
Player 1	C	2 , 0	1 , 2
	B	1 , 5	0 , 6

■

Exercise 6.9 Consider the following two-player game, where $o_1, o_2, \ldots, o_6$ are basic outcomes.

		Player 2	
		d	e
	a	o_1	o_2
Player 1	b	o_3	o_4
	c	o_5	o_6

The players rank the outcomes as indicated below (as usual, if outcome o is above outcome o' then o is strictly preferred to o' and if o and o' are on the same row then the player is indifferent between the two):

$$\text{Player 1}: \begin{pmatrix} o_1 \\ o_6 \\ o_4, o_2 \\ o_5 \\ o_3 \end{pmatrix} \qquad \text{Player 2}: \begin{pmatrix} o_3, o_4 \\ o_2 \\ o_1, o_5 \\ o_6 \end{pmatrix}$$

(a) One player has a strategy that is strictly dominated. Identify the player and the strategy.

[Note: in order to answer the following questions, you can make your life a lot easier if you simplify the game on the basis of your answer to part **(a)**.]

Player 1 satisfies the axioms of Expected Utility Theory and is indifferent between o_6 and the lottery $\begin{pmatrix} o_1 & o_5 \\ \frac{4}{5} & \frac{1}{5} \end{pmatrix}$ and is indifferent between o_2 and the lottery $\begin{pmatrix} o_6 & o_5 \\ \frac{1}{2} & \frac{1}{2} \end{pmatrix}$.

(b) Suppose that Player 1 believes that Player 2 is going to play d with probability $\frac{1}{2}$ and e with probability $\frac{1}{2}$. Which strategy should he play?

Player 2 satisfies the axioms of Expected Utility Theory and is indifferent between o_5 and the lottery $\begin{pmatrix} o_2 & o_6 \\ \frac{1}{4} & \frac{3}{4} \end{pmatrix}$.

(c) Suppose that Player 2 believes that Player 1 is going to play a with probability $\frac{1}{4}$ and c with probability $\frac{3}{4}$. Which strategy should she play?

(d) Find all the (pure- and mixed-strategy) Nash equilibria of this game.

■

Exercise 6.10 Consider the following game (where the payoffs are von Neumann-Morgenstern payoffs):

		Player 2	
		C	D
Player 1	A	x , y	3 , 0
	B	6 , 2	0 , 4

(a) Suppose that $x = 2$ and $y = 2$. Find the mixed-strategy Nash equilibrium and calculate the payoffs of both players at the Nash equilibrium.

(b) (b) For what values of x and y is $\left[\begin{pmatrix} A & B \\ \frac{1}{5} & \frac{4}{5} \end{pmatrix}, \begin{pmatrix} C & D \\ \frac{3}{4} & \frac{1}{4} \end{pmatrix}\right]$ a Nash equilibrium?

■

Exercise 6.11 Find the mixed-strategy Nash equilibria of the game of Exercise 6.3. Calculate the payoffs of both players at every Nash equilibrium that you find. ■

6.5.4 Exercises for Section 6.4: Strict dominance and rationalizability

The answers to the following exercises are in Section 6.6 at the end of this chapter.

Exercise 6.12 In the following game, for each player, find all the rationalizable pure strategies (that is, apply the cardinal IDSDS procedure).

		Player 2		
		L	M	R
	A	3 , 5	2 , 0	2 , 2
Player 1	B	5 , 2	1 , 2	2 , 1
	C	9 , 0	1 , 5	3 , 2

■

Note: The next three exercises are more difficult than the previous ones.

Exercise 6.13 Is the following statement true or false? Either prove that it is true or give a counterexample.

> "Consider a two-player strategic-form game with cardinal payoffs.
> - Let A and B be two pure strategies of Player 1.
> - Suppose that both A and B are rationalizable (that is, they survive the cardinal IDSDS procedure).
> - Then any mixed strategy that attaches positive probability to both A and B and zero to every other strategy is a best reply to some mixed strategy of Player 2."

■

Exercise 6.14 Consider the three-player game shown in Figure 6.13, where only the payoffs of Player 1 are recorded.

(a) Show that if Player 1 assigns probability $\frac{1}{2}$ to the event "Player 2 will play E and Player 3 will play G" and probability $\frac{1}{2}$ to the event "Player 2 will play F and Player 3 will play H", then playing D is a best reply.

Next we want to show that there is no mixed-strategy profile

$$\sigma_{-1} = \left(\begin{pmatrix} E & F \\ p & 1-p \end{pmatrix}, \begin{pmatrix} G & H \\ q & 1-q \end{pmatrix} \right)$$

of Players 2 and 3 against which D is a best reply for Player 1.

Define the following functions: $A(p,q) = \Pi_1(A, \sigma_{-1})$ (that is, $A(p,q)$ is Player 1's expected payoff if she plays the pure strategy A against σ_{-1}), $B(p,q) = \Pi_1(B, \sigma_{-1})$, $C(p,q) = \Pi_1(C, \sigma_{-1})$ and $D(p,q) = \Pi_1(D, \sigma_{-1})$.

(b) In the (p,q) plane (with $0 \le p \le 1$ and $0 \le q \le 1$) draw the curve corresponding to the equation $A(p,q) = D(p,q)$ and identify the region where $A(p,q) > D(p,q)$.

(c) In the (p,q) plane draw the curve corresponding to the equation $C(p,q) = D(p,q)$ and identify the region where $C(p,q) > D(p,q)$.

(d) In the (p,q) plane draw the two curves corresponding to the equation $B(p,q) = D(p,q)$ and identify the region where $B(p,q) > D(p,q)$.

(e) Infer from parts **(b)-(c)** that there is no mixed-strategy profile of Players 2 and 3 against which D is a best reply for Player 1.

■

Player 3: G

Player 1 \ Player 2	E	F
A	3	0
B	0	3
C	0	0
D	2	0

Player 3: H

Player 1 \ Player 2	E	F
A	0	0
B	3	0
C	0	3
D	0	2

Figure 6.13: A three-player game where only the payoffs of Player 1 are shown

Exercise 6.15 — ⋆⋆⋆ **Challenging Question** ⋆⋆⋆. A team of n professional swimmers ($n \geq 2$) – from now on called players – are partying on the bank of the Sacramento river on a cold day in January. Suddenly a passerby shouts "Help! My dog fell into the water!" Each of the swimmers has to decide whether or not to jump into the icy cold water to rescue the dog. One rescuer is sufficient: the dog will be saved if at least one player jumps into the water; if nobody does, then the dog will die. Each player prefers somebody else to jump in, but each player prefers to jump in himself if nobody else does.
Let us formulate this as a game. The strategy set of each player $i = 1,\dots,n$ is $S_i = \{J, \neg J\}$, where J stands for 'jump in' and $\neg J$ for 'not jump in'.
The possible basic outcomes can be expressed as subsets of the set $I = \{1,...,n\}$ of players: outcome $N \subseteq I$ is interpreted as 'the players in the set N jump into the water'; if $N = \emptyset$ the dog dies, while if $N \neq \emptyset$ the dog is saved.
Player i has the following ordinal ranking of the outcomes:
(1) $N \sim N'$, for every $N \neq \emptyset$, $N' \neq \emptyset$ with $i \notin N$ and $i \notin N'$,
(2) $N \succ N'$ for every $N \neq \emptyset$, $N' \neq \emptyset$ with $i \notin N$ and $i \in N'$,
(3) $\{i\} \succ \emptyset$.

(a) Find all the pure-strategy Nash equilibria.

(b) Suppose that each player i has the following von Neumann-Morgenstern payoff function (which is consistent with the above ordinal ranking):

$$\pi_i(N) = \begin{cases} v & \text{if } N \neq \emptyset \text{ and } i \notin N \\ v - c & \text{if } N \neq \emptyset \text{ and } i \in N \\ 0 & \text{if } N = \emptyset \end{cases} \quad \text{with } 0 < c < v.$$

Find the symmetric mixed-strategy Nash equilibrium (symmetric means that all the players use the same mixed strategy).

(c) Assuming that the players behave according to the symmetric mixed-strategy Nash equilibrium of Part **(b)**, is it better for the dog if n (the number of players) is large or if n is small? Calculate the probability that the dog is saved at the mixed-strategy Nash equilibrium as a function of n, for all possible values of c and v (subject to $0 < c < v$), and plot it for the case where $c = 10$ and $v = 12$. ■

6.6 Solutions to exercises

Solution to Exercise 6.1. The normalized von Neumann-Morgenstern utility functions are:

Player 1:

outcome	U_1
o_3	1
o_2	0.25
o_1, o_4	0

Player 2:

outcome	U_2
o_1, o_4	1
o_2	0.4
o_3	0

The reduced-form game is shown in Figure 6.14. □

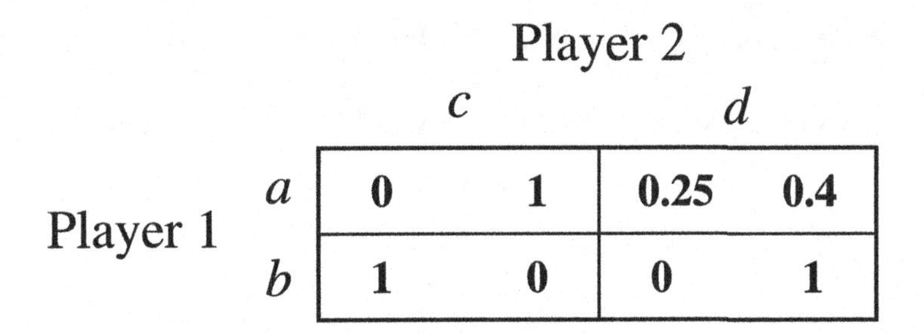

		Player 2	
		c	d
Player 1	a	**0** **1**	**0.25** **0.4**
	b	**1** **0**	**0** **1**

Figure 6.14: The reduced-form game for Exercise 6.1

Solution to Exercise 6.2.

The expected utility of the lottery $\begin{pmatrix} o_1 & o_4 \\ \frac{1}{4} & \frac{3}{4} \end{pmatrix}$ is $\frac{1}{4}(12)+\frac{3}{4}(16)=15$ for Player 1 and $\frac{1}{4}(6)+\frac{3}{4}(10)=9$ for Player 2.

The expected utility of the lottery $\begin{pmatrix} o_1 & o_2 \\ \frac{1}{2} & \frac{1}{2} \end{pmatrix}$ is 11 for Player 1 and 10 for Player 2.

The expected utility of the lottery $\begin{pmatrix} o_3 & o_4 \\ \frac{2}{5} & \frac{3}{5} \end{pmatrix}$ is 12 for Player 1 and 9.2 for Player 2.

The reduced-form game is shown in Figure 6.15. □

		Player 2	
		C	D
Player 1	A	15 9	11 10
	B	6 8	12 9.2

Figure 6.15: The reduced-form game for Exercise 6.2

Solution to Exercise 6.3.

(a) $\Pi_1=\frac{1}{8}(0)+\frac{1}{8}(6)+\frac{3}{8}(4)+\frac{3}{8}(2)=3$ and $\Pi_2=\frac{1}{8}(1)+\frac{1}{8}(3)+\frac{3}{8}(4)+\frac{3}{8}(0)=2$.

(b) No, because if Player 1 switched to the pure strategy C then her payoff would be $\frac{1}{2}(3)+\frac{1}{2}(4)=3.5>3$. □

Solution to Exercise 6.4. Player 1's payoff is $\Pi_1 = \frac{2}{6}(2) + \frac{2}{6}(8) + \frac{1}{6}(6) + \frac{1}{6}(4) = 5$.

If Player 1 switches to any other mixed strategy $\begin{pmatrix} A & B \\ p & 1-p \end{pmatrix}$, while Player 2's strategy is kept fixed at $\begin{pmatrix} C & D \\ \frac{1}{2} & \frac{1}{2} \end{pmatrix}$, then her payoff is $\Pi_1 = \frac{1}{2}(p)(2) + \frac{1}{2}(p)(8) + \frac{1}{2}(1-p)(6) + \frac{1}{2}(1-p)(4) = 5$.

Thus any mixed strategy of Player 1 is a best response to $\begin{pmatrix} C & D \\ \frac{1}{2} & \frac{1}{2} \end{pmatrix}$.

Similarly, Player 2's payoff is $\Pi_2 = \frac{2}{6}3 + \frac{2}{6}(5) + \frac{1}{6}(6) + \frac{1}{6}(2) = 4$. If Player 2 switches to any other mixed strategy $\begin{pmatrix} C & D \\ q & 1-q \end{pmatrix}$, while Player 1's strategy is kept fixed at $\begin{pmatrix} A & B \\ \frac{2}{3} & \frac{1}{3} \end{pmatrix}$, then her payoff is $\Pi_2 = \frac{2}{3}(q)(3) + \frac{2}{3}(1-q)(5) + \frac{1}{3}(q)(6) + \frac{1}{3}(1-q)(2) = 4$.

Thus any mixed strategy of Player 2 is a best response to $\begin{pmatrix} A & B \\ \frac{2}{3} & \frac{1}{3} \end{pmatrix}$.

Hence $\begin{pmatrix} A & B \\ \frac{2}{3} & \frac{1}{3} \end{pmatrix}$ is a best reply to $\begin{pmatrix} C & D \\ \frac{1}{2} & \frac{1}{2} \end{pmatrix}$ and $\begin{pmatrix} C & D \\ \frac{1}{2} & \frac{1}{2} \end{pmatrix}$ is a best reply to $\begin{pmatrix} A & B \\ \frac{2}{3} & \frac{1}{3} \end{pmatrix}$, that is, $\left[\begin{pmatrix} A & B \\ \frac{2}{3} & \frac{1}{3} \end{pmatrix}, \begin{pmatrix} C & D \\ \frac{1}{2} & \frac{1}{2} \end{pmatrix}\right]$ is a Nash equilibrium. □

Solution to Exercise 6.5.

(a) We have to find the Nash equilibrium of the following game:

		Player 2	
		c	d
Player 1	a	0 , 1	0.25 , 0.4
	b	1 , 0	0 , 1

To make calculations easier, let us multiply all the payoffs by 100 (that is, we re-scale the von Neumann-Morgenstern utility functions by a factor of 100):

		Player 2	
		c	d
Player 1	a	0 , 100	25 , 40
	b	100 , 0	0 , 100

There are no pure-strategy Nash equilibria. To find the mixed-strategy Nash equilibrium, let p be the probability with which Player 1 chooses a and q be the probability with which Player 2 chooses c.

Then, for Player 1, the payoff from playing a against $\begin{pmatrix} c & d \\ q & 1-q \end{pmatrix}$ must be equal to the payoff from playing b against $\begin{pmatrix} c & d \\ q & 1-q \end{pmatrix}$. That is, it must be that $25(1-q) = 100q$, which yields $q = \frac{1}{5}$. Similarly, for Player 2, the payoff from playing c against $\begin{pmatrix} a & b \\ p & 1-p \end{pmatrix}$ must be equal to the payoff from playing d against $\begin{pmatrix} a & b \\ p & 1-p \end{pmatrix}$. This requires $100p = 40p + 100(1-p)$, that is, $p = \frac{5}{8}$. Thus the Nash equilibrium is $\left[\begin{pmatrix} a & b \\ \frac{5}{8} & \frac{3}{8} \end{pmatrix}, \begin{pmatrix} c & d \\ \frac{1}{5} & \frac{4}{5} \end{pmatrix}\right]$.

(b) At the Nash equilibrium the payoffs are 20 for Player 1 and 62.5 for Player 2. (If you worked with the original payoffs, then the Nash equilibrium payoffs would be 0.2 for Player 1 and 0.625 for Player 2.) □

Solution to Exercise 6.6. We have to find the Nash equilibria of the following game.

		Player 2	
		C	D
Player	A	15 , 9	11 , 10
1	B	6 , 8	12 , 9.2

For Player 2 D is a strictly dominant strategy, thus at a Nash equilibrium Player 2 must play D with probability 1. For Player 1, the unique best reply to D is B. Thus the pure-strategy profile (B,D) is the only Nash equilibrium. □

Solution to Exercise 6.7. We have to find the Nash equilibria of the following game.

		Player 2	
		D	E
Player	A	2 , 3	8 , 5
1	B	6 , 6	4 , 2

(B,D) (with payoffs (6,6)) and (A,E) (with payoffs (8,5)) are both Nash equilibria. To see if there is also a mixed-strategy equilibrium we need to solve the following equations, where p is the probability of A and q is the probability of D: $2q + 8(1-q) = 6q + 4(1-q)$ and $3p + 6(1-p) = 5p + 2(1-p)$. The solution is $p = \frac{2}{3}$ and $q = \frac{1}{2}$ so that

$$\left[\begin{pmatrix} A & B \\ \frac{2}{3} & \frac{1}{3} \end{pmatrix}, \begin{pmatrix} D & E \\ \frac{1}{2} & \frac{1}{2} \end{pmatrix}\right]$$

is a Nash equilibrium. The payoffs at this Nash equilibrium are 5 for Player 1 and 4 for Player 2. □

Solution to Exercise 6.8. Since B is strictly dominated (by C), it cannot be assigned positive probability at a Nash equilibrium. Let p be the probability of T and q the probability of L. Then p must be such that $4p+0(1-p)=3p+2(1-p)$ and q must be such that $q+4(1-q)=2q+(1-q)$. Thus $p=\frac{2}{3}$ and $q=\frac{3}{4}$. Hence there is only one mixed-strategy equilibrium, namely

$$\left[\begin{pmatrix} T & C & B \\ \frac{2}{3} & \frac{1}{3} & 0 \end{pmatrix}, \begin{pmatrix} L & R \\ \frac{3}{4} & \frac{1}{4} \end{pmatrix}\right].$$

□

Solution to Exercise 6.9.

(a) Since Player 1 prefers o_5 to o_3 and prefers o_6 to o_4, strategy b is strictly dominated by strategy c.

Thus, at a Nash equilibrium, Player 1 will not play b with positive probability and we can simplify the game to

		Player 2	
		d	e
Player	a	o_1	o_2
1	c	o_5	o_6

Of the remaining outcomes, for Player 1 o_1 is the best outcome (we can assign utility 1 to it) and o_5 is the worst (we can assign utility 0 to it). Since he is indifferent between o_6 and the lottery $\begin{pmatrix} o_1 & o_5 \\ \frac{4}{5} & \frac{1}{5} \end{pmatrix}$, the utility of o_6 is $\frac{4}{5}$. Hence the expected utility of $\begin{pmatrix} o_5 & o_6 \\ \frac{1}{2} & \frac{1}{2} \end{pmatrix}$ is $\frac{1}{2}(0)+\frac{1}{2}\left(\frac{4}{5}\right)=\frac{2}{5}$ and thus the utility of o_2 is also $\frac{2}{5}$.

(b) If Player 2 plays d with probability $\frac{1}{2}$ and e with probability $\frac{1}{2}$, then for Player 1 playing a gives a payoff of $\frac{1}{2}(1)+\frac{1}{2}\left(\frac{2}{5}\right)=\frac{7}{10}$, while playing c gives a payoff of $\frac{1}{2}(0)+\frac{1}{2}\left(\frac{4}{5}\right)=\frac{4}{10}$. Hence he should play a.

If you did not follow the suggestion to simplify the analysis as was done above, then you can still reach the same answer, although in a lengthier way. You would still set $U(o_1)=1$. Then the expected payoff from playing a is

$$\Pi_1(a)=\tfrac{1}{2}U(o_1)+\tfrac{1}{2}U(o_2)=\tfrac{1}{2}+\tfrac{1}{2}U(o_2) \qquad (\star)$$

Since o_2 is as good as $\begin{pmatrix} o_5 & o_6 \\ \frac{1}{2} & \frac{1}{2} \end{pmatrix}$,

$$U(o_2)=\tfrac{1}{2}U(o_5)+\tfrac{1}{2}U(o_6). \qquad (\diamond)$$

Since o_6 is as good as $\begin{pmatrix} o_1 & o_5 \\ \frac{4}{5} & \frac{1}{5} \end{pmatrix}$,

$$U(o_6) = \tfrac{4}{5} + \tfrac{1}{5}U(o_5). \qquad (\dagger)$$

Replacing ($\dagger$) in ($\diamond$)(we get $U(o_2) = \frac{2}{5} + \frac{3}{5}U(o_5)$ and replacing this expression in ($\star$) we get $\Pi_1(a) = \frac{7}{10} + \frac{3}{10}U(o_5)$. Similarly,

$$\Pi_1(c) = \tfrac{1}{2}U(o_5) + \tfrac{1}{2}U(o_6) = \tfrac{1}{2}U(o_5) + \tfrac{1}{2}\left(\tfrac{4}{5} + \tfrac{1}{5}U(o_5)\right) = \tfrac{4}{10} + \tfrac{6}{10}U(o_5)$$

Now, $\Pi_1(a) > \Pi_1(c)$ if and only if $\frac{7}{10} + \frac{3}{10}U(z_5) > \frac{4}{10} + \frac{6}{10}U(z_5)$ if and only if $3 > 3U(o_5)$ if and only if $U(o_5) < 1$, which is the case because o_5 is worse than o_1 and $U(o_1) = 1$. Similar steps would be taken to answer parts **(c)** and **(d)**.

(c) In the reduced game, for Player 2 o_2 is the best outcome (we can assign utility 1 to it) and o_6 is the worst (we can assign utility 0 to it). Thus, since she is indifferent between o_5 and the lottery $\begin{pmatrix} o_2 & o_6 \\ \frac{1}{4} & \frac{3}{4} \end{pmatrix}$, the utility of o_5 is $\frac{1}{4}$ and so is the utility of o_1. Thus playing d gives an expected payoff of $\frac{1}{4}(\frac{1}{4}) + \frac{3}{4}(\frac{1}{4}) = \frac{1}{4}$ and playing e gives an expected utility of $\frac{1}{4}(1) + \frac{3}{4}(0) = \frac{1}{4}$. Thus she is indifferent between playing d and playing e (and any mixture of d and e).

(d) Using the calculations of parts **(b)** and **(c)** the game is as follows:

		Player 2	
		d	e
Player 1	a	$1\ ,\ \frac{1}{4}$	$\frac{2}{5}\ ,\ 1$
	c	$0\ ,\ \frac{1}{4}$	$\frac{4}{5}\ ,\ 0$

There is no pure-strategy Nash equilibrium. At a mixed-strategy Nash equilibrium, each player must be indifferent between his/her two strategies. From part **(c)** we already know that Player 2 is indifferent if Player 1 plays a with probability $\frac{1}{4}$ and c with probability $\frac{3}{4}$. Now let q be the probability with which Player 2 plays d. Then we need $q + \frac{2}{5}(1-q) = \frac{4}{5}(1-q)$, hence $q = \frac{2}{7}$. Thus the Nash equilibrium is $\left(\begin{array}{ccc|cc} a & b & c & d & e \\ \frac{1}{4} & 0 & \frac{3}{4} & \frac{2}{7} & \frac{5}{7} \end{array}\right)$ which can be written more succinctly as $\left(\begin{array}{cc|cc} a & c & d & e \\ \frac{1}{4} & \frac{3}{4} & \frac{2}{7} & \frac{5}{7} \end{array}\right)$. □

Solution to Exercise 6.10.

(a) Let p be the probability of A and q the probability of B. Player 1 must be indifferent between playing A and playing B: $2q + 3(1-q) = 6q$; this gives $q = \frac{3}{7}$. Similarly, Player 2 must be indifferent between playing C and playing D: $2 = 4(1-p)$; this gives $p = \frac{1}{2}$. Thus the Nash equilibrium is given by

$$\left[\begin{pmatrix} A & B \\ \frac{1}{2} & \frac{1}{2} \end{pmatrix}, \begin{pmatrix} C & D \\ \frac{3}{7} & \frac{4}{7} \end{pmatrix}\right]$$

The equilibrium payoffs are $\frac{18}{7} = 2.57$ for Player 1 and 2 for Player 2.

(b) Player 1 must be indifferent between playing A and playing B: $\frac{3}{4}(x) + \frac{1}{4}(3) = \frac{3}{4}(6)$. Thus $x = 5$. Similarly, Player 2 must be indifferent between playing C and playing D: $\frac{1}{5}(y) + \frac{4}{5}(2) = \frac{4}{5}(4)$. Thus $y = 8$. □

Solution to Exercise 6.11. We have to find the Nash equilibria of the following game:

		Player 2	
		D	E
	A	0 , 1	6 , 3
Player 1	B	4 , 4	2 , 0
	C	3 , 0	4 , 2

There are two pure-strategy equilibria, namely (B,D) and (A,E). To see if there is a mixed-strategy equilibrium we calculate the best response of Player 1 to every possible mixed strategy $\begin{pmatrix} D & E \\ q & 1-q \end{pmatrix}$ of Player 2 (with $q \in [0,1]$). For Player 1 the payoff from playing A against $\begin{pmatrix} D & E \\ q & 1-q \end{pmatrix}$ is $6 - 6q$, the payoff from playing B is $4q + 2(1-q) = 2 + 2q$ and the payoff from playing C is $3q + 4(1-q) = 4 - q$. These functions are shown in Figure 6.16, where the downward-sloping line plots the function where $A(q) = 6 - 6q$, the upward-sloping line plots the function $B(q) = 2 + 2q$ and the dotted line the function $C(q) = 4 - q$.

It can be seen from Figure 6.16 that

$$\text{Player1's best reply} = \begin{cases} A & \text{if } 0 \le q < \frac{2}{5} \\ \begin{pmatrix} A & C \\ p & 1-p \end{pmatrix} \text{ for any } p \in [0,1] & \text{if } q = \frac{2}{5} \\ C & \text{if } \frac{2}{5} < q < \frac{2}{3} \\ \begin{pmatrix} B & C \\ p & 1-p \end{pmatrix} \text{ for any } p \in [0,1] & \text{if } q = \frac{2}{3} \\ B & \text{if } \frac{2}{3} < q \le 1 \end{cases}$$

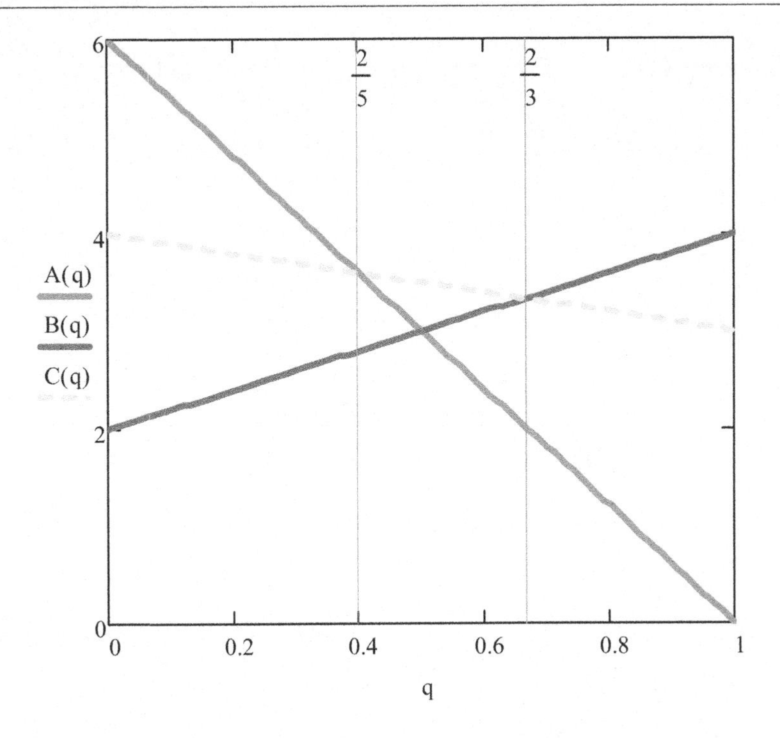

Figure 6.16: The best-reply diagram for Exercise 6.11

Thus if there is a mixed-strategy equilibrium it is either of the form

$$\left[\begin{pmatrix} A & C \\ p & 1-p \end{pmatrix}, \begin{pmatrix} D & E \\ \frac{2}{5} & \frac{3}{5} \end{pmatrix}\right] \quad \text{or of the form} \quad \left[\begin{pmatrix} B & C \\ p & 1-p \end{pmatrix}, \begin{pmatrix} D & E \\ \frac{2}{3} & \frac{1}{3} \end{pmatrix}\right].$$

In the first case, where Player 1 chooses B with probability zero, E strictly dominates D for Player 2 and thus $\begin{pmatrix} D & E \\ \frac{2}{5} & \frac{3}{5} \end{pmatrix}$ is not a best reply for Player 2, so that

$$\left[\begin{pmatrix} A & C \\ p & 1-p \end{pmatrix}, \begin{pmatrix} D & E \\ \frac{2}{5} & \frac{3}{5} \end{pmatrix}\right] \quad \text{is not a Nash equilibrium for any} p.$$

In the second case we need $D(p) = E(p)$, that is, $4p = 2(1-p)$, which yields $p = \frac{1}{3}$.

Thus the mixed-strategy Nash equilibrium is $\left[\begin{pmatrix} B & C \\ \frac{1}{3} & \frac{2}{3} \end{pmatrix}, \begin{pmatrix} D & E \\ \frac{2}{3} & \frac{1}{3} \end{pmatrix}\right]$ with payoffs of $\frac{10}{3}$ for Player 1 and $\frac{4}{3}$ for Player 2. □

Solution to Exercise 6.12. For Player 1, B is strictly dominated by $\begin{pmatrix} A & C \\ \frac{1}{2} & \frac{1}{2} \end{pmatrix}$; for Player 2, R is strictly dominated by $\begin{pmatrix} L & M \\ \frac{1}{2} & \frac{1}{2} \end{pmatrix}$. Eliminating B and R we are left with

		Player 2	
		L	M
Player	A	3 , 5	2 , 0
1	C	9 , 0	1 , 5

In this game no player has a strictly dominated strategy. Thus for Player 1 both A and C are rationalizable and for Player 2 both L and M are rationalizable. □

Solution to Exercise 6.13. The statement is false. Consider, for example, the following game:

		Player 2	
		L	R
	A	3 , 1	0 , 0
Player 1	B	0 , 0	3 , 1
	C	2 , 1	2 , 1

Here both A and B are rationalizable (indeed, they are both part of a Nash equilibrium; note that the cardinal IDSDS procedure leaves the game unchanged). However, the mixture $\begin{pmatrix} A & B \\ \frac{1}{2} & \frac{1}{2} \end{pmatrix}$ (which gives Player 1 a payoff of 1.5, no matter what Player 2 does) *cannot* be a best reply to any mixed strategy of Player 2, since it is strictly dominated by C. □

Solution to Exercise 6.14.

(a) If Player 1 assigns probability $\frac{1}{2}$ to the event "Player 2 will play E and Player 3 will play G" and probability $\frac{1}{2}$ to the event "Player 2 will play F and Player will play H", then A gives Player 1 an expected payoff of 1.5, B an expected payoff of 0, C an expected payoff of 1.5 and D an expected payoff of 2.
Thus D is a best reply to those beliefs.
The functions are as follows: $A(p,q) = 3pq$, $B(p,q) = 3(1-p)q + 3p(1-q)$, $C(p,q) = 3(1-p)(1-q)$, $D(p,q) = 2pq + 2(1-p)(1-q)$.

(b) $A(p,q) = D(p,q)$ at those points (p,q) such that $q = \frac{2-2p}{2-p}$. The set of such points is the continuous curve in the Figure 6.17. The region where $A(p,q) > D(p,q)$ is the region **above** the continuous curve.

(c) $C(p,q) = D(p,q)$ at those points (p,q) such that $q = \frac{1-p}{1+p}$. The set of such points is the dotted curve in the diagram shown in Figure 6.17. The region where $C(p,q) > D(p,q)$ is the region **below** the dotted curve.

(d) $B(p,q) = D(p,q)$ at those points (p,q) such that $q = \frac{2-5p}{5-10p}$ (for $p \neq \frac{1}{2}$). The set of such points is given by the two dashed curves in the diagram below. The region where $B(p,q) > D(p,q)$ is the region between the two dashed curves.

Thus

- in the region strictly above the continuous curve, A is better than D,
- in the region strictly below the dotted curve, C is better than D and
- in the region on and between the continuous curve and the dotted curve, B is better that D.

Hence, at every point in the (p,q) square there is a pure strategy of Player 1 which is strictly better than D. It follows that there is no mixed-strategy σ_{-1} against which D is a best reply. □

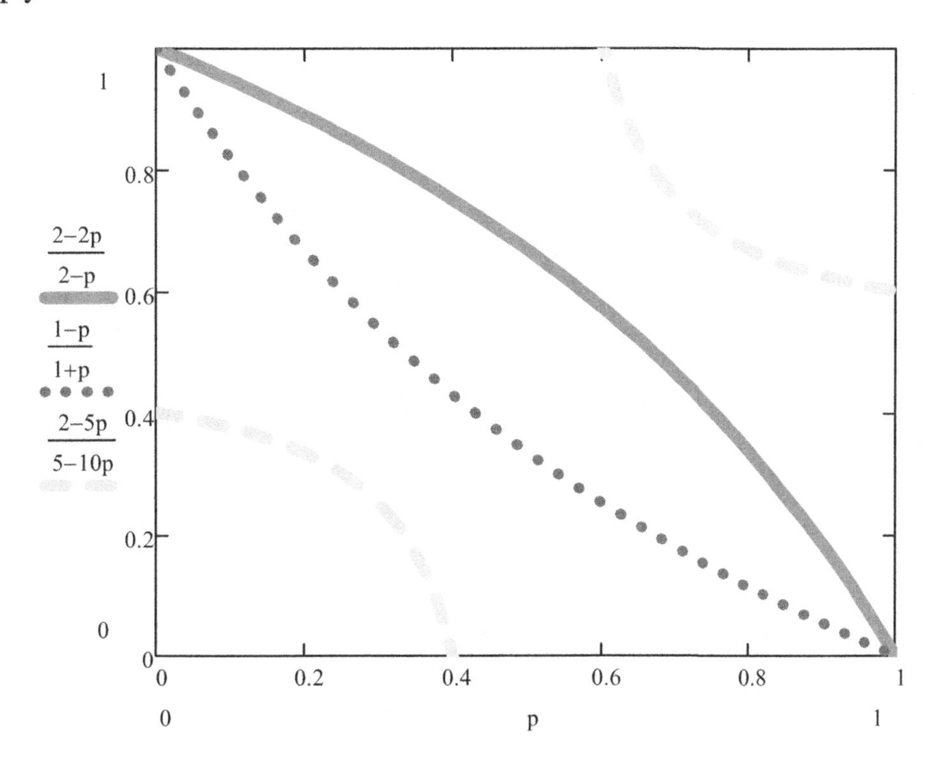

Figure 6.17: The diagram for Exercise 6.14

Solution to Exercise 6.15.

(a) There are n pure-strategy Nash equilibria: at each equilibrium exactly one player jumps in.

(b) Let p be the probability with which each player jumps into the water. Consider a Player i. The probability that none of the other players jump in is $(1-p)^{n-1}$ and thus the probability that somebody else jumps in is $\left[1-(1-p)^{n-1}\right]$.
Player i's payoff if he jumps in is $v-c$, while his expected payoff if he does not jump in is $v\left[1-(1-p)^{n-1}\right]+0(1-p)^{n-1}=v\left[1-(1-p)^{n-1}\right]$.

Thus we need $v-c=v\left[1-(1-p)^{n-1}\right]$, that is, $\boxed{p=1-\left(\frac{c}{v}\right)^{\frac{1}{n-1}}}$, which is strictly between 0 and 1 because $c<v$.

(c) At the Nash equilibrium the probability that nobody jumps in is $(1-p)^n = \left(\frac{c}{v}\right)^{\frac{n}{n-1}}$; thus this is the probability that the dog dies.
Hence, the dog is rescued with the remaining probability $1 - \left(\frac{c}{v}\right)^{\frac{n}{n-1}}$.
This is a decreasing function of n. The larger the number of swimmers who are present, the more likely it is that the dog dies.
The plot of this function when $c = 10$ and $v = 12$ is shown in Figure 6.18. □

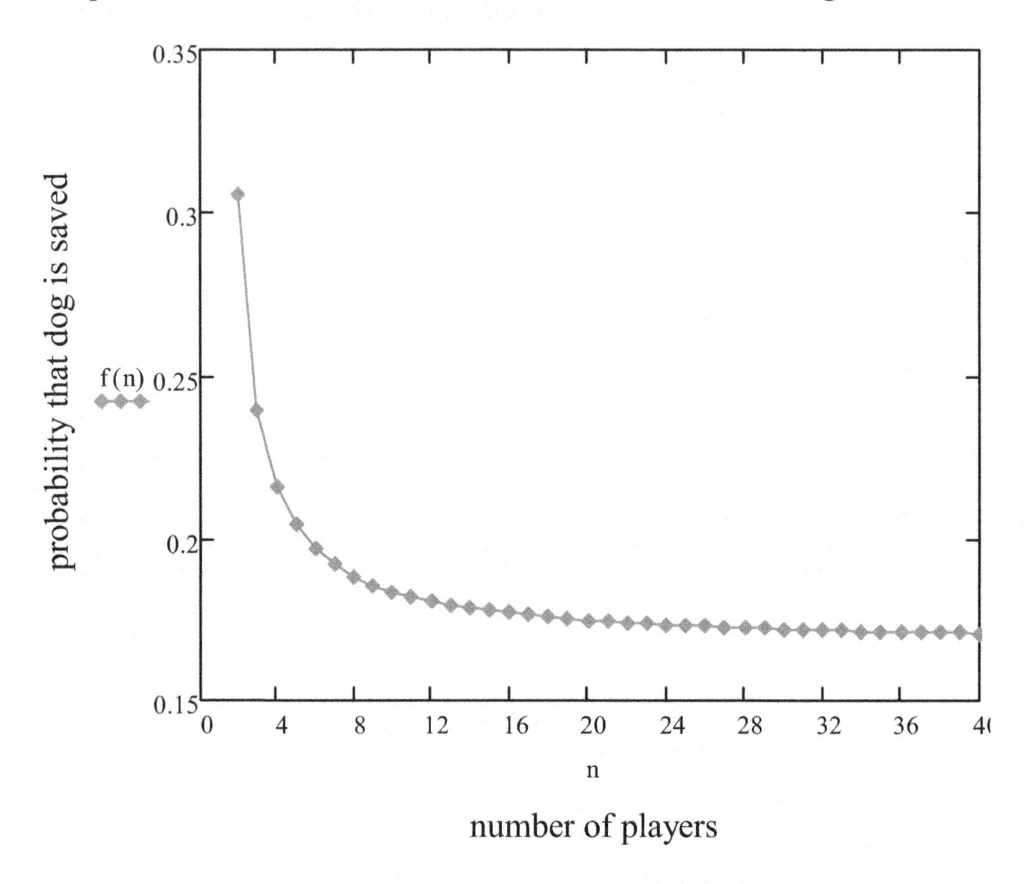

Figure 6.18: The probability that the dog is saved as a function of the number of potential rescuers

7. Extensive-form Games

7.1 Behavioral strategies in dynamic games

The definition of dynamic (or extensive-form) game-frame with cardinal payoffs is just like the definition of extensive-form game-frame with ordinal payoffs (Definition 4.1.1, Chapter 4), the only difference being that we postulate von Neumann-Morgenstern preferences instead of merely ordinal preferences.

In Chapter 6 we generalized the notion of strategic-form game-frame by allowing for lotteries (rather than just simple outcomes) to be associated with strategy profiles. One can do the same with extensive-form game-frames. For example, Figure 7.1 shows an extensive-form game-frame where associated with each terminal node (denoted by z_i, $i = 1,2,...,5$) is either a basic outcome (denoted by o_j, $j = 1,2,...,5$) or a lottery (probability distribution) over the set of basic outcomes $\{o_1, o_2, o_3, o_4, o_5\}$.

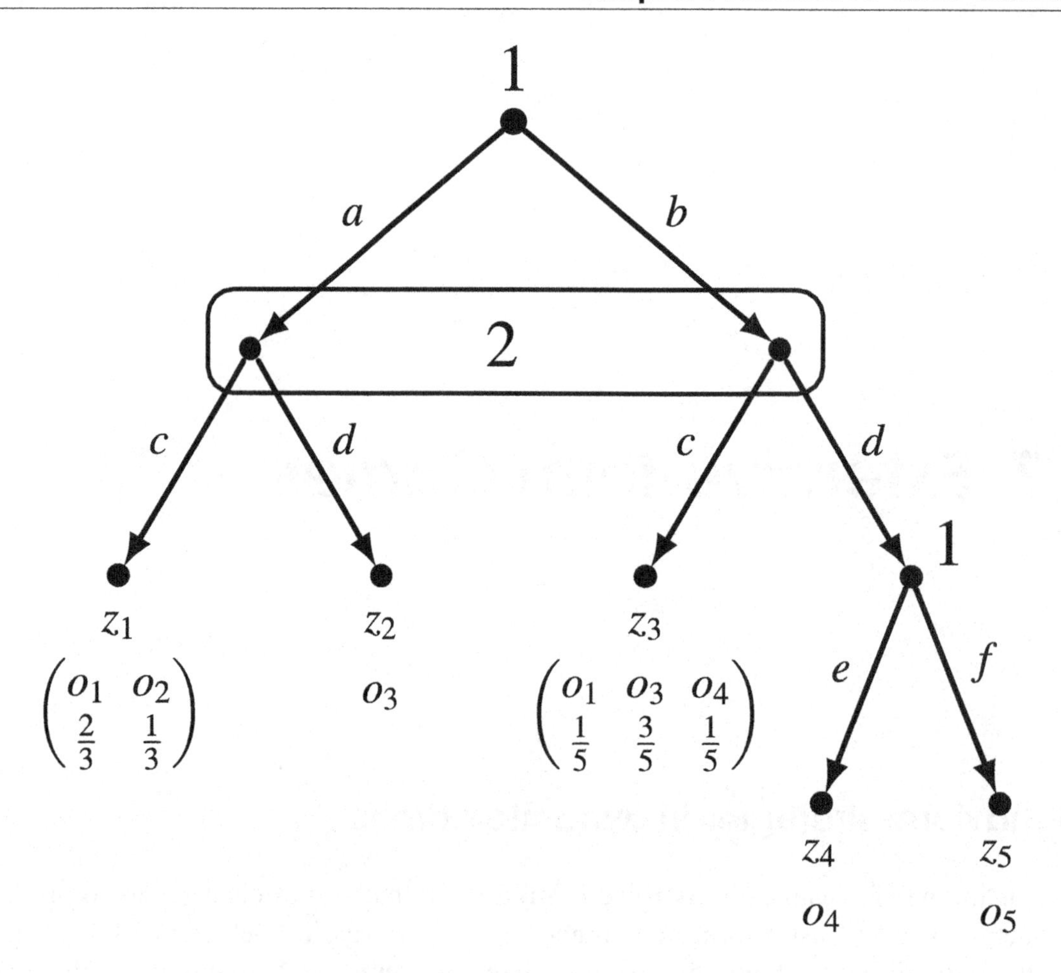

Figure 7.1: An extensive-form frame with probabilistic outcomes. The z_i's are terminal nodes and the o_i's are basic outcomes

In Figure 7.1 $\{z_1, z_2, ..., z_5\}$ is the set of terminal nodes and $\{o_1, o_2, ..., o_5\}$ is the set of basic outcomes.

Associated with z_1 is the lottery $\begin{pmatrix} o_1 & o_2 \\ \frac{2}{3} & \frac{1}{3} \end{pmatrix}$, while the lottery associated with z_3 is $\begin{pmatrix} o_1 & o_3 & o_4 \\ \frac{1}{5} & \frac{3}{5} & \frac{1}{5} \end{pmatrix}$, etc.

However, as we saw at the end of Chapter 4, in extensive forms one can explicitly represent random events by means of chance moves (also called moves of Nature). Thus an alternative representation of the extensive-form frame of Figure 7.1 is the extensive form shown in Figure 7.2.

We can continue to use the definition of extensive-form frame given in Chapter 4, but from now on we will allow for the possibility of chance moves.

The notion of strategy remains, of course, unchanged: a strategy for a player is a list of choices, one for every information set of that player (Definition 4.2.1, Chapter 4). For example, the set of strategies for Player 1 in the extensive frame of Figure 7.2 is $S_1 = \{(a,e),(a,f),(b,e),(b,f)\}$. Thus mixed strategies can easily be introduced also in extensive frames. For example, in the extensive frame of Figure 7.2, the set of mixed strategies for Player 1, denoted by Σ_1, is the set of probability distributions over S_1:

$$\Sigma_1 = \left\{ \begin{pmatrix} (a,e) & (a,f) & (b,e) & (b,f) \\ p & q & r & 1-p-q-r \end{pmatrix} : p,q,r \in [0,1] \text{ and } p+q+r \leqslant 1 \right\}.$$

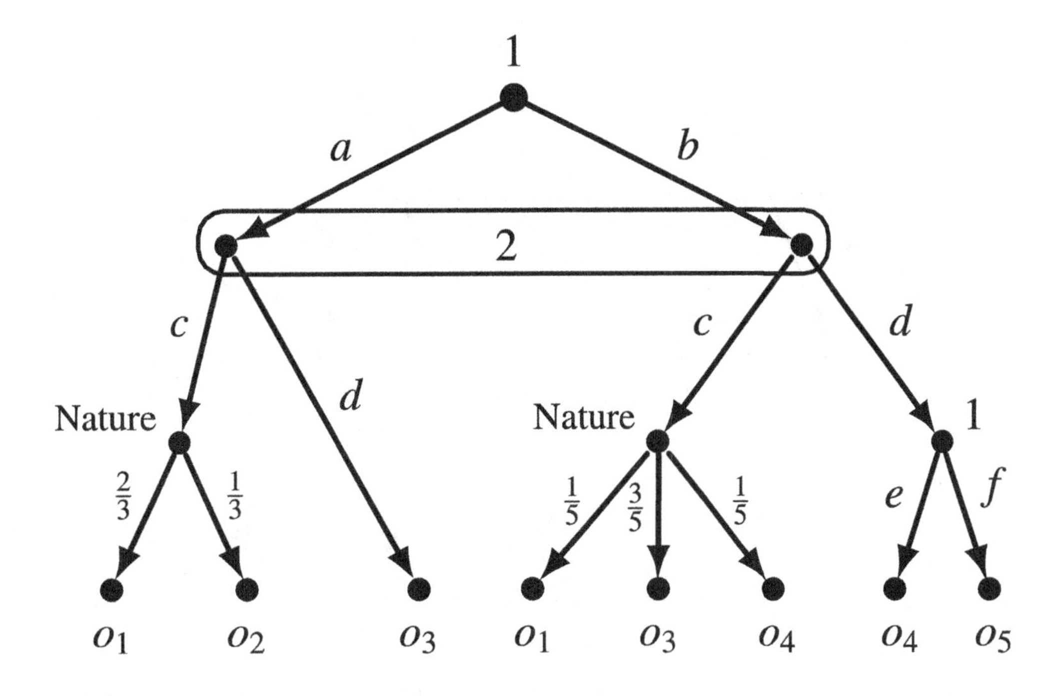

Figure 7.2: An alternative representation of the extensive frame of Figure 7.1. The terminal nodes have not been labeled. The o_i's are basic outcomes.

However, it turns out that in extensive forms with perfect recall one can use simpler objects than mixed strategies, namely behavioral strategies.

Definition 7.1.1 A *behavioral strategy* for a player in an extensive form is a list of probability distributions, one for every information set of that player; each probability distribution is over the set of choices at the corresponding information set.

For example, the set of behavioral strategies for Player 1 in the extensive frame of Figure 7.2 is:

$$\left\{ \left(\begin{array}{cc|cc} a & b & e & f \\ p & 1-p & q & 1-q \end{array} \right) : p,q \in [0,1] \right\}$$

.

A behavioral strategy is a simpler object than a mixed strategy: in this example, specifying a *behavioral strategy* for Player 1 requires specifying the values of two parameters (p and q), while specifying a *mixed strategy* requires specifying the values of three parameters (p, r and q). Can one then use behavioral strategies rather than mixed strategies? The answer is affirmative, as Theorem 7.1.1 below states.

First we illustrate with an example based on the extensive form of Figure 7.3 (the z_i's are terminal nodes and the outcomes have been omitted).

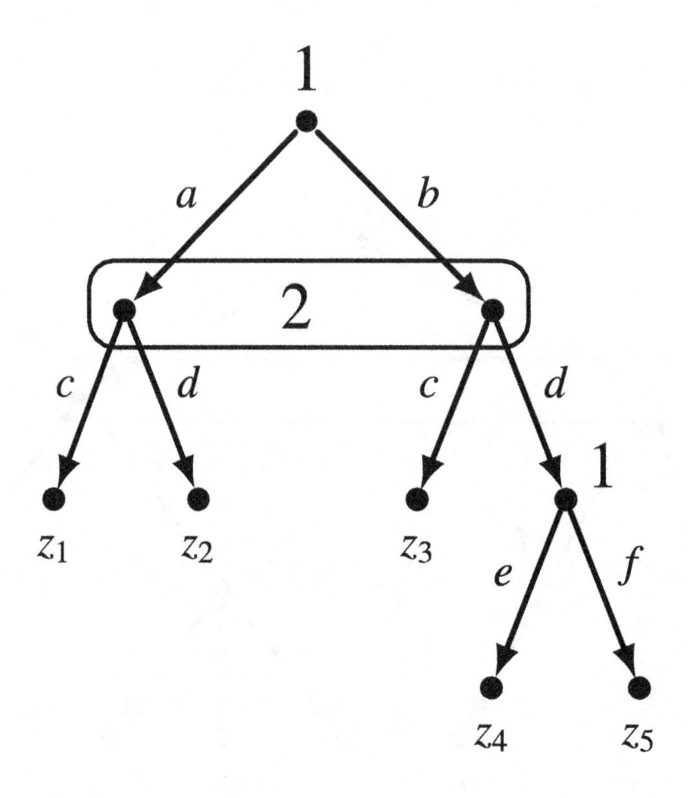

Figure 7.3: An extensive frame with the outcomes omitted. The z_i's are terminal nodes.

Consider the mixed-strategy profile $\sigma = (\sigma_1, \sigma_2)$ with

$$\sigma_1 = \begin{pmatrix} (a,e) & (a,f) & (b,e) & (b,f) \\ \frac{1}{12} & \frac{4}{12} & \frac{2}{12} & \frac{5}{12} \end{pmatrix} \quad \text{and} \quad \sigma_2 = \begin{pmatrix} c & d \\ \frac{1}{3} & \frac{2}{3} \end{pmatrix}.$$

We can compute the probability of reaching terminal node z_i, denoted by $P(z_i)$, as follows:

$$P(z_1) = \sigma_1((a,e))\,\sigma_2(c) + \sigma_1((a,f))\,\sigma_2(c) = \tfrac{1}{12}\left(\tfrac{1}{3}\right) + \tfrac{4}{12}\left(\tfrac{1}{3}\right) = \tfrac{5}{36}$$
$$P(z_2) = \sigma_1((a,e))\,\sigma_2(d) + \sigma_1((a,f))\,\sigma_2(d) = \tfrac{1}{12}\left(\tfrac{2}{3}\right) + \tfrac{4}{12}\left(\tfrac{2}{3}\right) = \tfrac{10}{36}$$
$$P(z_3) = \sigma_1((b,e))\,\sigma_2(c) + \sigma_1((b,f))\,\sigma_2(c) = \tfrac{2}{12}\left(\tfrac{1}{3}\right) + \tfrac{5}{12}\left(\tfrac{1}{3}\right) = \tfrac{7}{36}$$
$$P(z_4) = \sigma_1((b,e))\,\sigma_2(d) = \tfrac{2}{12}\left(\tfrac{2}{3}\right) = \tfrac{4}{36}$$
$$P(z_5) = \sigma_1((b,f))\,\sigma_2(d) = \tfrac{5}{12}\left(\tfrac{2}{3}\right) = \tfrac{10}{36}.$$

That is, the mixed-strategy profile $\sigma = (\sigma_1, \sigma_2)$ gives rise to the following probability distribution over terminal nodes:

$$\begin{pmatrix} z_1 & z_2 & z_3 & z_4 & z_5 \\ \frac{5}{36} & \frac{10}{36} & \frac{7}{36} & \frac{4}{36} & \frac{10}{36} \end{pmatrix}.$$

Now consider the following behavioral strategy of Player 1:

$$\left(\begin{array}{cc|cc} a & b & e & f \\ \frac{5}{12} & \frac{7}{12} & \frac{2}{7} & \frac{5}{7} \end{array}\right).$$

What probability distribution over the set of terminal nodes would it induce in conjunction with Player 2's mixed strategy $\sigma_2 = \begin{pmatrix} c & d \\ \frac{1}{3} & \frac{2}{3} \end{pmatrix}$? The calculations are simple:[1]

$$\begin{aligned}
P(z_1) &= P(a)\,\sigma_2(c) = \tfrac{5}{12}\left(\tfrac{1}{3}\right) = \tfrac{5}{36}, \\
P(z_2) &= P(a)\,\sigma_2(d) = \tfrac{5}{12}\left(\tfrac{2}{3}\right) = \tfrac{10}{36}, \\
P(z_3) &= P(b)\,\sigma_2(c) = \tfrac{7}{12}\left(\tfrac{1}{3}\right) = \tfrac{7}{36}, \\
P(z_4) &= P(b)\,\sigma_2(d)\,P(e) = \tfrac{7}{12}\left(\tfrac{2}{3}\right)\left(\tfrac{2}{7}\right) = \tfrac{4}{36}, \\
P(z_5) &= P(b)\,\sigma_2(d)\,P(f) = \tfrac{7}{12}\left(\tfrac{2}{3}\right)\left(\tfrac{5}{7}\right) = \tfrac{10}{36}.
\end{aligned}$$

Thus, against $\sigma_2 = \begin{pmatrix} c & d \\ \frac{1}{3} & \frac{2}{3} \end{pmatrix}$, Player 1's behavioral strategy $\left(\begin{array}{cc|cc} a & b & e & f \\ \frac{5}{12} & \frac{7}{12} & \frac{2}{7} & \frac{5}{7} \end{array}\right)$ and her mixed strategy $\begin{pmatrix} (a,e) & (a,f) & (b,e) & (b,f) \\ \frac{1}{12} & \frac{4}{12} & \frac{2}{12} & \frac{5}{12} \end{pmatrix}$ are equivalent, in the sense that they give rise to the same probability distribution over terminal nodes, namely

$$\begin{pmatrix} z_1 & z_2 & z_3 & z_4 & z_5 \\ \frac{5}{36} & \frac{10}{36} & \frac{7}{36} & \frac{4}{36} & \frac{10}{36} \end{pmatrix}.$$

[1] $P(x)$ denotes the probability of choice x for Player 1, according to the given behavioral strategy

Theorem 7.1.1 — Kuhn, 1953. In extensive forms *with perfect recall*, behavioral strategies and mixed strategies are equivalent, in the sense that, for every mixed strategy there is a behavioral strategy that gives rise to the same probability distribution over terminal nodes.[a]

[a] A more precise statement is as follows. Consider an extensive form with perfect recall and a Player i. Let x_{-i} be an arbitrary profile of strategies of the players other than i, where, for every $j \neq i$, x_j is either a mixed or a behavioral strategy of Player j. Then, for every mixed strategy σ_i of Player i there is a behavioral strategy b_i of Player i such that (σ_i, x_{-i}) and (b_i, x_{-i}) give rise to the same probability distribution over the set of terminal nodes.

Without perfect recall, Theorem 7.1.1 does not hold. To see this, consider the one-player extensive form shown in Figure 7.4 and the mixed strategy

$$\begin{pmatrix} (a,c) & (a,d) & (b,c) & (b,d) \\ \frac{1}{2} & 0 & 0 & \frac{1}{2} \end{pmatrix}$$

which induces the probability distribution $\begin{pmatrix} z_1 & z_2 & z_3 & z_4 \\ \frac{1}{2} & 0 & 0 & \frac{1}{2} \end{pmatrix}$ on the set of terminal nodes.

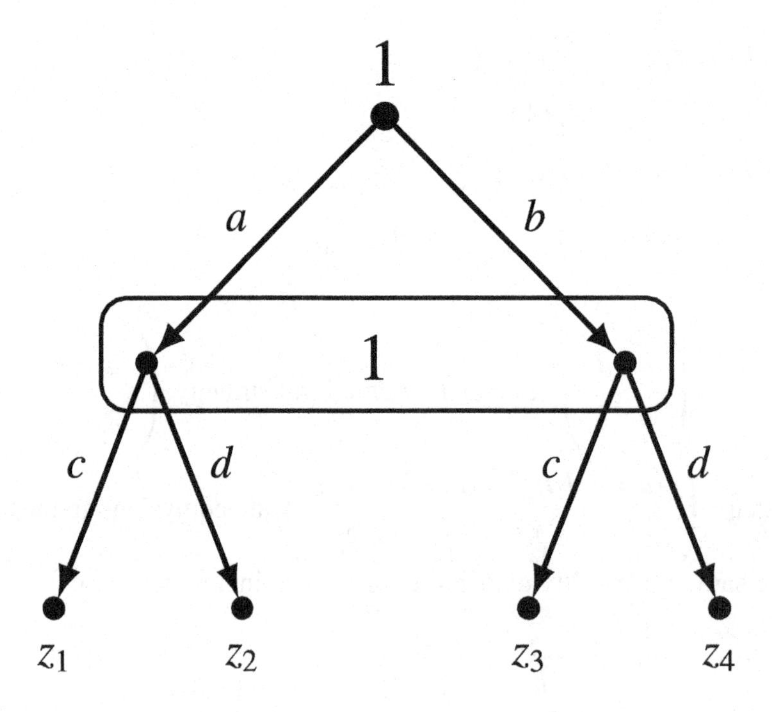

Figure 7.4: A one-player extensive frame without perfect recall.

Consider an arbitrary behavioral strategy

$$\left(\begin{array}{cc|cc} a & b & c & d \\ p & 1-p & q & 1-q \end{array}\right),$$

whose corresponding probability distribution over the set of terminal nodes is

$$\begin{pmatrix} z_1 & z_2 & z_3 & z_4 \\ pq & p(1-q) & (1-p)q & (1-p)(1-q) \end{pmatrix}.$$

In order to have $P(z_2) = 0$ it must be that either $p = 0$ or $q = 1$.

If $p = 0$ then $P(z_1) = 0$ and if $q = 1$ then $P(z_4) = 0$.

Thus the probability distribution

$$\begin{pmatrix} z_1 & z_2 & z_3 & z_4 \\ \frac{1}{2} & 0 & 0 & \frac{1}{2} \end{pmatrix}$$

cannot be achieved with a behavioral strategy.

Since the focus of this book is on extensive-form games with perfect recall, by appealing to Theorem 7.1.1, from now on we can restrict attention to behavioral strategies.

As usual, one goes from a frame to a game by adding preferences over outcomes. Let O be the set of basic outcomes (recall that with every terminal node is associated a basic outcome) and $\mathscr{L}(O)$ the set of lotteries (probability distributions) over O.

Definition 7.1.2 An *extensive-form game with cardinal payoffs* is an extensive frame (with, possibly, chance moves) together with a von Neumann-Morgenstern ranking $\succsim_i$ of the set of lotteries $\mathscr{L}(O)$, for every Player i.

As usual, it is convenient to represent a von Neumann-Morgenstern ranking by means of a von Neumann-Morgenstern utility function and replace the outcomes with a vector of utilities, one for every player.

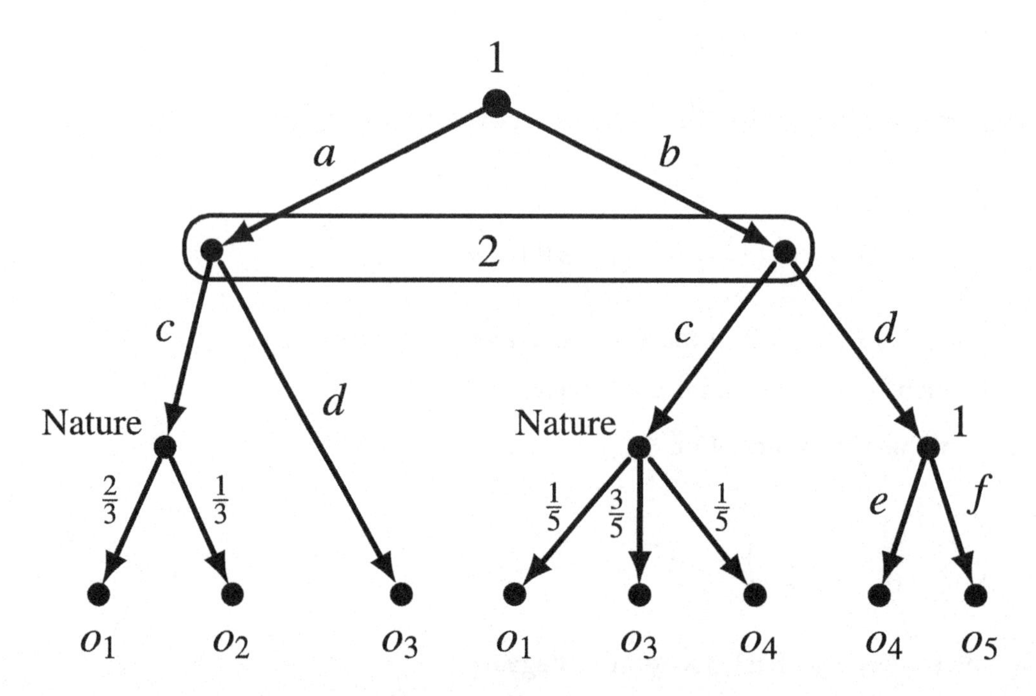

For example, consider the extensive form above, which reproduces Figure 7.2, where the set of basic outcomes is $O = \{o_1, o_2, o_3, o_4, o_5\}$ and suppose that Player 1 has a von Neumann-Morgenstern ranking of $\mathscr{L}(O)$ that is represented by the following von Neumann-Morgenstern utility function:

outcome:	o_1	o_2	o_3	o_4	o_5
U_1 :	5	2	0	1	3

Suppose also that Player 2 has preferences represented by the von Neumann-Morgenstern

utility function

outcome :	o_1	o_2	o_3	o_4	o_5
U_2 :	3	6	4	5	0

Then from the extensive frame of Figure 7.2 we obtain the extensive-form game with cardinal payoffs shown in Figure 7.5.

Since the expected utility of lottery $\begin{pmatrix} o_1 & o_2 \\ \frac{2}{3} & \frac{1}{3} \end{pmatrix}$ is 4 for both players, and the expected utility of lottery $\begin{pmatrix} o_1 & o_3 & o_4 \\ \frac{1}{5} & \frac{3}{5} & \frac{1}{5} \end{pmatrix}$ is 1.2 for Player 1 and 4 for Player 2, we can simplify the game by replacing the first move of Nature with the payoff vector (4,4) and the second move of Nature with the payoff vector (1.2, 4). The simplified game is shown in Figure 7.6.

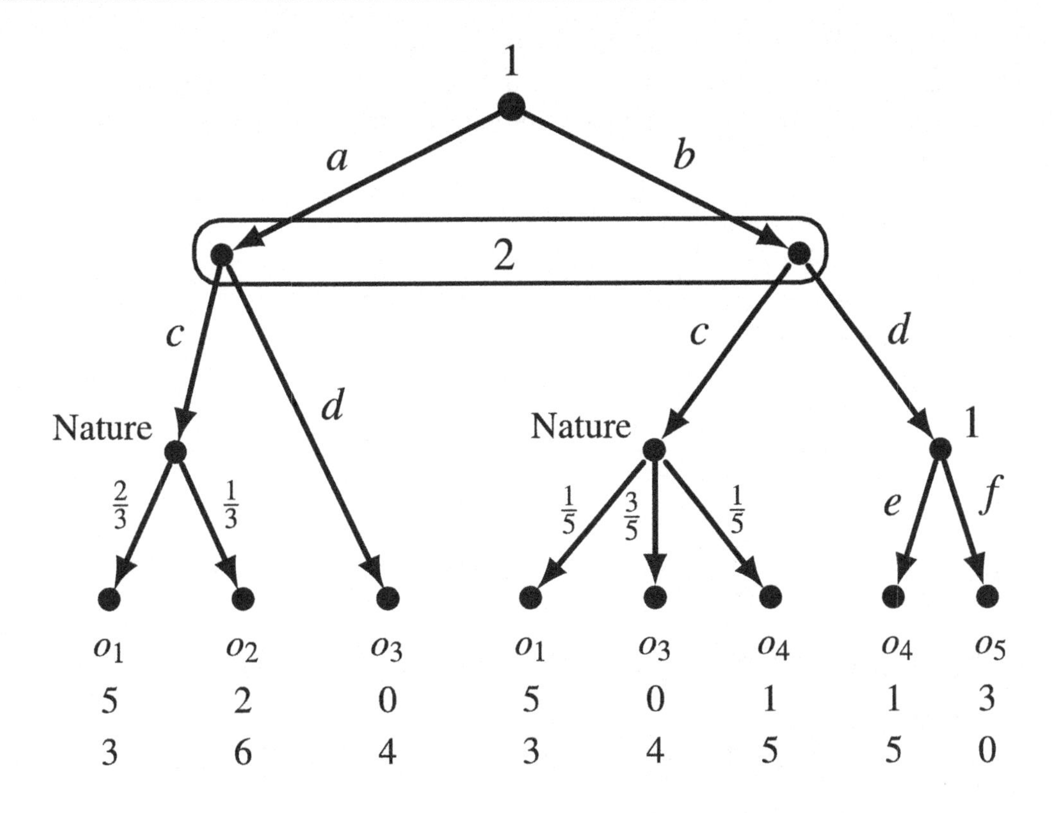

Figure 7.5: An extensive game based on the frame of Figure 7.2. The terminal nodes have not been labeled. The o_i's are basic outcomes.

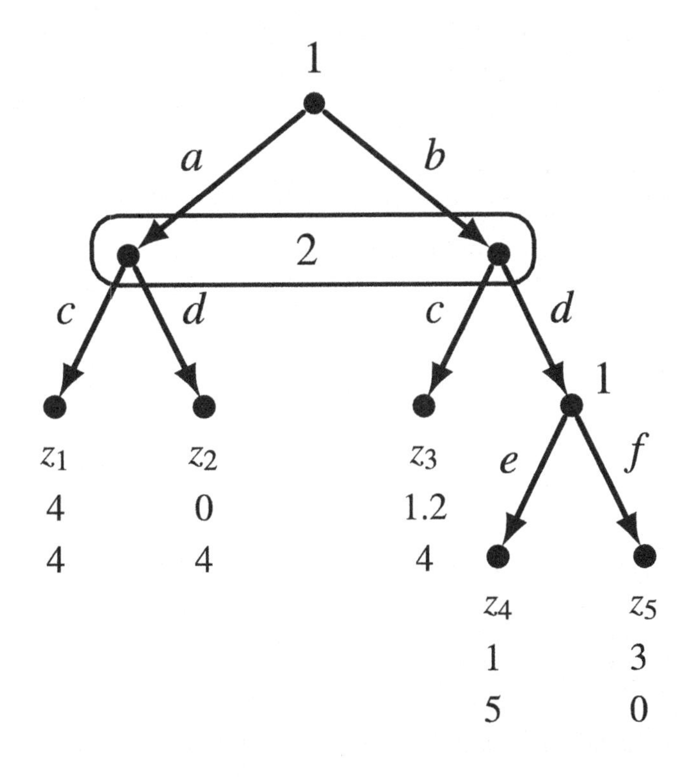

Figure 7.6: A simplified version of the game of Figure 7.5. The z_i's are terminal nodes. Note that this is a game based on the frame of Figure 7.3.

Given an extensive game with cardinal payoffs, associated with every behavioral strategy profile is a lottery over basic outcomes and thus – using a von Neumann-Morgenstern utility function for each player – a payoff for each player. For example, the behavioral strategy profile

$$\left[\left(\begin{array}{cc|cc} a & b & e & f \\ \frac{5}{12} & \frac{7}{12} & \frac{2}{7} & \frac{5}{7} \end{array}\right), \left(\begin{array}{cc} c & d \\ \frac{1}{3} & \frac{2}{3} \end{array}\right)\right]$$

for the extensive game of Figure 7.5 gives rise to the lottery

$$\left(\begin{array}{ccccc} o_1 & o_2 & o_3 & o_4 & o_5 \\ \frac{71}{540} & \frac{25}{540} & \frac{213}{540} & \frac{81}{540} & \frac{150}{540} \end{array}\right)$$

(for instance, the probability of basic outcome o_1 is calculated as follows:

$$P(o_1) = P(a)P(c)\tfrac{2}{3} + P(b)P(c)\tfrac{1}{5} = \tfrac{5}{12}\tfrac{1}{3}\tfrac{2}{3} + \tfrac{7}{12}\tfrac{1}{3}\tfrac{1}{5} = \tfrac{71}{540}).$$

Using the utility function postulated above for Player 1, namely

$$\begin{array}{rccccc} outcome: & o_1 & o_2 & o_3 & o_4 & o_5 \\ U_1: & 5 & 2 & 0 & 1 & 3 \end{array}$$

we get a corresponding payoff for Player 1equal to

$$\tfrac{71}{540}5 + \tfrac{25}{540}2 + \tfrac{213}{540}0 + \tfrac{81}{540}1 + \tfrac{150}{540}3 = \tfrac{936}{540} = 1.733.$$

An alternative way of computing this payoff is by using the simplified game of Figure 7.6 where the behavioral strategy profile

$$\left[\left(\begin{array}{cc|cc} a & b & e & f \\ \frac{5}{12} & \frac{7}{12} & \frac{2}{7} & \frac{5}{7} \end{array}\right), \left(\begin{array}{cc} c & d \\ \frac{1}{3} & \frac{2}{3} \end{array}\right)\right]$$

yields the probability distribution over terminal nodes

$$\left(\begin{array}{ccccc} z_1 & z_2 & z_3 & z_4 & z_5 \\ \frac{5}{36} & \frac{10}{36} & \frac{7}{36} & \frac{4}{36} & \frac{10}{36} \end{array}\right),$$

which, in turn, yields the probability distribution

$$\left(\begin{array}{ccccc} 4 & 0 & 1.2 & 1 & 3 \\ \frac{5}{36} & \frac{10}{36} & \frac{7}{36} & \frac{4}{36} & \frac{10}{36} \end{array}\right)$$

over utilities for Player 1. From the latter we get that the expected payoff for Player 1 is

$$\tfrac{5}{36}4 + \tfrac{10}{36}0 + \tfrac{7}{36}1.2 + \tfrac{4}{36}1 + \tfrac{10}{36}3 = \tfrac{936}{540} = 1.733.$$

The calculations for Player 2 are similar (see Exercise 7.3).

Test your understanding of the concepts introduced in this section, by going through the exercises in Section 7.4.1 at the end of this chapter.

7.2 Subgame-perfect equilibrium revisited

The notion of subgame-perfect equilibrium was introduced in Chapter 4 (Definition 4.4.1) for extensive-form games with ordinal payoffs.

When payoffs are *ordinal*, a subgame-perfect equilibrium may fail to exist because either the entire game or a proper subgame does not have any Nash equilibria.

In the case of finite extensive-form games with *cardinal* payoffs, a subgame-perfect equilibrium always exists, because – by Nash's theorem (Theorem 6.2.1, Chapter 6) – every finite game has at least one Nash equilibrium in mixed strategies.

Thus, in the case of cardinal payoffs, the subgame-perfect equilibrium algorithm (Definition 4.4.2, Chapter 4) never halts and the output of the algorithm is a subgame-perfect equilibrium.

We shall illustrate this with the extensive-form game shown in Figure 7.7.

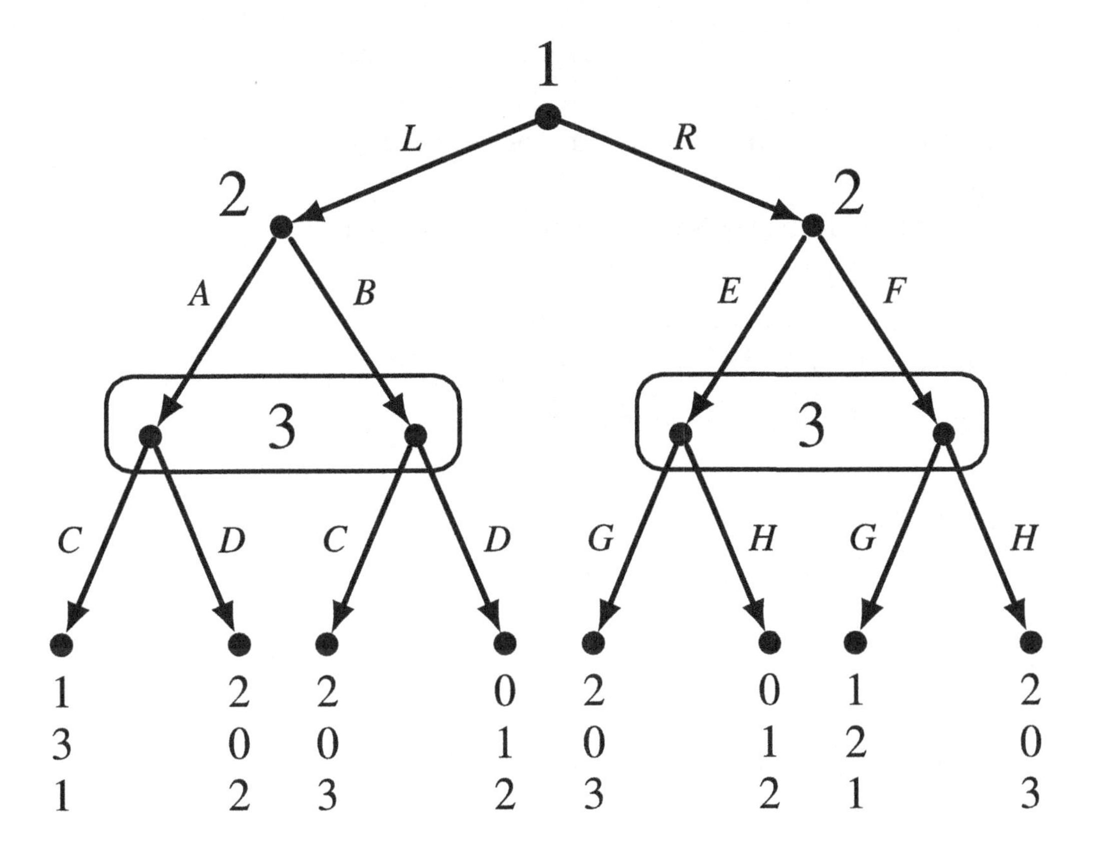

Figure 7.7: An extensive-form game with cardinal payoffs.

Let us apply the subgame-perfect equilibrium algorithm to this game. We start with the proper subgame that begins at Player 2's decision node on the left, whose strategic form is shown in Figure 7.8. Note that this subgame has no pure-strategy Nash equilibria. Thus if payoffs were merely ordinal payoffs the algorithm would halt and we would conclude that the game of Figure 7.7 has no subgame-perfect equilibria. However, we will assume that payoffs are cardinal (that is, that they are von Neumann-Morgenstern utilities).

		Player 3	
		C	D
Player 2	A	3 1	0 2
	B	0 3	1 2

Figure 7.8: The strategic form of the proper subgame on the left in the game of Figure 7.7.

To find the mixed-strategy Nash equilibrium of the game of Figure 7.8, let p be the probability of A and q the probability of C.
- Then we need q to be such that $3q = 1 - q$, that is, $q = \frac{1}{4}$,
- and p to be such that $p + 3(1 - p) = 2$, that is, $p = \frac{1}{2}$.
Thus the Nash equilibrium of this proper subgame is:

$$\left[\begin{pmatrix} A & B \\ \frac{1}{2} & \frac{1}{2} \end{pmatrix}, \begin{pmatrix} C & D \\ \frac{1}{4} & \frac{3}{4} \end{pmatrix}\right],$$

yielding the following payoffs:

$$\text{for Player 1: } \tfrac{1}{2}\tfrac{1}{4}1 + \tfrac{1}{2}\tfrac{3}{4}2 + \tfrac{1}{2}\tfrac{1}{4}2 + \tfrac{1}{2}\tfrac{3}{4}0 = 1.125$$

$$\text{for Player 2: } \tfrac{1}{2}\tfrac{1}{4}3 + \tfrac{1}{2}\tfrac{3}{4}0 + \tfrac{1}{2}\tfrac{1}{4}0 + \tfrac{1}{2}\tfrac{3}{4}1 = 0.75$$

$$\text{for Player 3: } \tfrac{1}{2}\tfrac{1}{4}1 + \tfrac{1}{2}\tfrac{3}{4}2 + \tfrac{1}{2}\tfrac{1}{4}3 + \tfrac{1}{2}\tfrac{3}{4}2 = 2.$$

Thus we can simplify the game of Figure 7.7 as shown in Figure 7.9.

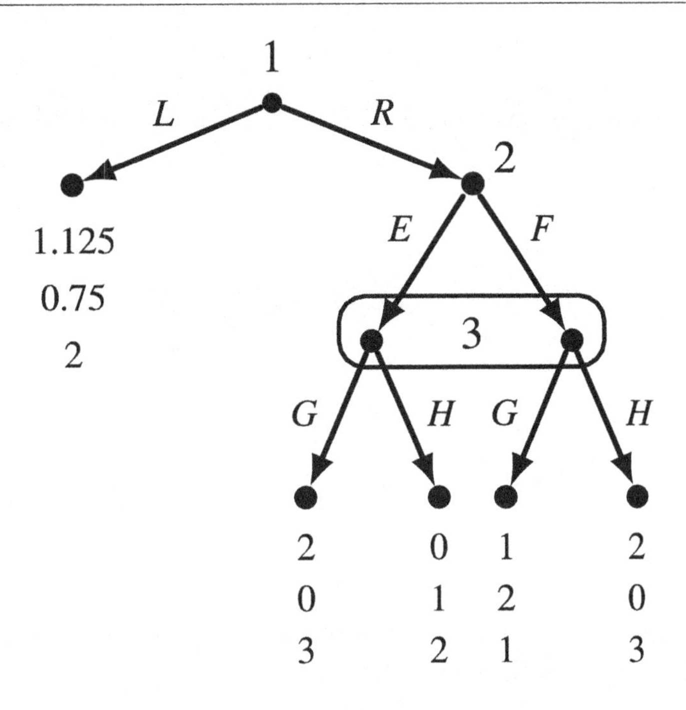

Figure 7.9: The game of Figure 7.7 after replacing the proper subgame on the left with the payoffs associated with its Nash equilibrium.

Now consider the proper subgame of the game of Figure 7.9 (the subgame that starts at Player 2's node). Its strategic form is shown in Figure 7.10.

		Player 3	
		G	H
Player 2	E	0 3	1 2
	F	2 1	0 3

Figure 7.10: The strategic form of the proper subgame of the game of Figure 7.9.

Again, there is no pure-strategy Nash equilibrium. To find the mixed-strategy equilibrium let p be the probability of E and q the probability of G.

- Then we need q to be such that $1-q=2q$, that is, $q=\frac{1}{3}$,
- and p to be such that $3p+1-p=2p+3(1-p)$, that is, $p=\frac{2}{3}$.
- Hence the Nash equilibrium is $\left[\begin{pmatrix} E & F \\ \frac{2}{3} & \frac{1}{3} \end{pmatrix}, \begin{pmatrix} G & H \\ \frac{1}{3} & \frac{2}{3} \end{pmatrix}\right]$ yielding the following payoffs:

for Player 1: $\frac{2}{3}\left(\frac{1}{3}\right)(2)+\frac{2}{3}\left(\frac{2}{3}\right)(0)+\frac{1}{3}\left(\frac{1}{3}\right)(1)+\frac{1}{3}\left(\frac{2}{3}\right)(2)=1.$

for Player 2: $\frac{2}{3}\left(\frac{1}{3}\right)(0)+\frac{2}{3}\left(\frac{2}{3}\right)(1)+\frac{1}{3}\left(\frac{1}{3}\right)(2)+\frac{1}{3}\left(\frac{2}{3}\right)(0)=0.67.$

for Player 3: $\frac{2}{3}\left(\frac{1}{3}\right)(3)+\frac{2}{3}\left(\frac{2}{3}\right)(2)+\frac{1}{3}\left(\frac{1}{3}\right)(1)+\frac{1}{3}\left(\frac{2}{3}\right)(3)=2.33.$

Thus we can simplify the game of Figure 7.9 as shown in Figure 7.11, where the optimal choice for Player 1 is L.

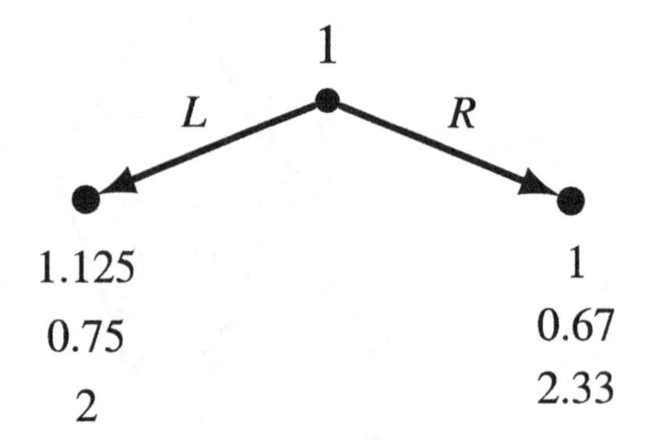

Figure 7.11: The game of Figure 7.9 after replacing the proper subgame with the payoffs associated with the Nash equilibrium.

Hence the subgame-perfect equilibrium of the game of Figure 7.7 (expressed in terms of behavioral strategies) is:

$$\left[\begin{pmatrix} L & R \\ 1 & 0 \end{pmatrix}, \left(\begin{array}{cc|cc} A & B & E & F \\ \frac{1}{2} & \frac{1}{2} & \frac{2}{3} & \frac{1}{3} \end{array}\right), \left(\begin{array}{cc|cc} C & D & G & H \\ \frac{1}{4} & \frac{3}{4} & \frac{1}{3} & \frac{2}{3} \end{array}\right)\right]$$

We conclude this section with the following theorem, which is a corollary of Theorem 6.2.1 (Chapter 6).

Theorem 7.2.1 Every finite extensive-form game with cardinal payoffs has at least one subgame-perfect equilibrium in mixed strategies.

Test your understanding of the concepts introduced in this section, by going through the exercises in Section 7.4.2 at the end of this chapter.

7.3 Problems with the notion of subgame-perfect equilibrium

The notion of subgame-perfect equilibrium is a refinement of Nash equilibrium. As explained in Chapter 3, in the context of perfect-information games, the notion of subgame-perfect equilibrium eliminates some "unreasonable" Nash equilibria that involve incredible threats. However, not every subgame-perfect equilibrium can be viewed as a "rational solution". To see this, consider the extensive-form game shown in Figure 7.12. This game has no proper subgames and thus the set of subgame-perfect equilibria coincides with the set of Nash equilibria. The pure-strategy Nash equilibria of this game are (a,f,c), (a,e,c), (b,e,c) and (b,f,d). It can be argued that neither (a,f,c) nor (b,f,d) can be considered "rational solutions".

Consider first the Nash equilibrium (a, f, c). Player 2's plan to play f is rational only in the very limited sense that, given that Player 1 plays a, what Player 2 plans to do is irrelevant because it cannot affect anybody's payoff; thus f is as good as e. However, if we take Player 2's strategy as a "serious" plan specifying what Player 2 would actually do if she had to move, then – given that Player 3 plays c – e would give Player 2 a payoff of 2, while f would only give a payoff of 1. Thus e seems to be a better strategy than f, if Player 2 takes the contingency "seriously".

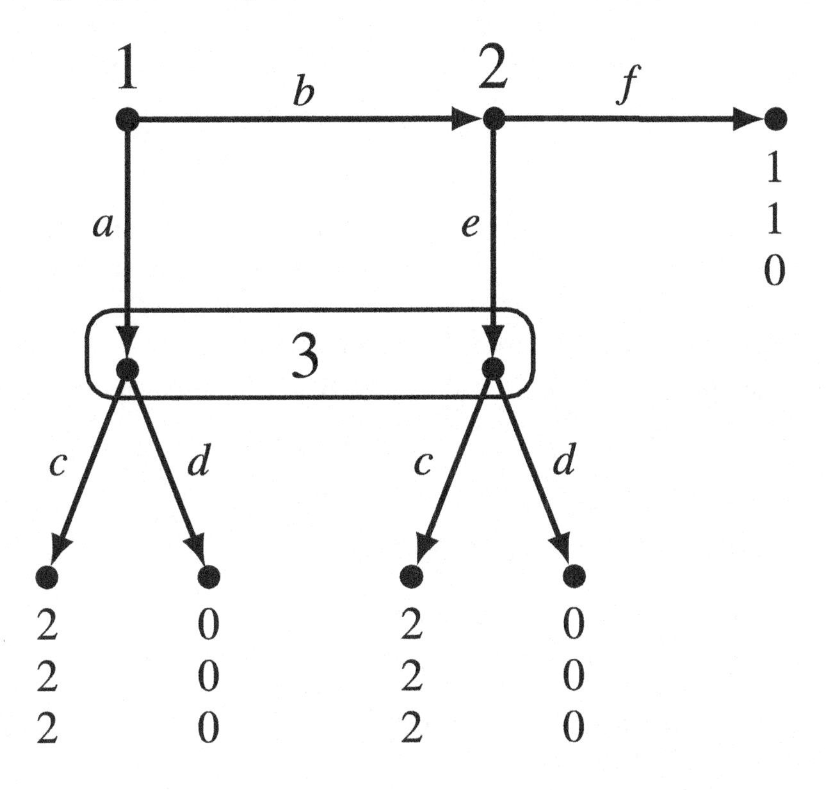

Figure 7.12: An extensive-form game showing the insufficiency of the notion of subgame-perfect equilibrium.

Consider now the Nash equilibrium (b, f, d) and focus on Player 3. As before, Player 3's plan to play d is rational only in the very limited sense that, given that Player 1 plays a and Player 2 plays f, what Player 3 plans to do is irrelevant, so that c is as good as d. However, if Player 3 did find himself having to play, it would not be rational for him to play d, since d is a strictly dominated choice: no matter whether he is making his choice at the left node or at the right node of his information set, c gives him a higher payoff than d. How can it be then that d can be part of a Nash equilibrium? The answer is that d is strictly dominated *conditional on Player 3's information set being reached* but not as a plan formulated before the play of the game starts. In other words, d is strictly dominated *as a choice but not as a strategy.*

The notion of subgame-perfect equilibrium is not strong enough to eliminate "unreasonable" Nash equilibria such as (a, f, c) and (b, f, d) in the game of Figure 7.12 In order to do that we will need a stronger notion. This issue is postponed to Part IV (Chapters 11-13).

7.4 Exercises

7.4.1 Exercises for section 7.1: Behavioral strategies in dynamic games

The answers to the following exercises are in Section 7.5 at the end of this chapter.

Exercise 7.1 What properties must an extensive-form frame satisfy in order for it to be the case that, for a given player, the set of mixed strategies coincides with the set of behavioral strategies? [Assume that there are at least two choices at every information set.] ■

Exercise 7.2 Suppose that, in a given extensive-form frame, Player 1 has four information sets: at one of them she has two choices and at each of the other three she has three choices.

(a) How many parameters are needed to specify a mixed strategy of Player 1?

(b) How many parameters are needed to specify a behavioral strategy of Player 1? ■

Exercise 7.3 From the behavioral strategy profile

$$\left[\begin{pmatrix} a & b & | & e & f \\ \frac{5}{12} & \frac{7}{12} & | & \frac{2}{7} & \frac{5}{7} \end{pmatrix}, \begin{pmatrix} c & d \\ \frac{1}{3} & \frac{2}{3} \end{pmatrix}\right]$$

calculate the payoff of Player 2 in two ways:
(1) using the game of Figure 7.5 and
(2) using the simplified game of Figure 7.6 ■

Exercise 7.4 Consider the extensive form of Figure 7.13, where $o_1, \ldots, o_5$ are basic outcomes. Player 1's ranking of O is

$$o_1 \succ_1 o_5 \succ_1 o_4 \succ_1 o_2 \sim_1 o_3;$$

furthermore, she is indifferent between o_5 and the lottery

$$\begin{pmatrix} o_1 & o_2 & o_3 \\ \frac{6}{8} & \frac{1}{8} & \frac{1}{8} \end{pmatrix}$$

and is also indifferent between o_4 and the lottery

$$\begin{pmatrix} o_2 & o_5 \\ \frac{2}{3} & \frac{1}{3} \end{pmatrix}.$$

Player 2's ranking of O is

$$o_1 \sim_2 o_2 \sim_2 o_4 \succ_2 o_3 \succ_2 o_5;$$

furthermore, he is indifferent between o_3 and the lottery

$$\begin{pmatrix} o_1 & o_2 & o_5 \\ \frac{1}{10} & \frac{1}{10} & \frac{8}{10} \end{pmatrix}.$$

Finally, Player 3's ranking of O is

$$o_2 \succ_3 o_4 \succ_3 o_3 \sim_3 o_5 \succ_3 o_1;$$

furthermore, she is indifferent between o_4 and the lottery

$$\begin{pmatrix} o_1 & o_2 & o_3 \\ \frac{1}{4} & \frac{1}{2} & \frac{1}{4} \end{pmatrix}$$

and is also indifferent between o_3 and the lottery

$$\begin{pmatrix} o_1 & o_2 \\ \frac{3}{5} & \frac{2}{5} \end{pmatrix},$$

Write the corresponding extensive-form game. ■

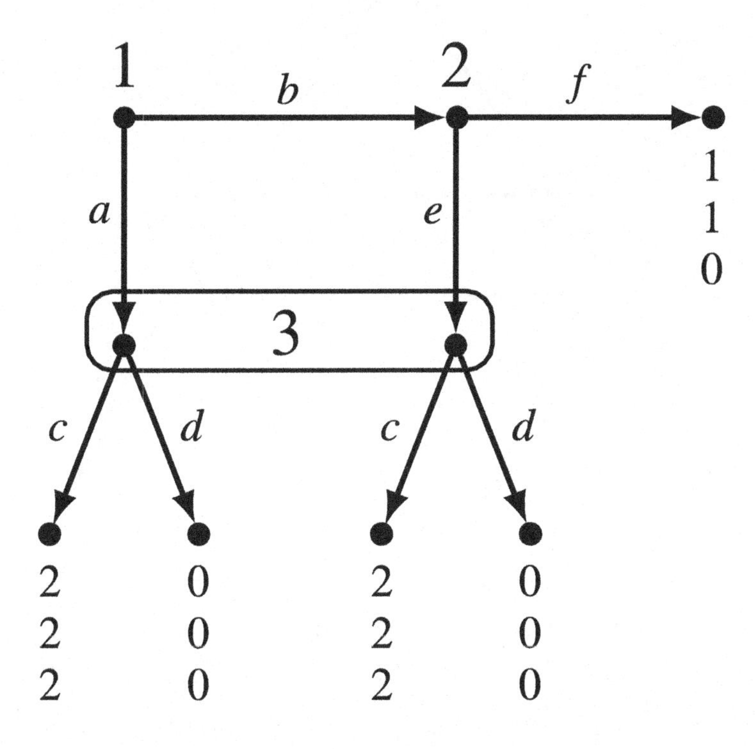

Figure 7.13: The game for Exercise 7.4

7.4.2 Exercises for section 7.2: Subgame-perfect equilibrium revisited

The answers to the following exercises are in Section 7.5 at the end of this chapter.

Exercise 7.5 Consider the extensive-form game with cardinal payoffs shown in Figure 7.14 .

(a) Write the corresponding strategic-form game and find all the pure-strategy Nash equilibria.

(b) Find the subgame-perfect equilibrium. ■

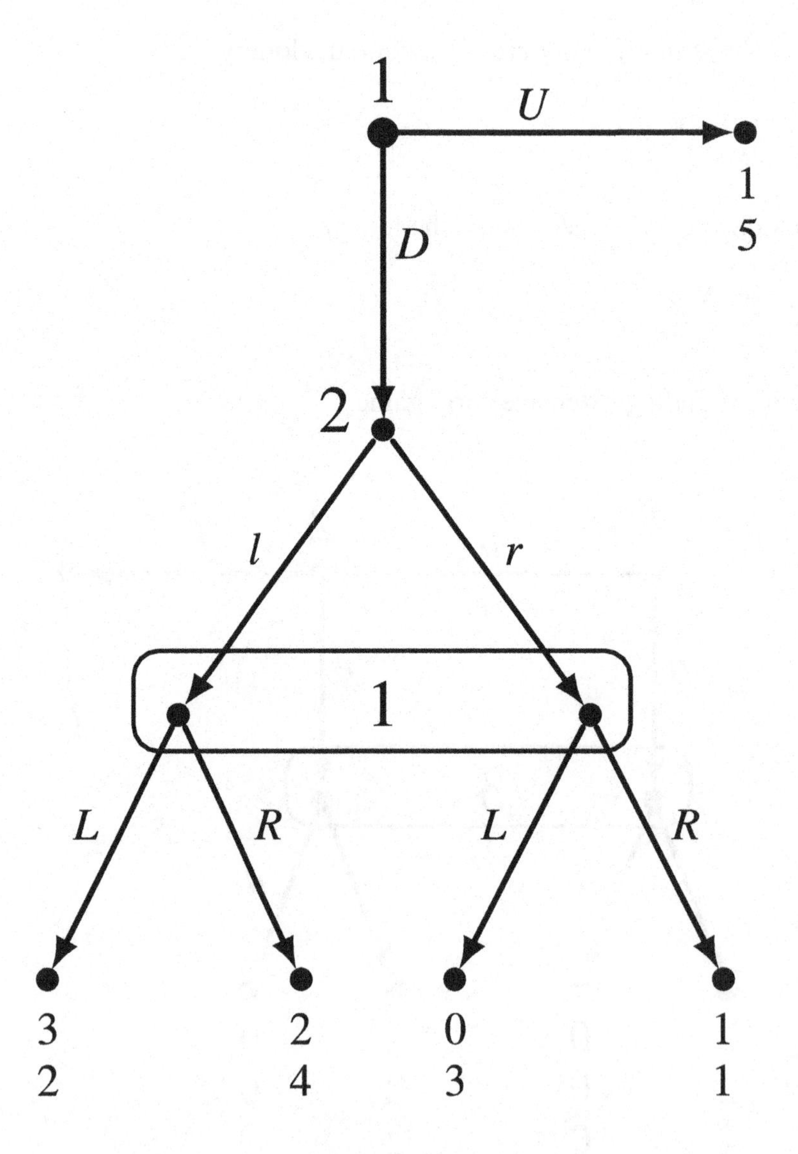

Figure 7.14: The game for Exercise 7.5

Exercise 7.6 Consider the extensive form shown in Figure 7.15 (where the basic outcomes are denoted by xj instead of o_j, $j = 1, \dots 10$). All the players satisfy the axioms of expected utility. They rank the outcomes as indicated below (as usual, if outcome w is above outcome y then w is strictly preferred to y, and if w and y are written next to each other then the player is indifferent between the two):

$$Player\ 1: \begin{pmatrix} x_7, x_9 \\ x_1, x_2, x_4, x_5 \\ x_{10} \\ x_3, x_6, x_8 \end{pmatrix} \quad Player\ 2: \begin{pmatrix} x_1, x_3 \\ x_4, x_5 \\ x_2, x_7, x_8 \\ x_6 \\ x_9 \end{pmatrix} \quad Player\ 3: \begin{pmatrix} x_2, x_7 \\ x_8 \\ x_1, x_4, x_9 \\ x_3, x_5, x_6 \end{pmatrix}$$

Furthermore, Player 2 is indifferent between x_4 and the lottery $\begin{pmatrix} x_1 & x_2 \\ \frac{1}{2} & \frac{1}{2} \end{pmatrix}$

and Player 3 is indifferent between x_1 and the lottery $\begin{pmatrix} x_2 & x_5 \\ \frac{1}{2} & \frac{1}{2} \end{pmatrix}$.

Although the above information is not sufficient to determine the von Neumann-Morgenstern utility functions of the players, it *is* sufficient to compute the subgame-perfect equilibrium. [Hint: apply the IDSDS procedure to the subgame.]
Find the subgame-perfect equilibrium. ■

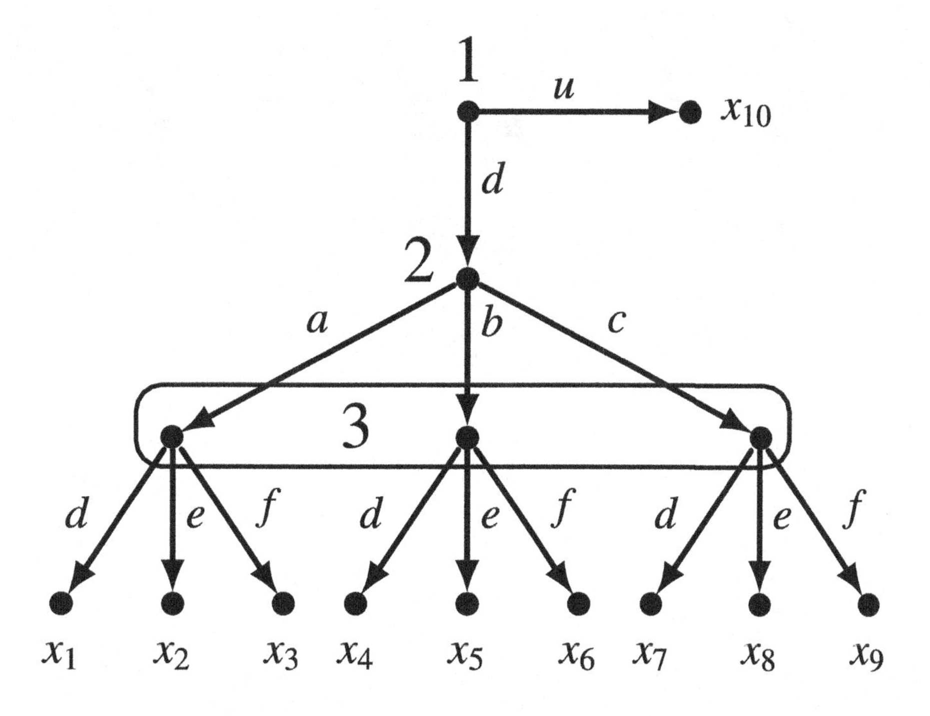

Figure 7.15: The game for Exercise 7.6

Exercise 7.7 Consider the extensive-form game shown in Figure 7.16.

(a) Write the corresponding strategic-form game.

(b) Find all the pure-strategy Nash equilibria.

(c) Find the mixed-strategy subgame-perfect equilibrium.

■

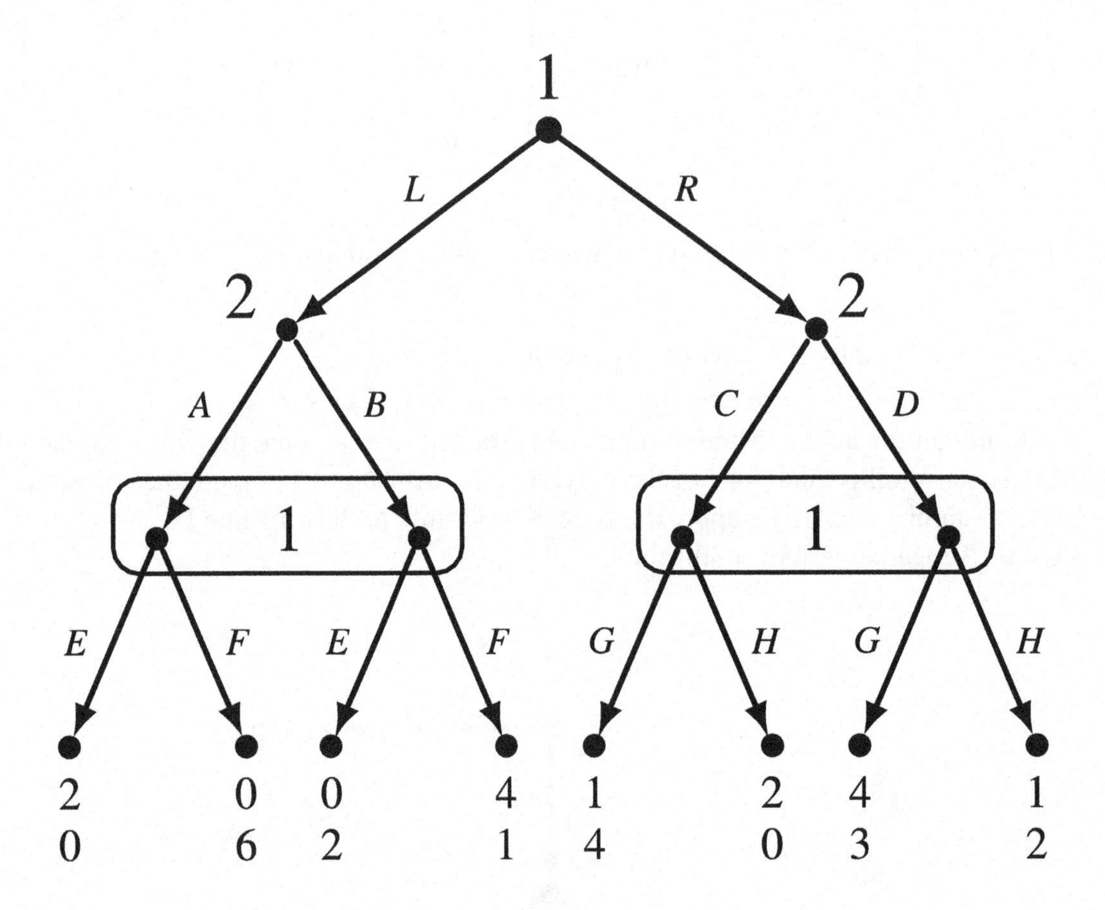

Figure 7.16: The game for Exercise 7.7

Exercise 7.8 Consider the extensive-form game shown in Figure 7.17.

(a) Find all the pure-strategy Nash equilibria. Which ones are also subgame perfect?

(b) (This is a more challenging question) Prove that there is no mixed-strategy Nash equilibrium where Player 1 plays M with probability strictly between 0 and 1. ■

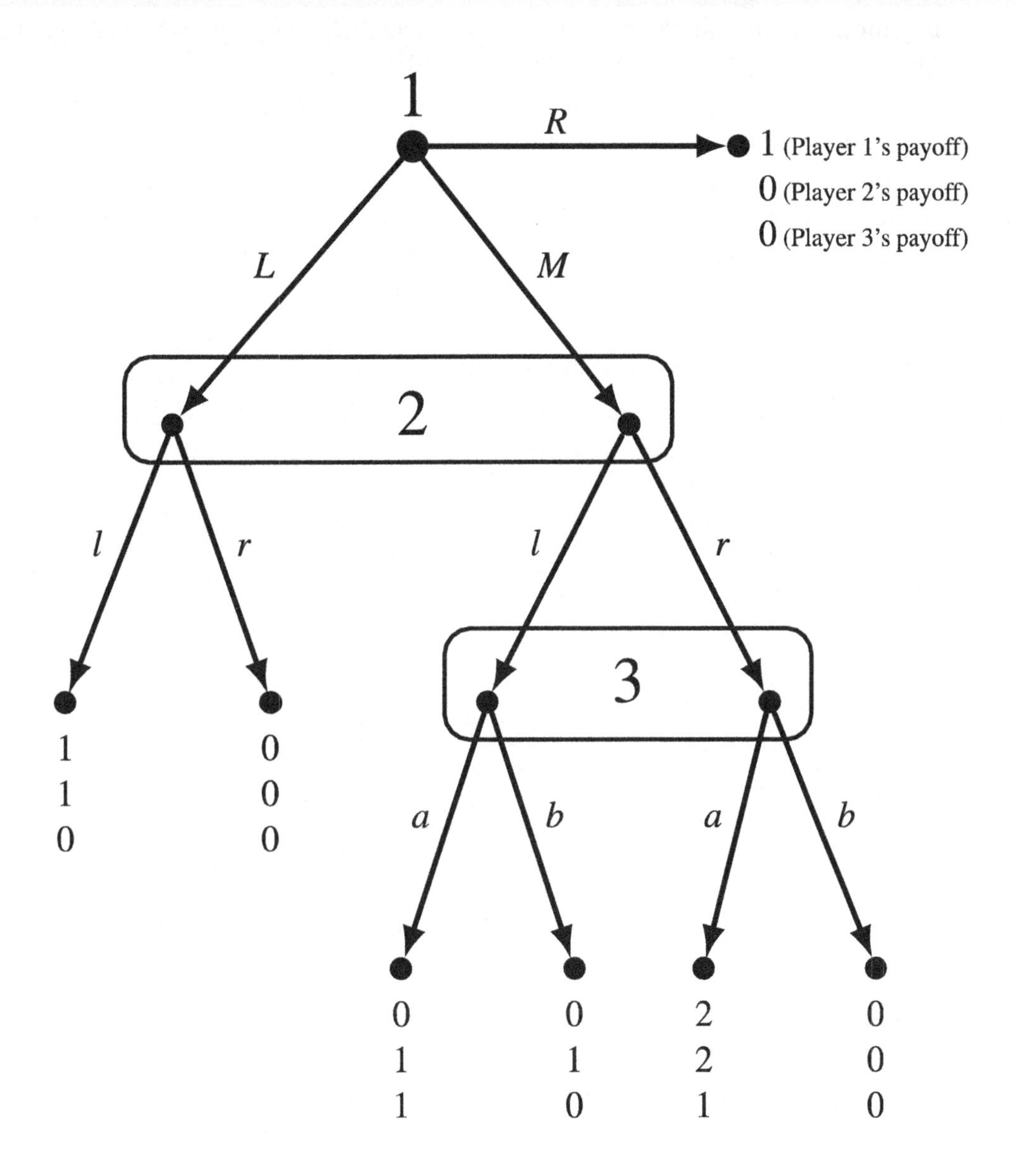

Figure 7.17: The game for Exercise 7.8

Exercise 7.9 — ⋆⋆⋆ **Challenging Question** ⋆⋆⋆.

You have to go to a part of town where many people have been mugged recently. You consider whether you should leave your wallet at home or carry it with you. Of the four possible outcomes, your most preferred one is having your wallet with you and not being mugged. Being mugged is a very unpleasant experience, so your second favorite alternative is not carrying your wallet and not being mugged (although not having any money with you can be very inconvenient). If, sadly enough, your destiny is to be mugged, then you prefer to have your wallet with you (possibly with not too much money in it!) because you don't want to have to deal with a frustrated mugger. A typical potential mugger's favorite outcome is the one where you have your wallet with you and he mugs you. His least preferred outcome is the one where he attempts to mug you and you don't have your wallet with you (he risks being caught for nothing). He is indifferent to whether or not you are carrying your wallet if he decides not to mug you. Denote the possible outcomes as shown in Figure 7.18.

(a) What is the ordinal ranking of the outcomes for each player?

Suppose that both players have von Neumann-Morgenstern utility functions. You are indifferent between the following lotteries:

$$L_1 = \begin{pmatrix} z_1 & z_2 & z_3 & z_4 \\ \frac{3}{20} & \frac{14}{20} & \frac{3}{20} & 0 \end{pmatrix} \quad \text{and} \quad L_2 = \begin{pmatrix} z_1 & z_2 & z_3 & z_4 \\ 0 & \frac{1}{2} & 0 & \frac{1}{2} \end{pmatrix};$$

furthermore, you are indifferent between

$$L_3 = \begin{pmatrix} z_1 & z_2 & z_3 & z_4 \\ 0 & \frac{2}{3} & \frac{1}{3} & 0 \end{pmatrix} \quad \text{and} \quad L_4 = \begin{pmatrix} z_1 & z_2 & z_3 & z_4 \\ \frac{1}{2} & \frac{1}{2} & 0 & 0 \end{pmatrix}.$$

The potential mugger is indifferent between the two lotteries

$$L_5 = \begin{pmatrix} z_1 & z_2 & z_3 & z_4 \\ \frac{1}{4} & \frac{1}{4} & \frac{1}{4} & \frac{1}{4} \end{pmatrix} \quad \text{and} \quad L_6 = \begin{pmatrix} z_1 & z_2 & z_3 & z_4 \\ \frac{8}{128} & \frac{67}{128} & \frac{16}{128} & \frac{37}{128} \end{pmatrix}.$$

(b) For each player find the normalized von Neumann-Morgenstern utility function.

You have to decide whether or not to leave your wallet at home. Suppose that, if you leave your wallet at home, with probability p (with $0 < p < 1$) the potential mugger will notice that your pockets are empty and with probability $(1 - p)$ he will not notice; in the latter case he will be uncertain as to whether you have your wallet with you or you don't. He will be in the same state of uncertainty if you did take your wallet with you.

(c) Represent this situation as an extensive game with imperfect information.

(d) Write the corresponding normal form.

(e) Find all the subgame-perfect equilibria (including the mixed-strategy ones, if any). (Hint: your answer should distinguish between different values of p).

■

		Potential mugger: Not mug	Potential mugger: Mug
You	Leave wallet at home	z_1	z_2
You	Take wallet with you	z_3	z_4

Figure 7.18: The outcomes for Exercise 7.9

7.5 Solutions to exercises

Solution to Exercise 7.1 It must be the case that the player under consideration has only one information set. □

Solution to Exercise 7.2

(a) 53. The number of pure strategies is $2 \times 3 \times 3 \times 3 = 54$ and thus 53 probabilities are needed to specify a mixed strategy.

(b) 7: one probability for the information set where she has two choices and two probabilities for each of the other three information sets. □

Solution to Exercise 7.3

1. The induced probability distribution on basic outcomes is

$$\begin{pmatrix} o_1 & o_2 & o_3 & o_4 & o_5 \\ \frac{71}{540} & \frac{25}{540} & \frac{213}{540} & \frac{81}{540} & \frac{150}{540} \end{pmatrix}.$$

Thus Player 2's expected utility is

$$\tfrac{71}{540}3 + \tfrac{25}{540}6 + \tfrac{213}{540}4 + \tfrac{81}{540}5 + \tfrac{150}{540}0 = \tfrac{1620}{540} = 3.$$

2. The induced probability distribution on terminal nodes is

$$\begin{pmatrix} z_1 & z_2 & z_3 & z_4 & z_5 \\ \frac{5}{36} & \frac{10}{36} & \frac{7}{36} & \frac{4}{36} & \frac{10}{36} \end{pmatrix}.$$

Thus Player 2's expected payoff is

$$\tfrac{5}{36}4 + \tfrac{10}{36}4 + \tfrac{7}{36}4 + \tfrac{4}{36}5 + \tfrac{10}{36}0 = \tfrac{108}{36} = 3.$$

Not surprisingly, the same number as in Part 1. □

Solution to Exercise 7.4 The normalized von Neumann-Morgenstern utility functions are

	o_1	o_2	o_3	o_4	o_5
U_1	1	0	0	0.25	0.75
U_2	1	1	0.2	1	0
U_3	0	1	0.4	0.6	0.4

Thus the extensive-form game is shown in Figure 7.19.

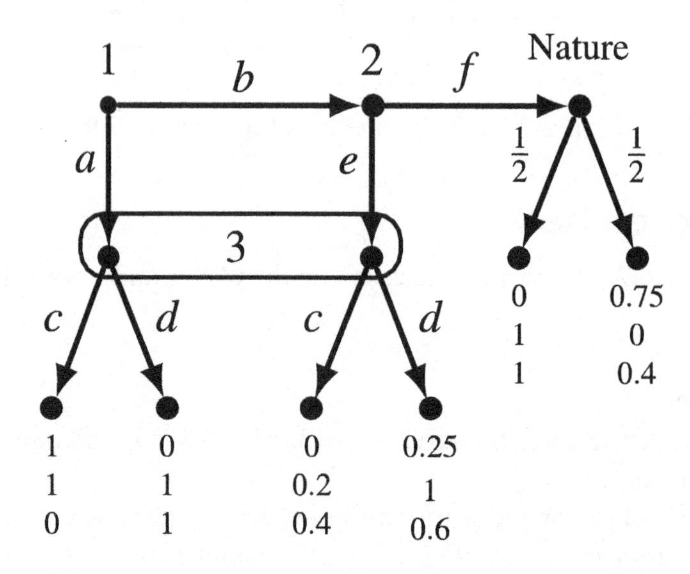

Figure 7.19: The game for Exercise 7.4

Or, in a simplified form obtained by removing the move of Nature, as shown in Figure 7.20. □

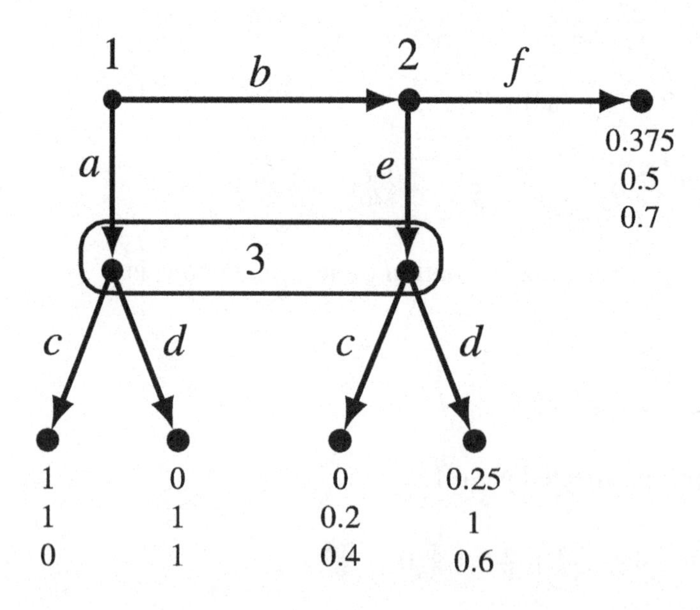

Figure 7.20: The simplified game for Exercise 7.4

Solution to Exercise 7.5

(a) The strategic form is shown in Figure 7.21. The pure-strategy Nash equilibria are (UL, r) and (UR, r).

Player 2

Player 1	l		r	
UL	1	5	1	5
UR	1	5	1	5
DL	3	2	0	3
DR	2	4	1	1

Figure 7.21: The strategic form for Part **(a)** of Exercise 7.5

(b) The strategic form of the proper subgame that starts at Player 2's node is as follows:

		Player 2 l	Player 2 r
Player 1	L	3 , 2	0 , 3
	R	2 , 4	1 , 1

This game has a unique mixed-strategy Nash equilibrium given by

$$\left[\begin{pmatrix} L & R \\ \frac{3}{4} & \frac{1}{4} \end{pmatrix}, \begin{pmatrix} l & r \\ \frac{1}{2} & \frac{1}{2} \end{pmatrix}\right], \text{ yielding Player 1 an expected payoff of 1.5.}$$

Thus the unique subgame-perfect equilibrium, expressed as a behavioral-strategy profile, is

$$\left[\begin{pmatrix} U & D & \Big| & L & R \\ 0 & 1 & \Big| & \frac{3}{4} & \frac{1}{4} \end{pmatrix}, \begin{pmatrix} l & r \\ \frac{1}{2} & \frac{1}{2} \end{pmatrix}\right]$$

or, expressed as a mixed-strategy profile,

$$\left[\begin{pmatrix} UL & UR & DL & DR \\ 0 & 0 & \frac{3}{4} & \frac{1}{4} \end{pmatrix}, \begin{pmatrix} l & r \\ \frac{1}{2} & \frac{1}{2} \end{pmatrix}\right].$$

□

Solution to Exercise 7.6 There is only one proper subgame starting from Player 2's node; its strategic-form frame is as follows:

		Player 3		
		d	e	f
	a	x_1	x_2	x_3
Player 2	b	x_4	x_5	x_6
	c	x_7	x_8	x_9

For Player 2 strategy c is strictly dominated by strategy b (she prefers x_4 to x_7, and x_5 to x_8 and x_6 to x_9) and for Player 3 strategy f is strictly dominated by strategy d (she prefers x_1 to x_3, and x_4 to x_6 and x_7 to x_9). Thus we can simplify the game as follows:

		Player 3	
		d	e
Player	a	x_1	x_2
2	b	x_4	x_5

Restricted to these outcomes the payers' rankings are:

$$Player\ 2: \begin{pmatrix} x_1 \\ x_4, x_5 \\ x_2 \end{pmatrix} \qquad Player\ 3: \begin{pmatrix} x_2 \\ x_1, x_4 \\ x_5 \end{pmatrix}.$$

Let U be Player 2's von Neumann-Morgenstern utility function. We can set $U(x_1) = 1$ and $U(x_2) = 0$. Thus, since she is indifferent between x_4 and x_5 and also between x_4 and the lottery $\begin{pmatrix} x_1 & x_2 \\ \frac{1}{2} & \frac{1}{2} \end{pmatrix}$, $U(x_4) = U(x_5) = \frac{1}{2}$.

Let V be Player 3's von Neumann-Morgenstern utility function. We can set $V(x_2) = 1$ and $V(x_5) = 0$. Thus, since she is indifferent between x_1 and x_4 and also between x_1 and the lottery $\begin{pmatrix} x_2 & x_5 \\ \frac{1}{2} & \frac{1}{2} \end{pmatrix}$, $V(x1) = V(x4) = \frac{1}{2}$.

Hence the above game-frame becomes the following game:

		Player 3	
		d	e
Player	a	$1\ ,\ \frac{1}{2}$	$0\ ,\ 1$
2	b	$\frac{1}{2}\ ,\ \frac{1}{2}$	$\frac{1}{2}\ ,\ 0$

There is no pure-strategy Nash equilibrium. Let p be the probability of a and q the probability of d. Then for a Nash equilibrium we need $q = \frac{1}{2}$ and $p = \frac{1}{2}$.

Hence in the subgame the outcome will be $\begin{pmatrix} x_1 & x_2 & x_4 & x_5 \\ \frac{1}{4} & \frac{1}{4} & \frac{1}{4} & \frac{1}{4} \end{pmatrix}$.

Since all of these outcomes are better than x_{10} for Player 1, Player 1 will play d. Thus the subgame-perfect equilibrium is

$$\left[\begin{pmatrix} d & u \\ 1 & 0 \end{pmatrix}, \begin{pmatrix} a & b & c \\ \frac{1}{2} & \frac{1}{2} & 0 \end{pmatrix}, \begin{pmatrix} d & e & f \\ \frac{1}{2} & \frac{1}{2} & 0 \end{pmatrix}\right].$$

□

Solution to Exercise 7.7

(a) The strategic form is shown in Figure 7.22.

Player 2 (columns), Player 1 (rows)

	AC		AD		BC		BD	
LEG	2	0	2	0	0	2	0	2
LEH	2	0	2	0	0	2	0	2
LFG	0	6	0	6	4	1	4	1
LFH	0	6	0	6	4	1	4	1
REG	1	4	4	3	1	4	4	3
REH	2	0	1	2	2	0	1	2
RFG	1	4	4	3	1	4	4	3
RFH	2	0	1	2	2	0	1	2

Figure 7.22: The strategic form for Exercise 7.7

(b) There are no pure-strategy Nash equilibria.

(c) First let us solve the subgame on the left, whose strategic form is as follows:

Player 2 (columns), Player 1 (rows)

	A	B
E	2 , 0	0 , 2
F	0 , 6	4 , 1

There is no pure-strategy Nash equilibrium. Let us find the mixed-strategy equilibrium. Let p be the probability assigned to E and q the probability assigned to A.
- Then p must be the solution to $6(1-p) = 2p + (1-p)$ and q must be the solution to $2q = 4(1-q)$.
- Thus $p = \frac{5}{7}$ and $q = \frac{2}{3}$.
- The expected payoff of Player 1 is $\frac{4}{3} = 1.33$, while the expected payoff of player 2 is $\frac{12}{7} = 1.714$.

Next we solve the subgame on the right, whose strategic form is as follows:

		Player 2	
		C	D
Player 1	G	1 , 4	4 , 3
	H	2 , 0	1 , 2

There is no pure-strategy Nash equilibrium. Let us find the mixed-strategy equilibrium.

- Let p be the probability assigned to G and q the probability assigned to C.
- Then p must be the solution to $4p = 3p + 2(1-p)$ and q must be the solution to $q + 4(1-q) = 2q + (1-q)$.
- Thus $p = \frac{2}{3}$ and $q = \frac{3}{4}$.
- The expected payoff of Player 1 is $\frac{7}{4} = 1.75$. Thus the game reduces to the the one shown in Figure 7.23, where the optimal choice is R. Hence the subgame-perfect equilibrium is:

$$\left[\left(\begin{array}{cc|cc|cc} L & R & E & F & G & H \\ 0 & 1 & \frac{5}{7} & \frac{2}{7} & \frac{2}{3} & \frac{1}{3} \end{array}\right), \left(\begin{array}{cc|cc} A & B & C & D \\ \frac{2}{3} & \frac{1}{3} & \frac{3}{4} & \frac{1}{4} \end{array}\right)\right]$$ □

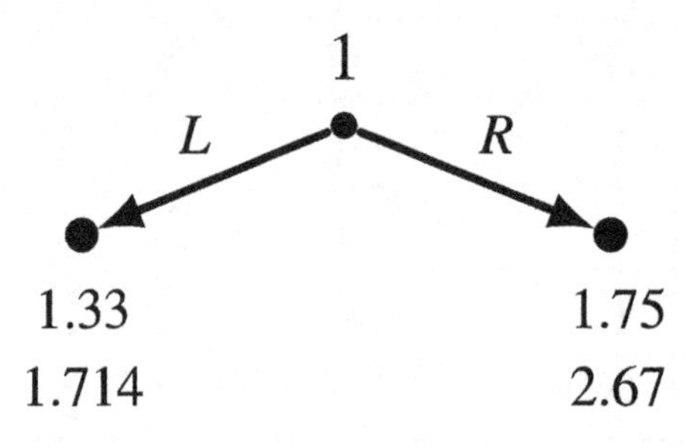

Figure 7.23: The reduced game after eliminating the proper subgames

Solution to Exercise 7.8

(a) The strategic form is shown in Figure 7.24.

Player 3: a

		Player 2: l			Player 2: r		
Player 1	R	1	0	0	1	0	0
	M	0	1	1	2	2	1
	L	1	1	0	0	0	0

Player 3: b

		Player 2: l			Player 2: r		
Player 1	R	1	0	0	1	0	0
	M	0	1	0	0	0	0
	L	1	1	0	0	0	0

Figure 7.24: The strategic form for Exercise 7.8

The pure-strategy Nash equilibria are highlighted: (R,l,a), (M,r,a), (L,l,a), (R,l,b), (R,r,b) and (L,l,b). They are all subgame perfect because there are no proper subgames.

(b) Since, for Player 3, a strictly dominates b, conditional on his information set being reached, he will have to play a if his information set is reached with positive probability. Now, Player 3's information set is reached with positive probability if and only if Player 1 plays M with positive probability. Thus when $P(M) > 0$ the game essentially reduces to the one shown in Figure 7.25.

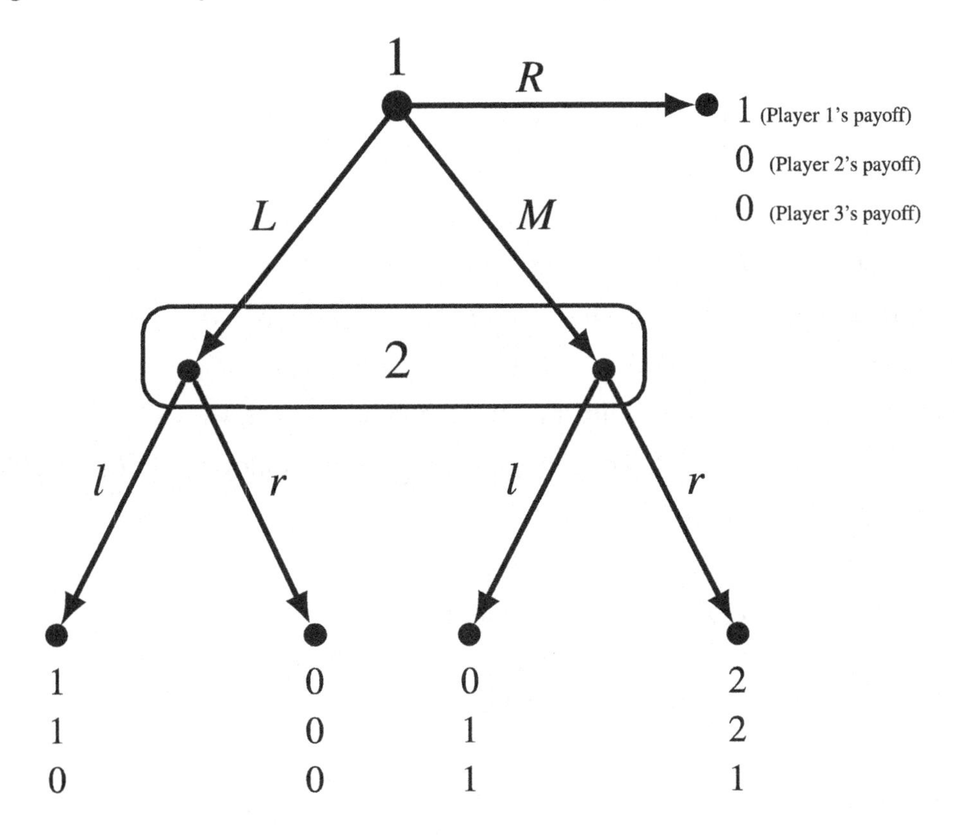

Figure 7.25: The extensive-form game for Part (b) of Exercise 7.8

Now, in order for Player 1 to be willing to assign positive probability to M he must expect a payoff of at least 1 (otherwise R would be better) and the only way he can expect a payoff of at least 1 is if Player 2 plays r with probability at least $\frac{1}{2}$.

- If Player 2 plays r with probability greater than $\frac{1}{2}$, then M gives Player 1 a higher payoff than both L and R and thus he will choose M with probability 1, in which case Player 2 will choose r with probability 1 (and Player 3 will choose a with probability 1) and so we get the pure strategy equilibrium (M,r,a).
- If Player 2 plays r with probability exactly $\frac{1}{2}$ then Player 1 is indifferent between M and R (and can mix between the two), but finds L inferior and must give it probability 0. But then Player 2's best reply to a mixed strategy of Player 1 that assigns positive probability to M and R and zero probability to L is to play r with probability 1 (if his information set is reached it can only be reached at node x_2).
- Thus there cannot be a mixed-strategy equilibrium where Player 1 assigns to M probability p with $0 < p < 1$: it must be either $Pr(M) = 0$ or $Pr(M) = 1$. □

Solution to Exercise 7.9

(a) The rankings are as follows:

$$\text{You: } \begin{pmatrix} \text{best} & z_3 \\ & z_1 \\ & z_4 \\ \text{worst} & z_2 \end{pmatrix}, \quad \text{Potential Mugger: } \begin{pmatrix} \text{best} & z_4 \\ & z_1, z_3 \\ \text{worst} & z_2 \end{pmatrix}$$

(b) Let U be your utility function. Let $U(z_3) = 1, U(z_1) = a, U(z_4) = b$ and $U(z_2) = 0$, with $0 < b < a < 1$. The expected utilities are as follows: $EU(L_1) = \frac{3}{20}a + \frac{3}{20}$, $EU(L_2) = \frac{1}{2}b$, $EU(L_3) = \frac{1}{3}$ and $EU(L_4) = \frac{1}{2}a$.
From $EU(L_3) = E(L_4)$ we get that $a = \frac{2}{3}$.
Substituting this into the equation $EU(L_1) = EU(L_2)$ gives $b = \frac{1}{2}$.
Thus $U(z_3) = 1$, $U(z_1) = \frac{2}{3}$, $U(z_4) = \frac{1}{2}$ and $U(z_2) = 0$.
Let V be the mugger's utility function. Let $V(z_4) = 1, V(z_1) = V(z_3) = c$ and $V(z_2) = 0$ with $0 < c < 1$. The expected utilities are as follows: $EV(L_5) = \frac{1}{4}(2c+1)$ and $EV(L_6) = \frac{1}{128}(24c+37)$.
Solving $EV(L_5) = EV(L_6)$ gives $c = \frac{1}{8}$.
Thus, $V(z_4) = 1, V(z_1) = V(z_3) = \frac{1}{8}$ and $V(z_2) = 0$.

(c) The extensive game is shown in Figure:7.26.

(d) The strategic form is shown in Figure 7.27 (for the mugger's strategy the first item refers to the left node, the second item to the information set on the right).

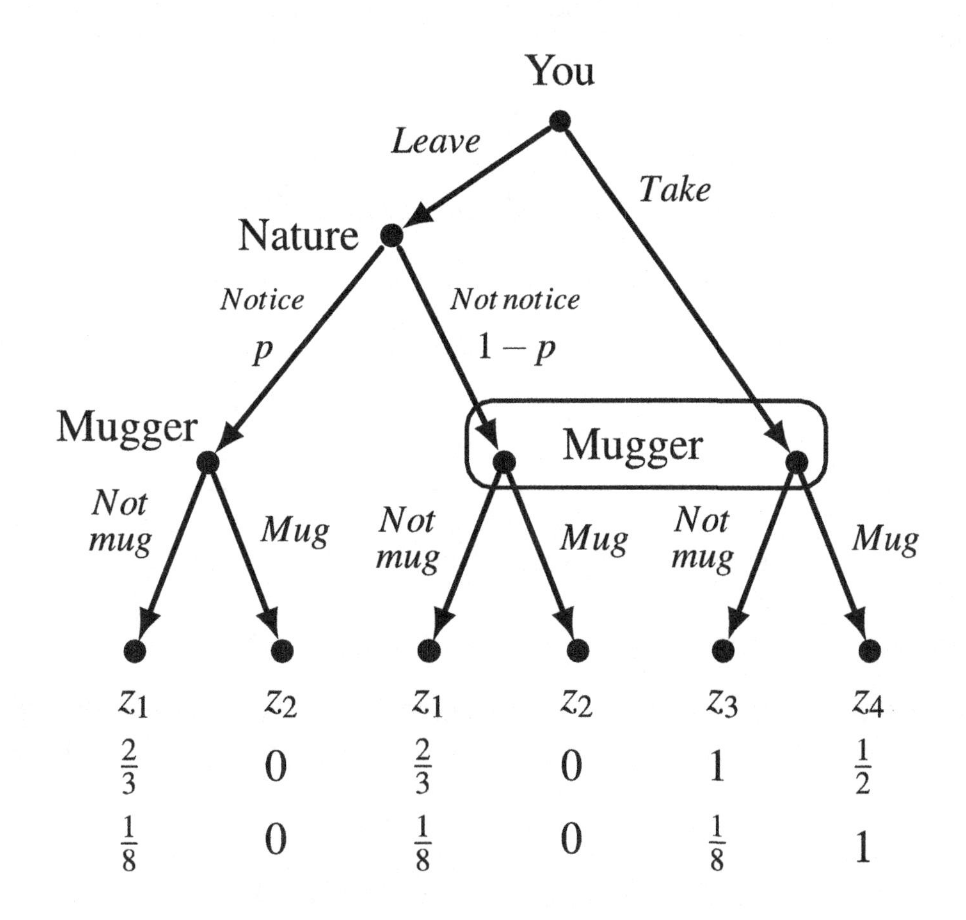

Figure 7.26: The extensive game for the Exercise 7.9

Potential Mugger

You	*NN*		*NM*		*MN*		*MM*	
L	$\frac{2}{3}$	$\frac{1}{8}$	$\frac{2}{3}p$	$\frac{1}{8}p$	$\frac{2}{3}(1-p)$	$\frac{1}{8}(1-p)$	0	0
T	1	$\frac{1}{8}$	$\frac{1}{2}$	1	1	$\frac{1}{8}$	$\frac{1}{2}$	1

Figure 7.27: The strategic form for the game of Figure 7.26

(e) At a subgame-perfect equilibrium the mugger will choose not to mug when he notices your empty pockets. Thus the normal form can be simplified as shown in Figure 7.28.

Potential Mugger

		NN		NM	
You	L	$\frac{2}{3}$	$\frac{1}{8}$	$\frac{2}{3}p$	$\frac{1}{8}p$
	T	1	$\frac{1}{8}$	$\frac{1}{2}$	1

Figure 7.28: The reduced game for Exercise 7.9

Thus,

- If $p < \frac{3}{4}$ then Take is a strictly dominant strategy for you and therefore there is a unique subgame-perfect equilibrium given by (Take, (Not mug,Mug)).
- If $p = \frac{3}{4}$ then there is a continuum of equilibria where the Mugger chooses (Not mug, Mug) with probability 1 and you choose L with probability q and T with probability $(1-q)$ for any q with $0 \leq q \leq \frac{28}{29}$, obtained from the following condition about the Potential Mugger:

$$\underbrace{\tfrac{3}{32}q + 1 - q}_{\substack{\text{expected payoff} \\ \text{from playing } NM}} \geq \underbrace{\tfrac{1}{8}}_{\substack{\text{payoff from} \\ \text{playing } NN}}$$

- If $p > \frac{3}{4}$ then there is no pure-strategy subgame-perfect equilibrium. Let q be the probability that you choose L and r the probability that the mugger chooses NN. Then the unique mixed strategy equilibrium is given by the solution to:

$$\tfrac{2}{3}r + \tfrac{2}{3}p(1-r) = r + \tfrac{1}{2}(1-r) \quad \text{and} \quad \tfrac{1}{8} = \tfrac{1}{8}pq + (1-q)$$

which is $q = \frac{7}{8-p}$ and $r = \frac{4p-3}{4p-1}$. Thus the unique subgame-perfect equilibrium is:

$$\left(\begin{array}{cc|cccc} L & T & NN & NM & MN & MM \\ \frac{7}{8-p} & \frac{1-p}{8-p} & \frac{4p-3}{4p-1} & \frac{2}{4p-1} & 0 & 0 \end{array}\right).$$

□

Index

I

K

L

M

N

O

P

R

S

T

U

V

W

Z

Made in the USA
Coppell, TX
26 August 2023